AF342301

Interaction of Mechanics and Mathematics

Series editor

Lev Truskinovsky, Laboratoire de Mechanique des Solid, Palaiseau, France
e-mail: trusk@lms.polytechnique.fr

About this Series

The Interaction of Mechanics and Mathematics (IMM) series publishes advanced textbooks and introductory scientific monographs devoted to modern research in the wide area of mechanics. The authors are distinguished specialists with international reputation in their field of expertise. The books are intended to serve as modern guides in their fields and anticipated to be accessible to advanced graduate students. IMM books are planned to be comprehensive reviews developed to the cutting edge of their respective field and to list the major references.

Advisory Board

D. Colton, USA
R. Knops, UK
G. DelPiero, Italy
Z. Mroz, Poland
M. Slemrod, USA
S. Seelecke, USA
L. Truskinovsky, France

IMM is promoted under the auspices of ISIMM (International Society for the Interaction of Mechanics and Mathematics).

More information about this series at http://www.springer.com/series/5395

Serge Preston

Non-commuting Variations in Mathematics and Physics

A Survey

 Springer

Serge Preston
Portland, OR
USA

ISSN 1860-6245   ISSN 1860-6253 (electronic)
Interaction of Mechanics and Mathematics
ISBN 978-3-319-28321-0   ISBN 978-3-319-28323-4 (eBook)
DOI 10.1007/978-3-319-28323-4

Library of Congress Control Number: 2015959935

© Springer International Publishing Switzerland 2016
This work is subject to copyright. All rights are reserved by the Publisher, whether the whole or part
of the material is concerned, specifically the rights of translation, reprinting, reuse of illustrations,
recitation, broadcasting, reproduction on microfilms or in any other physical way, and transmission
or information storage and retrieval, electronic adaptation, computer software, or by similar or dissimilar
methodology now known or hereafter developed.
The use of general descriptive names, registered names, trademarks, service marks, etc. in this
publication does not imply, even in the absence of a specific statement, that such names are exempt from
the relevant protective laws and regulations and therefore free for general use.
The publisher, the authors and the editors are safe to assume that the advice and information in this
book are believed to be true and accurate at the date of publication. Neither the publisher nor the
authors or the editors give a warranty, express or implied, with respect to the material contained herein or
for any errors or omissions that may have been made.

Printed on acid-free paper

This Springer imprint is published by SpringerNature
The registered company is Springer International Publishing AG Switzerland

NOTES ON THE NONCOMMUTING VARIATIONS.

SERGE PRESTON

Content by Chapters:

(1) Chapter 1. Basics of the Lagrangian Field Theory - p.3,

(2) Chapter 2. Lagrangian Field Theory with the non-commuting (NC) variations - p.17,

(3) Chapter 3. Vertical connections in the Configurational bundle and the NC-variations - p.53,

(4) Chapter 4. K-twisted prolongations and μ-symmetries (by works of Muriel, Romero, Gaeta, Morando, etc.) - p.79 ,

(5) Chapter 5. Applications: Holonomic and non-Holonomic Mechanics,H.Kleinert Action principle, Uniform Materials,
and the Dissipative potentials - p.111,

(6) Chapter 6. Material time, NC-variations and the Material Aging - p.139,

(7) Appendix I. Fiber bundles and their geometrical structures, absolute parallelism - p.187

(8) Appendix II. Jet bundles, contact structures and connections on Jet bundles- p.205,

(9) Symmetry groups of systems of Differential Equations and the Noether balance laws - p   223

CONTENTS

1.	Preface	x
2.	Introduction.	xi
3.	Configurational bundle , 1-jet bundle and the Lagrangian action.	3
3.1.	Configurational bundle (Y, π, X).	3
3.2.	First orders Lagrangians and the Action functional.	4
4.	First Variation and the Euler-Lagrange system.	5
4.1.	First Variations.	5
4.2.	Euler-Lagrange Equations and natural boundary conditions.	7
4.3.	Symmetries and Noether Theorem.	8
4.4.	Energy-Momentum balance law.	9

4.5.   General variations, group of automorphisms of the configurational
              bundle: case when $\phi_t \in Aut(\pi)$.   10
5.   Introduction.   17
6.   **Euler-Lagrange equations with K-twisted variations.**   17
7.   Noether Theorem, Energy-Momentum balance law.   20
7.1.   Stress-Energy-Momentum balance law.   21
8.   On the non-unicity of NC-representation (6.5).   22
9.   Case of natural (tensor-like) bundles: Canonical NC-tensor $K^i_{\mu j}$ 24
10.   **Examples.**   27
11.   Weak and strong minimizers for the systems with NC-variations.   31
11.1.   Minimizers for systems with NC-variations.   32
12.   Second variation.   36
12.1.   Jacobi equation.   37
13.   **Hamiltonian systems and the NC-variations.**   38
13.1.   Comparison with the metriplectic model   40
14.   Hamilton-Jacobi Equation.   42
14.1.   Case $n = 1$. Formula for variations.   42
14.2.   Case n=1, Hamilton-Jacoby equation and failure of Second Jacoby
              Theorem.   43
14.3.   Basic Field Equation for non-conservative dynamical systems.   45
14.4.   Basic Field Equation for a Hamiltonian system with nonconservative
              forces.   45
14.5.   Complete solution of BFE and related conservation laws.   46
14.6.   General solution of modified Hamiltonian system (14.16) from the
              general solution of BFE.   47
15.   Introduction   53
16. K-twisted prolongation $\xi \to Pr^1_K(\xi)$.   54
16.1. K-twisted total derivative.   54
17.   Non-conservation of Cartan distribution by a K-twisted prolongations
              of vector fields.   55
18.   Obstruction for a K-twisted prolongation $\xi \to \xi^{(1)}_K$ to be the Lie
              algebra morphism.   57
19.   Curvature and the sources f_j.   61
19.1.   Special cases.   62
20.   Case: Zero order tensor K: $K^\alpha_{\beta i} \in C^\infty(Y)$.   64
20.1.   Curvature $\tilde{R}$ and the potential forces.   65
21.   NC variations and the "Dynamical connections" in Hamiltonian
              systems.   67
21.1.   Case of Field Theory   68
22.   Kinematical connection (Γ, S), dynamical connection K, and the
              energy-momentum balance law.   70
23.   Infinitesimal Variational Calculus and the non-commutative variations.   72
23.1.   Poincare-Cartan formalism, case of the first order.   72
24.   **Lifted Poincare-Cartan form of a balance system and its
              contact source modification.**   73
25.   **Higher order Lagrangian systems with NC variations.**   75
25.1.   Higher order Lagrangian field theory.   75
25.2.   Degenerate Lagrangian and higher order dissipation.   77

26. Introduction. 79
27. Twisted prolongation of vector fields in Y to the jet bundles. 79
27.1. K-twisted prolongations of order one. 80
28. λ-twisted prolongations - case $m = n = 1$ 80
28.1. λ-prolongations of higher order. 81
28.2. Characteristic of λ-prolongation in terms of contact structure. 81
29. λ-symmetries. 82
30. **μ-prolongations and μ-symmetries.Case of one PDE (m=1).** 83
31. **μ-prolongation and μ-symmetries for the systems of PDE.** 87
31.1. Compatibility condition for the 1-form $\mu = \Lambda_i dx^i$. 87
31.2. Twisted invariance condition of the contact structure preservation at the μ-prolongation: general case. 89
32. μ-symmetries and reduction of PDE systems. 90
33. μ-conservation laws. 91
34. Noether Theorem for μ-symmetries. 92
34.1. Conservation laws for μ-symmetries. 93
35. Gauge transformations and comparison of μ- and conventional prolongations. 94
35.1. Exponential vector fields and the μ-symmetries. 94
35.2. Gauge transformations and the action of Linear Group . 95
35.3. Darboux derivative and the local presentation of the form μ. 96
35.4. Comparison of flow- and μ-prolongations. 98
36. Applications and Examples. 100
37. Deformation of exterior differential, Lie derivatives and the μ-prolongation (by P.Morando,[98]). 102
37.1. Witten's gauging and the deformed operators d and $\mathcal{L}_\xi$. 102
37.2. Derformation of exterior differential and Lie derivative. 103
37.3. Modifications for the case of the 1-jet bundle. 104
38. Variational λ- and μ-symmetries and the Noether Theorem for λ and μ-symmetries. 106
38.1. Case of ODE 106
38.2. The case of PDE. 107
39. Appendix:Darboux derivatives. 108
40. Nonconservative forces in Holonomic Mechanics. 111
40.1. Geometrization via affine connection. 111
40.2. Description of a mechanical system with non-conservative forces via NC-variations. 113
40.3. Fundamental quadrilateral and non-commuting variations, see more. 115
41. **Variational methods in the Nonholonomic Mechanics and Boltzmann connection.** 117
42. **Gauge transformations, torsion and H.Kleinert's Action Principle.** 123
42.1. Connection defined by an automorphism. 123
42.2. Automorphism and the prolongation of variations. 123
42.3. **Kustaanheimo-Stiefel transformation carries the Kepler-Coulomb problem to the Harmonic Oscillator** 126
42.4. **Motion of a point in Cartan space-time (by H.Kleinert, A.Pelster, P.Fiziev).** 128

43.   **Elastic deformation of uniform materials.**   129
43.1. Elasticity in material coordinates.   129
43.2. Uniform materials, material connections.   130
44.   **Dissipative Potentials in Continuum Thermodynamics and
       the NC-variations.**   133
44.1. Rayleigh dissipative functions in Mechanics and the Dissipative
       potentials in continuum thermodynamics (see [140]).   133
44.2. Dissipative potentials in Field Theory.   133
44.3. Dissipative potentials versus NC-variations.   135
44.4. EL-systems with NC variations defining the dissipative potential. 135
44.5. Tensor K of NC-variations defined by a dissipative potential.   136
45.   **Introduction.**   139
46.   **Introduction:4-dim material space-time.**   139
47.   **Thermasy and the entropy balance as an Euler-Lagrange
       Equation.**   140
47.1. **Internal (Lagrangian) space-time picture and the principle
       of "material relativity".**   143
47.2. Energy balance law   144
48.   Interlude: History of the "material-time".   145
49.   **Introduction to the material space-time.**   147
50.   **4D kinematics of media with an inner (material) metric.**   148
50.1. Physical and Material Space-Time   148
50.2. Deformation History   149
50.3. ADM-decomposition of Material Metric, Lapse and Shift.   151
51.   Mass conservation law   153
52.   Elastic, Inelastic and Total Strain Tensors   154
52.1. Elastic Strain Tensor   155
52.2. Inelastic Strain Tensor   156
52.3. Strain Rate Tensor   157
53.   **Parameters of Material Evolution, metric Lagrangian.**   158
54.   Action, boundary term, Hooke's law.   161
55.   Euler-Lagrange Equations.   162
56.   Special cases and examples.   164
56.1. Block-diagonal metric G 164
56.2. Spacial subsystem.   165
56.3. Statical case.   165
56.4. Almost flat case.   166
56.5. Homogeneous media   166
57.   **Two examples.**   166
57.1. Modeling of Necking Phenomena in Polymers.   166
57.2. Example: Variation of Material Metric g due to the Chemical
       Degradation   167
58.   Physical and Material Balance Laws   169
59.   Energy-Momentum Balance Law and the Eshelby Tensor.   170
60.   **Aging of a homogeneous rod.**   173
60.1. Deformation, strain tensors and tensor K 173
60.2. **Unconstrained aging.**   174
60.3. **Stress relaxation.**   176

60.4.   **Creep**. 177
61.   Appendix A. Strain energy as a perturbation of the "ground state energy". 179
62.   Appendix B. Variations. 182
63.   **Conclusion.** 183
64.   **Differentialbe manifolds.** 187
64.1.   Differentible mappings of manifolds. 188
65.   **Fibre bundles.** 188
65.1.   Tangent and Cotangent bundles. 190
66.   Vector and affine bundles. 191
66.1.   Vertical bundle $V(\pi)$. 191
66.2.   Natural bundles 192
67.   Mappings (morphisms) and Automorphisms of bundles. 192
67.1.   One parametrical groups of automorphisms and the infinitesimnal automorphisms of fiber bundles. 193
68.   **Connections on the fibre bundles** (For more details, see [46, 69, 70]). 194
68.1.   Connections in the bundle tower. 196
69.   Linear connections. 196
69.1.   Linear connections. 197
69.2.   Curvature and Torsion. 199
69.3.   Metric connections and the Nonmetricity. 200
70.   **Absolute parallelism**. 200
70.1.   Non-holonomic frame and absolute parallelism. 201
70.2.   Non-holonomic (pure gauge) transformations and induced connections. 202
71.   **Automorphisms of the vertical bundle and their prolongation.** 203
72.   Introduction. 205
73.   Jet bundle $J^1(\pi)$. 205
73.1.   **The 1-jet bundle of a fibre bundle $\pi : Y \to X$.** 205
74.   Higher order jet bundles $J^k(\pi)$. 207
74.1.   Infinite jet-bundle $J^\infty(\pi)$. 209
74.2.   **Total derivatives**. 209
75.   Contact structure on the k-jet bundles. 210
76.   Prolongation of vector fields to the jet bundles. 213
77.   Connections on the 1-jet bundle $\pi_{10} : J^1(\pi) \to Y$. 217
77.1.   Vertical connections. 220
77.2.   Connections in the infinite tower $J^\infty / \ldots / X$ 221
78.   **Lie groups actions on the jet bundles and the symmetry groups of (systems of) differential equations.** 224
79.   **Symmetries of Lagrangian and the first Noether Theorem.** 225
79.1.   Symmetries and infinitesimal symmetries of the Lagrangian Action. 225
79.2.   Generalized vector fields and the symmetries of systems of differential equations. 227
80.   **Symmetries and Noether conservation laws.** 227
81.   Noether balance laws. 229
82.   Conclusion. 231
References 231

1. Preface

The aim of this text is to present and study the method of so-called "**non-commuting variations (shortly, NC-variations)**" in Variational Calculus. To present this method we recall one of the basic rules of Variational Calculus - the rule defining the variations of derivatives $\frac{\partial y^{\mu}}{\partial x^{i}}$ of dynamical variables $y^{\mu}(x)$ (fields in the Field Theory) corresponding to a variation ξ of dynamical variables (fields): "variation of a derivative equals to the derivative of variation". In Mechanics, this rule takes the form $\delta \dot{y}^{\mu} = \frac{d}{dt}\delta y^{\mu}$. In Classical Field theory this rule takes the form

$$\delta \frac{\partial y^{\mu}}{\partial x^{i}} = \frac{\partial}{\partial x^{i}}\delta y^{\mu}.$$

This rule can be formulated as follows: "taking of variations of dynamical variables $y^{i}(x)$ commute with the taking of derivatives."

This rule was universally adopted in the XVIII and XIX centuries but, as early as in 1887, this rule was questioned by Vito Volterra, see [130, 131]. Studying non-holonomic mechanical systems, V.Volterra noticed that the use of the conventional rule of defining variations of derivatives does not allows us to obtain equations of motion for non-holonomic systems by variational methods. Further developments including works of L.Boltzman,[8] G.Hamel,[57], T.Levi Civita and U.Amaldi,[83] led to the conflicting points of view at the range of applicability of the conventional rule of defining variations (see a Historical Review between Chapters 1 and 2 below). Finally, the status of this, conventional, rule and its relation to the alternative rules - the use of "non-commuting variations" in Non-Holonomic Mechanics were clarified in works of J.Neimark and his coauthors in the 1950s of XX century ([104, 105]) and by A.Lurie in 1961, [88].

Later on, the non-commuting variations were used in the works of B.Vujanovich and T.Atanackovic on dynamical systems with non-conservative forces ([133, 134, 4, 5]), in Elasticity Theory, and in works of H.Kleinert, P.Fiziev and A.Pelster on the dynamics in Cartan-Riemann spaces ([35, 65]).

While studying the application of non-commuting variations in classical field theory we noticed that the usage of non-commuting rules to define variations of derivatives **is equivalent to the use of a non-trivial vertical connections** to modify the procedure of flow prolongation of variational vector fields in the space Y of the configurational bundle $\pi : Y \to X$ of a physical system to the 1-jet bundle $J^{1}(\pi) \to Y$ over π, [112]. This led us to the study of the geometrical structures underlying the method of non-commuting variations of derivatives in Lagrangian formalism. In particular,a natural variety of questions that arises here is: which of the basic methods of Variational Calculus - Theory of second variations, Hamiltonian systems and Legendre transformation, conservation laws (including Noether theorems), Hamilton-Jacoby Equations, etc. - are preserved in this modified scheme and which parts require modifications to stay true. These and some other related questions are studied in the present work.

We will also show that any system of PDE of the form: "Euler- Lagrange equations with sources"

$$E_{j}(L) = f_{j} \tag{1.1}$$

can be realized by the Lagrangian formalism with a conventional action functional $\mathcal{A}(L)$ and the non-commuting variations defined by an appropriately chosen (defined by the sources f_{j}) tensor K of NC-variations. We show that the basic methods of conventional Lagrangian formalism - Noether Theorem, second variation technique, Hamiltonian equations, Weyl fields preserve their form in Lagrangian formalism with NC-variations. We study the relations between the properties of sources f_j and the curvature $\tilde{R}$ of the vertical connection tensor K.

We demonstrate that a variety of geometrical structures that appeared in the study of dynamics in some physical systems - dissipative potentials, non-holonomic transformations, torsion of zero curvature connections (absolute parallelism), material time and thermasy (= heat displacement), introduced by H.Helmholtz and studied by D.van Dantzig ([135, 136, 110]) are special cases or are closely related to the use of non-commutative variations defined by a vertical connection in the conventional Lagrangian formalism.

Our perspective in this work, supported by the results of the geometrical (bundle) form of Variational Calculus, is that the conventional rules of taking variations of the derivatives of dynamical variables (fields) (underlying the flow prolongation method) have important mathematical advantages (preservation of Cartan distribution, preservation of Lie bracket, etc.) making them more fundamental. Yet, a more general approach allows inclusion into the framework of Variational Calculus, the physical systems that can not be described by the conventional Lagrangian formalism.

In that we adopt the point of view of B. Vujanovich and that of A.Lurie's that in difference to the variations of fields y^i, variations of their derivatives $y^i_{,\mu}$ *are not only kinematical, but dynamical notions* and should be dealt with as such. In particular, this allows us to introduce geometrical factors that have dynamical meaning into the definition of variations of derivatives that have dynamical meaning. This allows us to describe dissipative processes in the system.

These notes are based on the Lectures delivered by author at the 15th Summer School in Global Analysis at Masaruk University,Brno, CZ on August 8-12, 2011. I am using this case to thank participants of this school and, especially, its organizer Professor Demeter Krupka and Dr.Marcella Palese for useful discussions during the school.

In particular, during this school Marcella Palese informed me about a non-conventional procedures of the non-commuting variations introduced by C.Murial and J.Romero in Spain and used by the group of specialists in Spain and Italy with the goal to extend the range of symmetry groups of Lagrangian systems. Their goals were different from ours but their constructions (of λ and μ-prolongations) are similar, but not identical, to our approach of using vertical connections. We have included a condensed exposition of their work and the relations with our scheme in the present text (see Chapter 4).

Preliminary results on the NC-variations Lagrangian formalism where published in the Proceedings of the GCM2008, [112].

2. Introduction.

In Chapter I we give a short sketch of classical Lagrangian formalism. Here we tried to make a presentation as simple as possible. Yet, we introduce in the beginning of Chapter 1, some basic invariant notions, whose more detailed description reader can find in the Appendix (and, in more details, in the literature refered there to. We define the configuration bundle, $\pi : Y \to X$, one-jet bundle $J^1(\pi)$ as the domain of Lagrangian functions and the total derivatives d_μ on the space $J^1(\pi)$ used to write down the Euler-Lagrange equations in an invariant way.

In Chapter 2 we define the non-commutative variations for an action functional of a Lagrangian $L(x^\mu, y^i, y^i_{,x^\mu})$ of the first order. Non-commutative variations are defined using tensors $K^\mu_{i\nu}$ in the configurational space Y. Variations of derivatives defined with the help of tensor K will be called **K-twisted variations** and the Euler-Lagrange equations obtained using variations defined this way will be called K-twisted EL-equations.

We get the Euler-Lagrange equations

$$EL(L)_\mu = f_\mu, \ \mu = 1, \ldots, m$$

with the sources f_j defined by a "tensor K", formulate corresponding Noether Theorem (proved in Appendix III), present the canonical Energy-Momentum balance law.

A variety of examples of $EL + NV$ systems and classes of such systems are presented here.

Using Legendre transformation we construct corresponding Hamilton equations with sources and compare them with the "metriplectic or "double bracket" systems.

Then we show the form taken by the second variation formalism (sufficient conditions, Jacoby equation, etc.) in the case of NC-variations. At the end of this Chapter we show that this approach to the Lagrangian formalism can be readily extended to the higher order Lagrangian problems and to the case of "degenerate Lagrangians" where the source terms are of higher order then the Lagrangian itself.

In Chapter 3, we show that the procedure of K-twisted prolongation of a variation $\xi = \xi^i \partial_{y^\mu}$ of dynamical fields y^μ to the 1-jet bundle $J^1(\pi)$ is lacking two basic properties of the conventional flow prolongations of variational vector fields: conservation of Lie vector fields brackets and preservation of Cartan distribution in the 1-jet space. While the second property is valid only if tensor K vanishes, obstruction to the preservation of the Lie brackets is determined. It consists in two parts - curvature form tensor $\tilde{R}$ and the "skew-symmetric bracket" presenting the deformation of the Lie bracket of the vector fields.

Next, we show that the tensor K defining the Non-Commuting Lagrangian formalism has the form of the vertical component of an (Ehresmann) connection ω on the affine bundle $\pi_{10} : J^1(\pi) \to Y$ - the component responsible for the term of the form $a^i_\mu \partial_{y^i_\mu}$ of the K-vertical lift of a vector field $\xi = \xi^\mu \partial_{x^\mu} + \xi^i \partial_{y^i}$.

More then this, vertical/vertical component of the curvature $R(\omega)$ coincide with the tensor $\tilde{R}$ mentioned above. It is shown that one can define the covariant flow prolongation of vector fields from Y to $J^1(Y)$ so that the K-twisted prolongation of a vector field coincide with the modified by K flow prolongation.

In Sec.20, we consider the case where tensor K does not depend on the derivatives of dynamical fields $K^i_{\mu j} \in C^\infty(Y)$. We calculate different quantities characterizing K-twisted prolongations for this case and study the relation between the form of source terms, f_j, and the properties of the "curvature" $\tilde{R}$.

In Chapter 4 we present a short description of works of the group os spanish and italian mathematicians developed the Theory of twisted prolongations of vector fields to the jet bundles in many respects similar to our scheme. Their works had different goal - to construct, using the twisted prolongations of vector fields, alternative classes of symmetry groups of differential equations and systems of differential equations. their "theory of λ and μ -prolongations and symmetries has an important property - vector fields obtained by the prolongations to the jet bundles preserves, in some modified sense, the Cartan distributions and contact formes. This property has an elegant form and probably can be useful in further development of Geometrical Theory of Differential Equations.

In Chapter 5, we discuss several situations when the non-commuting variations were used explicitly or implicitly in the variational description of some physical systems. For some time a "geometrization" of a mechanical system, i.e., presentation equations of motion of such a system as the geodesic motion with respect to some linear connection in the configurational space Q of this system was a very popular problem in Mechanics. In Sec.40, it is shown, following the work of B.Vujanovich,[134, 133] that the same result can be achieved without changing the geometry of the space Q but, instead, by using conventional Lagrangian of this system and redefining the variations of velocities in the tangent space $T(Q)$. In Sec.41, the short review of variational approach to the non-holonomic mechanical systems is presented. Using the approach of L.Boltzman we construct the equations of motion in non-holonomic systems with line non-holonomic relations. We notice that the Bolzmann tensor defining the non-commutativity tensor K of the variations is defined here by the torsion of the zero curvature connection corresponding to the non-holonomic frame (see Appendix I, Sec.66).

In Sections 42,43 we present the use of non-holonomic (gauge) transformations for constructing Variational principle with non-commuting variations defined by the torsion of the (absolute parallelism) connection given by this transformation. First example of such scheme (see Section 42) is the one that was developed by H.Kleinert and his collaborators P.Fiziev and A.Pelster [35, 65] to describe Mechanics in spaces with metrics and connections (Cartan spaces). In Section 43 we present a short review of properties of **Uniform Materials**. Uniform materials were defined by K.Kondo (1955) and developed by a variety of specialists including E.Kroner, B.Bilby, C.C.Wang, C.Truesdell in 60th of XX century and by many specialists later on. We refer to the monographs [22, 137] for the detailed description of "Uniform materials"" theory.

In Sec.44, we discuss the relation between the method of non-commuting variations and the use of **dissipative potentials** (special case of which are Rayleigh dissipative function) in Lagrangian formalism.

In Chapter 6, we present an application of Lagrangian formalism with the NC-variations to the description of irreversible evolution of a continuous media with heating and structural changes.

In Sec. (46) we introduce thermasy, a scalar variable introduced by H. Helmholtz and later on, used by D.van Dantzig in his study of thermodynamics of moving matter, see [135], A.Green and P.Nahdi in thermoelasticity, see [53, 54] and G.Maugin and V.Kalpakides in Continuum Thermodynamics, [92]. Using thermasy, whose time derivative is absolute temperature one can formulate **entropy balance of a thermodynamical system as the Euler-Lagrange Equation**. We present modified and simplified version of this variational system and write down corresponding energy balance and the heat propagation equation that has the form of Cattaneo heat propagation law, see [100].

Then we introduce the model of material metric space-time (P, G) that was introduced by A.Chudnovsky and the author in order to model the aging processes in the materials, [15, 16]. In this model, evolution of the material is presented by smooth embedding of the material space-time into the Galilean space-time and the material metric G describes the structural properties of material. In particular, the rate S of the proper (=material) time τ relative to the physical type: $d\tau = Sdt$ is the characteristic of the entropy production in the material (if entropy production is zero, $S = 1$). We show that the entropy balance in a thermodynamical system obtained as the Euler-Lagrange for thermasy using the NC-variations defined by the rate of material time S coincide with the Euler-Lagrange Equation for thermasy **obtained using the material time τ instead of physical time t and conventional variations instead of NC-variations**.

This duality shows that the usage of NC-variations allows us to model complex irreversible phenomena that is impossible to do using conventional Lagrangian approach.

Appendix:

In the Appendix I we present a short review of geometrical notions used in the text: manifolds, fiber bundles, connections and their curvature, linear connection and its rorsion, prolongation of vector fields from Y to the jet bundles, absolute parallelism. In appendix II we define jet bundles, their mappings, total derivatives, contact structure of jet bundles, connections in jet bundles, Lie vector fields, properties of vertical connections. In Appendix III we define the symmetries and infinitesimal symmetries of the differential systems and the Lagrangian action, define the Noether formal;ism and probe the the First Noether Theorem. In the case od Euler-lagrange equation with sources, Noether equations corresponding to the symmetry Lie groups are **balance equations** rather then the conservation laws. This referees, in particular, to the energy-momentum balance laws.

Part I. Non-commuting variations - elementary topics.

Chapter 1. Basics of the Lagrangian Field Theory.

In this Chapter, we introduce the basic notions of Classical Lagrangian Field Theory of the first order - configurational bundle, action functional, Euler-Lagrange equations in the volume necessary in the main text. We will keep this presentation as short as possible for two reasons: the first reason is that this, classical material is well presented in a variety of well known sources - see [44, 118] for a classical introduction to the Variational Calculus. The second reason is that we prefer to introduce some notions (second variation, Hamilton-Jacobi equation) in the sections where we can compare them with the form they take in a case of non-commuting variations. For more advanced exposition of geometrical structure of classical field theory, including Lagrangian Field Theory we refer to the sources [33, 45, 46, 47, 106].

3. CONFIGURATIONAL BUNDLE , 1-JET BUNDLE AND THE LAGRANGIAN ACTION.

3.1. Configurational bundle (Y, π, X). Dynamical variables of Classical Field Theory typically appear to be tensor or tensor density fields defined in the domain of a physical, material or mezoscopic space-time (for the last one, see [96] and other works of W. Muschik).

To organize these fields and their derivatives into a natural geometrical picture it is convenient to introduce the **configurational fiber bundle** - a triple (Y, π, X) with the base space $X, \dim X = n$ and the total space $Y, \dim Y = n + m$, where the smooth mapping $\pi : Y^{n+m} \to X^n$ is onto and of constant rank (see Appendix I for a short introduction of geometrical and topological notions used here). For most situations studied in this book, it is sufficient to assume that the space Y is the product of the base X and the typical fiber F and that both X and F are either Euclidian vector spaces of dimension n and m respectively, or open domains in Euclidian spaces of corresponding dimension. In the general case, spaces X, Y, F are differentiable manifolds, sometimes, with the boundary. See Appendix I where these notions are defined and examples of manifolds are presented.

For a point $x \in X$, the sub-manifold $Y_x = \pi^{-1}(x)$ is called the fiber over x. Fibers Y_x of the bundle π are assumed to be connected m-dim manifolds diffeomorphic to the fixed manifold F^m (see Appendix I).

We will be using local coordinates in the bundle Y, adjusted to the bundle structure - "fibred charts" $(V, x^i, i = 1, \ldots, n; y^\alpha, \alpha = 1, \ldots, m)$ in Y defined in a domain $V \subset Y$. Here x^i are coordinates in the domain $\pi(V) \subset X$ and y^α are fiber coordinates in the fibers Y_x.

Below we will be using shortened notations for partial derivatives in the fiber charts: ∂_i for $\frac{\partial}{\partial x^i}$, and ∂_μ for $\frac{\partial}{\partial y^\mu}$.

A smooth mapping $s : U \to Y$ where U is an open subset of X is called the **section of the bundle** π if the composition of projector π and mapping s is the identity mapping of the domain U: $\pi \circ s = id_U$. In applications (in Physics and in Continuum Mechanics), functions $y^\alpha(x)$ represent the components of tensor fields which are the dynamical variables of the considered theory.

If the configurational bundle, π, is a *vector bundle* (see Appendix I), the fibers F_x are real vector spaces (isomorphic to the fixed vector space F^m) and the transition

© Springer International Publishing Switzerland 2016
S. Preston, *Non-commuting Variations in Mathematics and Physics*,
Interaction of Mechanics and Mathematics,
DOI 10.1007/978-3-319-28323-4_1

transformations (see Appendix I) are the linear mappings between the fibers at each point of the intersection of the domains of coordinate charts.

Quite often the configuration bundle is trivial so that the space Y is the product $Y = X \times F^m$ of the base X, and the standard fiber F^m and the bundle projection $\pi : Y = X \times F \to X$ is the projection onto the first factor.

Remark 1. Spaces (smooth manifolds) that appear in classical physics are *often* of one of the following types:

(1) An open subset $X \subset R^n$ or an open subset with the boundary ∂X,

(2) A k-dimensional submanifold $X^k \subset R^n$ with the boundary ∂X. A 2-dim surface with the boundary in the Euclidian space E^3 is an example.

(3) Compact manifolds - spheres, torus, etc.

Most base manifolds used in classical physics are of the first and second type. Manifolds of the third type usually appear in the problems of Geometry and of Gauge Field Theory, see [111]. Quite often X is the physical or material space-time and $n = 4$.

It is assumed that **the base manifold X is endowed with the Riemannian or pseudo-Riemannian metric** G. We denote by dv the volume n-form corresponding to the metric G. In local coordinates (U, x^i), metric G is presented by the non-degenerate symmetrical (0,2)-tensor field $G = G_{ij}(x)dx^i dx^j$ and the volume form dv has the form

$$dv = \sqrt{|det(G_{ij}(x))|}dx^1 \wedge \ldots \wedge dx^n.$$

3.2. **First orders Lagrangians and the Action functional.** A **Lagrangian Field Theory of order 1 with the configurational bundle** $\pi : Y^{n+m} \to X^n$ is defined by a Lagrangian - a function L on the **first jet bundle** $J^1\pi$ **of the configurational bundle** : $L \in C^\infty(J^1(\pi))$. The 1-jet bundle space $J^1\pi$ is fibred over Y and X $(J^1\pi \to Y \to X)$. Jet bundle $J^1\pi$ carries, in its fibers over Y, the information about the first derivatives $y_i^\alpha = \frac{\partial y^\alpha}{\partial x^i}$ of the components y^α of the sections $s = \{y^i(x)\}$ of configurational bundle $\pi : Y \to X$ (see Appendix II).

We refer to Appendix II for the definition and basic properties of the 1-jet bundle of a fiber bundle. We remark that a fibred chart (V, x^i, y^α) in the configurational bundle $\pi : Y \to X$ defines the "lifted fibred chart" $(V^1, x^i, y^\alpha, y_i^\alpha)$ in the domain $V^1 = \pi_{10}^{-1}(V)$ of the 1-jet bundle $J^1(\pi)$

Remark 2. Mathematically, it is more natural to define Lagrangian as an n-form $\lambda = L(x^i, y^\alpha, y_i^\alpha)dv$. Here we mostly use the simple definition that is closer to the applications in physics.

Let $L \in C^\infty(J^1(\pi))$ be a Lagrangian of the first order and let $D \subset X$ be a domain in the base space X.

The **action functional** $\mathcal{A}_D(s)$ on the sections $s : D \to Y$ $(D \subset X$ being a domain in X) is defined by the integral

$$\mathcal{A}_D(s) = \int_D L(j^1 s(x))dv = \int_D L(x^i, y^\alpha(x), y_{,x^i}^\alpha(x))dv. \tag{3.1}$$

The **main postulate of Lagrangian Field Theory** is that the configurations of dynamical fields $s(x) = \{y^\alpha(x)\}$, realized in real situations, are the **critical**

points of the action functional. In many cases, real solution $s(x)$ delivers an extremum to the action functional between all configurations (geometrically - sections s of the bundle π that satisfy some additional conditions (boundary conditions, initial conditions, Lagrange conditions, etc).

In the next section we sketch the basic components of the formalism of Calculus of Variations - we introduce first variations and Euler-Lagrange Equations for the first order theory.

4. First Variation and the Euler-Lagrange system.

Consider a Lagrangian Field Theory of the first order with a Lagrangian $L(x^i, y^\alpha, y_i^\alpha)$, where $L \in C^\infty(J^1(\pi))$ is an infinitely differentiable function on the 1-jet space. Let $s(x) = \{y^\alpha(x)\} : U \to Y$ be a section of the configurational bundle π that delivers an extremum (minimum or maximum) of the action functional $A_U(s)$ (see (3.1)). The extremal property of the section s is determined by comparing the value of the action functional at the section s with its action at neighborhood sections $s(x) + \epsilon\xi(x)$. Thus, we need to introduce variations of a section s in a convenient form.

4.1. **First Variations.** Let $s : U \to Y$ be a section of the the configurational bundle $\pi : Y \to X$ defined in a open subset $U \subset X$. In local fibred coordinates, section $s(x)$ is presented by its components $s(x) = (s^\alpha(x^k), \ \alpha = 1, \ldots, m)$.

Definition 1. (1) *A variation of section $s : D \to Y$ is the one-parameter family $t \to s_t(x)$ of local sections of the configurational bundle (defined in D) such that $s_0 = s$.*

(2) *Infinitesimal variation, corresponding to the variation $s_t(x)$ is the π-vertical vector field $\xi = \xi^\mu \partial_\mu$ along the image of section s (i.e. defined at the points $s(D) \subset Y$).*

Here and below, ∂_μ is the shortened notation for the partial derivative

$$\partial_\mu = \frac{\partial}{\partial y^\mu}.$$

$$\xi(s(x)) = \frac{d}{dt}\Big|_{t=0} s_t(x). \tag{4.1}$$

Vice versa, let $\xi = \xi^\mu(x, y)\frac{\partial}{\partial y^\mu}$ be a vertical vector field defined in the domain $V \subset Y$ of a fibred chart (V, x^i, y^μ) and let ϕ^t be the phase flow of this vector field. Then, the association $t \to \phi^t(s(x))$ is the variation of a section $s(x)$.

Yet, in order to define variations of action $A_D(s)$ (see (3.1)), we have to define values of derivatives $\frac{\partial s^\alpha}{\partial x^i}$ for variated sections $s_t(x)$. A natural way to do this (called the flow prolongation) is to define variations of sections $s(x)$ by the phase flow of a vertical vector field $\xi(x, y)$, then **extend (lift) this flow to the 1-jet space** $J^1\pi$ and apply obtained 1-parametrical (local) group of transformations to the 1-jet $j^1s(x)$ of section $s(x)$.

Let a variation of section s, given by the collection of components $s^\alpha(x)$ (dynamical fields), be defined by a π-vertical vector field $\xi = \xi^\alpha \partial_{y^\alpha} \in V(\pi)(U^0)$ (where $U^0 = \pi^{-1}U \subset Y$). In terms of components, an infinitesimal variation of section $s(x)$ has the form $s(x) = \{s^\alpha(x)\} \to \{s^\alpha(x) + \xi^\alpha(x, y)\epsilon\}$ for small ϵ.

Corresponding **variations of derivatives** entering the Lagrangian L and, therefore, the action functional

$$A_D(s) = \int_D L(j^1 s)dv \qquad (4.2)$$

are determined by a **prolongation procedure** lifting vector fields ξ in Y to the vector fields in the 1-jet space $J^1(\pi)$.

$$\xi = \xi^\alpha \partial_\alpha \to \xi^1 = \xi + \xi_i^\alpha \partial_{y_i^\alpha}.$$

Remark 3. Flow prolongation procedure $\xi \to Pr^1(\xi)$ has the form (see Appendix II for more details)

$$Pr^1(\xi) = \xi^\alpha \partial_\alpha + \xi_i^\alpha \partial_{y_i^\alpha} = \xi^\alpha \partial_\alpha + d_i \xi^\alpha \partial_{y_i^\alpha}. \qquad (4.3)$$

Here $\xi_i^\alpha = d_i \xi^\alpha = \partial_i \xi^\alpha + y_i^\beta \partial_\beta \xi^\alpha$ is the **total derivative** of the coefficients $\xi^\alpha \in C^\infty(J^1(\pi))$ by x^i. The second term in the expression (4.3) can be interpreted as the analog of *the derivative by x^i of the variation of the dynamical field y^α*, a component of $d\delta y^\alpha$.

The 1-jet component $\{\xi_i^\alpha\}$ of the prolonged vector field in $J^1(\pi)$ represent the variation of the derivative - $\delta y_{,i}^\alpha$.

The conventional rule "variation of the derivatives is equal to the derivative of variation" $\delta dy^\alpha = d\delta y^\alpha$ now takes the form of the following condition for a variational vector field $\xi = \xi^\alpha(x^i, y^\alpha, y_i^\alpha)\partial_{y^\alpha} + \xi_i^\alpha \partial_{y_i^\alpha}$:

$$\xi_i^\alpha(z) = \delta y_i^\alpha = d\delta y^\alpha = d_i \xi^\alpha. \qquad (4.4)$$

Thus, **the commutativity rule of prolongation of variations to the first jet bundle is equivalent to the statement that the π-vertical variations $\xi = \xi^\alpha \partial_\alpha$ of dynamical fields $y^\alpha(x)$ are prolonged to the 1-jet space by the flow prolongation**.

This rule of prolongation, basic in the Variational Calculus with one-dimensional base (Mechanics), is expressed by the relation $\delta \dot{y}^\alpha = \dot{\delta y}^\alpha$, [44]. The same property is basic in the Lagrangian Field Theory, [44, 45, 106].

It is this rule that has been challenged in the works cited in the Introduction. So, in order to present the methods of B.Vujanovic, H.Kleinert and their coauthors in the geometrical form we have to study a natural modifications of the prolongation procedure (4.3). This includes the prolongation of the vector fields on the base X to the jet spaces which is imperative, for instance, for the study of symmetries of Euler-Lagrange equations and corresponding Noether *balance laws*.

Remark 4. Unfortunately, modifying the rule of variations of jet bundle variables, one has, in general, to sacrifice some properties of these variations that are taken for granted in the Variational Calculus. In Ch.3, Sec.18 we show that the basic properties of the flow prolongation - *Lie algebra morphism* and the preservation of the Cartan distribution are generically lost in the modified picture of lifting $\xi \to Pr_K^1(\xi)$. We also introduce the geometrical structures responsible for this loss and, in some sense, characterizing it. In Chapter 4 we describe the modification of Cartan distribution preservation property suggested by E.Pucci and G.Saccomandi ([115],). Their condition can be applied to a large class of prolongation procedures defined by the C.Muriel and J.Romano for ordinary differential equations and extended by the group of italian mathematicians to the systems of partial differential equations (see Chapter 4 and references therein).

4.2. **Euler-Lagrange Equations and natural boundary conditions.** Form the variation of the action $A_D(s)$ using this **flow prolongation** (4.3) of a vertical variational vector field $\xi = \xi^\alpha(x,y)\partial_\alpha$. Infinitesimal variations of fields and their derivatives (independent variables x^μ are not varied) have the form

$$\begin{cases} y^\alpha \to y^\alpha + \epsilon\xi^\alpha(x,y), \\ y_i^\alpha \to y_i^\alpha + \epsilon d_i\xi^\alpha. \end{cases}$$

Calculate the first variation of action:

$$\Delta A_D(s)(\epsilon\xi) = A_D(s_\epsilon) - A_D(s) =$$

$$= \int_D [L(x, s^\alpha(x) + \epsilon\xi^\alpha(x, s(x)), s^\alpha_{,i}(x) + \epsilon(d_i\xi^\alpha(x, s(x)))) - L(x, s(x), s^\alpha_{,i}(x))]dv =$$

$$= \epsilon \int_D \left[\frac{\partial L}{\partial y^\alpha}\xi^\alpha(x, s(x)) + \frac{\partial L}{\partial y_i^\alpha}(d_i\xi^\alpha(x, s(x))) \right] dv + O(\epsilon^2) =$$

$$= \epsilon \int_D \left[\frac{\partial L}{\partial y^\alpha} - d_i\left(\frac{\partial L}{\partial y_i^\alpha} \right) \right] \xi^\alpha(x, s(x))dv + \epsilon \int_{\partial D} \frac{\partial L}{\partial y_i^\alpha}\xi^\alpha n_i dS + O(\epsilon^2) \quad (4.5)$$

First variation of the action $A_D(s)$ at a section $s(x) = (x^i, y^\mu(x))$ has the form

$$\delta A_D(s) = \int_D [\frac{\partial L}{\partial y^\alpha} - d_i\left(\frac{\partial L}{\partial y_i^\alpha} \right)]\xi^\alpha(x, s(x))\epsilon dv + \epsilon \int_{\partial D} \frac{\partial L}{\partial y_i^\alpha}\xi^\alpha n_i dS. \quad (4.6)$$

where the boundary term $\int_{\partial D} \frac{\partial L}{\partial y_i^\alpha}\xi^\alpha dS^{n-1}$ appears after integrating by parts.

If the variational vector field $\xi = \xi^\alpha\partial_\alpha$ vanishes on the boundary ∂D, the last term in the previous expression for the variation vanish and the arbitrariness of the variations ξ^α lead to the system of Euler-Lagrange equations in the form

$$E_\beta(L) = \frac{\partial L}{\partial y^\beta} - d_i\left(\frac{\partial L}{\partial y_i^\beta} \right) = 0, \; \beta = 1, \ldots, m. \quad (4.7)$$

Here $d_i = \frac{\partial}{\partial x^i} + y_i^\mu\frac{\partial}{\partial y^\mu}$ is the total derivative by x^i (see Appendix II).

If, having this equation, we omit the condition that the variational vector fields ξ vanish on the boundary ∂D of the domain D together with its normal derivatives, we extend the class of admissible variations. If, in such a case, we integrate by parts in the expression for variation of action (4.2), the boundary integral in the sum (4.6)

$$\int_{\partial D} L_{,y_i^\alpha}(s)\xi^\alpha dS_i$$

appears to be nonzero.

. As a result, using at first, the variational vector fields that vanishes on the boundary ∂D, and then the general variations we get, in addition to the **Euler-Lagrange equations (4.7)**, the **natural boundary conditions**

$$L_{,y_i^\alpha}(s) \cdot n_i = \pi_\mu^i n_i = 0, \; \alpha = 1, \ldots, m, \quad (4.8)$$

where $n_* = \{n_i\}$ is the covariant vector corresponding to the unit normal vector n^i on the boundary ∂D and $\pi_\mu^i = L_{,y_i^\alpha}(s)$ is the momentum (1,1)-tensor

. Thus, the natural boundary condition requires that the normal component of the momenta vanish.

4.3. Symmetries and Noether Theorem.

Let $L(x^i, y^\mu, y_i^\mu)$ be a first order Lagrangian and let $G \subset Diff(Y)$ be a finite-dimensional Lie group of diffeomorphisms of the space Y that is, at the same time, the group (geometrical) symmetries of the Lagrangian L (see Appendix III, Sec. 77) where definitions and properties of groups of **variational** and **divergent** symmetries are presented). In particular, A Lie group G of (diffeomorphic) transformations of the manifold Y is a **group of divergent symmetries of Lagrangian** L if the infinitesimal condition (78.7) (see Appendix III, Sec.78) is fulfilled.

Locally, in terms of fibred coordinates (x^i, y^μ), transformations ϕ of the space Y, corresponding to the elements $g \in G$, have the form

$$\phi : (x^i, y^\mu) \to (\bar{\phi}^i(x^j, y^\nu),), \phi^\mu(x^j, y^\nu)), \tag{4.9}$$

with $\bar{\phi}^i, \phi^\mu(x, y)$ being smooth functions of corresponding variables.

An important special case of geometrical transformations is the case of **projectable** transformations ("automorphisms of the bundle $\pi : Y \to X$"). In fibred coordinates, projectable transformations, are characterized by the condition $\bar{\phi}^i = \bar{\phi}^i(x^j)$ in (4.9).

Infinitesimally, Lie algebra $\mathfrak{g}$ of the group G of geometrical transformations is formed by the vector fields in Y, $\zeta = \zeta^i(x, y)\partial_{x^i} + \zeta^\mu(x, y)\partial_{y^\mu}$, while the Lie algebra of a group of automorphisms of π formed by the projectable vector fields $\zeta = \zeta^i(x)\partial_{x^i} + \zeta^\mu(x, y)\partial_{y^\mu}$. In the case of projective transformations, transformations $\phi \in G$ generate (by taking projections) the group $\bar{G}$ of transformations $\bar{\phi}$ of the base X.

Notice that the defining property of a one-parameter group ϕ^t to be the group of symmetries of Lagrangian L is that the action of fase field transformations ϕ^t, $t \in R$ transform solutions of Euler-Lagrange Equations to the other solutions of the same Euler-Lagrange system.

Therefore, infinitesimal (phase) fields of such one-parameter groups act as the infinitesimal variations of the action $A_D(s)$ corresponding to L. Thus, in order to realize this symmetry of the action $A_D(s)$, geometrical symmetries (acting in Y) should be lifted to the 1-jet bundle $J^1(\pi)$ by the same procedure $\xi \to Pr^1(\xi)$ as the variational vector fields.

Let $\zeta \in \mathfrak{g}$ be an arbitrary element of Lie algebra $\mathfrak{g}$. Let $Pr^1(\zeta)$ be the flow prolongation of vector field ζ to the 1-jet bundle $J^1(\pi)$. Then, the vector field ζ is the **infinitesimal divergent symmetry** of Lagrangian, that generates the (at least local) one-parameter group of symmetries of L - phase flow) if and only if there exists a horizontal 1-form $B = B_k(x^i, y^\mu, y_i^\mu)dx^i$ in $J^1(\pi)$ such that

$$pr^1(\zeta)L + Ldiv(\bar{\zeta}) = div(B), \tag{4.10}$$

see (80.90) or [106], Chapter 4 for more details.

Divergence here is defined using the volume form dv defined by the metric g in the base X.

Now we formulate the Noether Theorem for the variational and the divergent symmetry groups. See Appendix III for the proof of this Theorem.

Theorem 1. *Let L be a Lagrangian of order k and let*

$$E_\alpha(L) = 0, \ \alpha = 1,\ldots, m, \qquad (4.11)$$

be an Euler-Lagrange system with the Lagrangian L of order k. Let ξ be a vector field in Y i an infinitesimal variational symmetry of Lagrangian L.

Let $Pr^k(\xi)$ be the prolongation of vector field ξ of order k - to the k-jet bundle $J^k\pi$. Then there exist an n-tuple of the smooth functions P^i such that for some functions $A \in C^\infty(J^k\pi$ the following equality (Noether conservation law) is fulfilled

$$Div(A + L\xi) = -Q^\mu E_\mu(L). \qquad (4.12)$$

As a result, for all solutions y of the Euler Lagrange system of equations,

$$Div(P)(y) = 0. \qquad (4.13)$$

For the proof of this theorem, see [106], Ch.4 or here, Appendix III, Section 79. par

If a Lie group acting on the space Y is a group of *divergent symmetries* (see Appendix III), relation (4.12) is replaces by the relation

$$Div(A + L\xi) = +Q^\mu E_\mu(L) = Div(B), \qquad (4.14)$$

for the n-tuple of functions $B_i \in C^\infty(J^k\pi)$.

As a result, conservation law (4.13) in Theorem 1 for the solutions of Euler Lagrange system holds with the following modification:

$$P = B - A - L\xi. \qquad (4.15)$$

corresponding **Noether conservation law** for the solution of Euler-Lagrange equations has the form (comp. (79.7).)

$$Div(P) = 0, \text{where } P^i = \zeta^\mu L_{,y_i^\mu} + \zeta^i L - \zeta^j y_j^\nu L_{,y_i^\mu} - B. \qquad (4.16)$$

As an example, illustrating the Noether method, to associating conservation (or balance) laws to the symmetry Lie groups of transformations, we present the Stress-Energy-Momentum balance law for the Lagrangian Field Theory.

4.4. **Energy-Momentum balance law.**

Here we write down the canonical stress-energy-momentum (CEM) balance law corresponding to a Lagrangian $L \in C^\infty(J^k(\pi))$. We will be using the approach of [106], modified to produce the balance law, see Appendix III.

We take the base space $X = R^4$ to be the physical space-time endowed with standard Euclidian or standard Lorentz metric.

In this case $\xi = \xi_k = \partial_{x^k}$ lifted to Y by a connection Γ in the bundle π: $\partial_{x^k} \to \hat{\xi}_k = \partial_{x^k} + \Gamma_k^\mu \partial_\mu$. Then, the characteristic of the vector field $\hat{\xi}_k$ has the components $Q^\mu = \Gamma_k^\mu - y_k^\mu$ and we define the Energy-Momentum Tensor in its standard form [80].

$$T_k^i = L\delta_k^i - (\Gamma_k^\mu - y_k^\mu)L_{,y_i^\mu}. \qquad (4.17)$$

As a result the balance equation 81.4 corresponding to this vector field (*stress-energy-momentum "balance" law*) has the form

$$d_i T_k^i = d_i\left(L\delta_k^i - y_k^\mu L_{,y_i^\mu}\right) = -\frac{\partial L}{\partial x^k}\bigg|_{expl.}, \ k = 0, 1, 2, 3. \qquad (4.18)$$

If L does not depend explicitly on x^k, this balance law becomes the "conservation law".

In particular, for $k = 0$ we get the energy balance law (or conservation law if $\frac{\partial L}{\partial x^0}_{expl.} = 0$)

$$d_i T_0^i = d_i \left(L\delta_0^i - y_0^\sigma L_{,y_i^\sigma} \right) = -\frac{\partial L}{\partial x^0}_{expl.}. \qquad (4.19)$$

Notice the linear dependence of the second term in T_0^i on the *velocities* y_0^σ and on the *momenta* $-\pi_\sigma^i = L_{,y_i^\sigma}$.

4.5. General variations, group of automorphisms of the configurational bundle: case when $\phi_t \in Aut(\pi)$..

In order to define the Euler-Lagrange equations (4.7) and the natural boundary conditions (4.8) corresponding to an action (4.2), it is sufficient to use *m independent vertical variations* generated by one-parameter groups of automorphisms $\phi_t \in Aut(\pi)$ of the configurational bundle $\pi : Y \to X$ (See Appendix I) **acting along the fibers** Y_x. In local fibred coordinates (W, x^i, y^μ) such authomorphisms have the representation $(x, y) \to (x, \phi_t^\mu(x, y))$. Infinitesimal variations of this type have the form

$$\xi = \phi^\mu(x, y)\partial_\mu.$$

Such variations are sometimes called the "**outer variations**", [47], Ch.3.

On the other hand, in order to get conservation laws related to the Noether symmetries of the Lagrangian, one has to use vector fields of as general type as possible, therefore including variations of independent variables x^i - **inner variations**. In particular, the energy-momentum balance law (4.16) appears from applying the variations ∂_i generated by the translation of independent variables - space-time coordinates t, x^1, x^2, x^3.

Thus, it is interesting to see a result of variations of action (4.2) generated by the one-parameter groups ϕ_t of *general automorphisms* of the configurational bundle π:

$$(x, y) \to (\bar{\phi}^i(x), \phi^\mu(x, y)). \qquad (4.20)$$

Transformations of such one-parameter groups act on the sections as follows: $s(x) \to \phi_t s(\bar{\phi}_{-t})$.

In the infinitesimal form, the variational vector field corresponding to such 1-parameter subgroup is

$$\xi = -\xi^i(x)\partial_i + \xi^\mu(x, y)\partial_\mu, \qquad (4.21)$$

and its flow prolongation to the 1-jet bundle $J^1\pi$ is

$$Pr^1(\xi) = \xi + (d_i\xi^\mu - y_j^\mu d_i\xi^j)\partial_{y_i^\mu} = (\xi^i\partial_i - y_j^\mu d_i\xi^j \partial_{y_i^\mu}) + (\xi^\mu\partial_\mu + d_i\xi^\mu \partial_{y_i^\mu}). \qquad (4.22)$$

Using such variations of the action (1.1) and the standard flow prolongation (4.3) of vector field ξ to the 1-jet bundle $J^1(\pi)$, we obtain for the first variation of the action

$$\delta A_D(s) = \epsilon \int_D \left[\frac{\partial L}{\partial y^\alpha} - d_i(\frac{\partial L}{\partial y_i^\alpha}) \right]\xi^\alpha(x, s(x))dv + \epsilon \int_{\partial D} \frac{\partial L}{\partial y_i^\alpha}\xi^\alpha n_i dS +$$

$$+ \epsilon \int_D (L_{,x^k}(-\xi^k) + L_{,y_i^\mu} y_j^\mu d_i(\xi^j) + L(-\xi_{,x^i}^i))dv. \qquad (4.23)$$

Notice that the trace $\xi_{,x^i}^i$ of the Jacobi matrix $\xi_{,x^j}^i$ in the last term appears due to the action of transformation $\bar{\phi}^i(x)$ on the volume form dv.

Integrating by parts the terms containing derivatives of the components of the vector fields in the last integral we get the expression for the first variation of action A_D for an arbitrary variation $\phi_t \in Aut(\pi)$,

$$\delta A_D(s) = \epsilon \int_D [\frac{\partial L}{\partial y^\alpha} - d_i \left(\frac{\partial L}{\partial y_i^\alpha} \right)] \xi^\alpha(x, s(x)) dv + \epsilon \int_{\partial D} \frac{\partial L}{\partial y_i^\alpha} \xi^\alpha n_i dS +$$

$$+ \epsilon \int_D [L\delta_i^j - L_{,x^k}\delta_i^k + d_j \left(y_i^\mu L_{,y_j^\mu} \right)] \xi^i dv - \epsilon \int_D d_i(L_{,y_i^\mu} y_j^\mu \xi^j) dv. \quad (4.24)$$

Recall the form of the energy-momentum tensor (see 4.14) $T_j^i = L\delta_i^j - y_j^\mu L_{,y_i^\mu}$.

Requiring that the first variations of action $A_D(s)$ vanish at any variation of general type, and using the independence of variations ξ^μ and ξ^i, we get, in addition to the Euler-Lagrange equations (4.7) and the natural boundary conditions (4.8), one more equation - the **energy-momentum balance law-**

$$d_j \left(L\delta_i^j - y_i^\mu L_{,y_j^\mu} \right) = -L_{,x^i} \quad (4.25)$$

in the domain D, and the additional condition

$$L_{,y_i^\mu} y_j^\mu \xi^j \cdot n^i = 0 \quad (4.26)$$

on the boundary ∂D. This condition is the consequence of the natural boundary condition (4.8).

We refer to [47],Ch.3 for more details about the properties of **inner** and **outer** variations in the smooth situations, the notion of **inner extremals** and their relations to the usual (outer) extremals. We also refer to the article ([49]) and to the references therein for the exposition of the use of inner and outer variations in the important case of non-smooth (Lipschitz) variations and examples of non-smooth minimizers in applications.

[1]

---

[1] I would like to thank L.Truskinovsky who has sent to me this paper containing an example of non-commuting inner and outer variations in a situation with Lipschitz variations.

Interlude: Historical review.

The usefulness and the place of Lagrangian Variational methods in Mechanics, Geometry, and in the Classical Field Theory, is common knowledge between mathematicians and physicists.

The advantage in having all structural and dynamical properties of a system concentrated in one function, Lagrangian L, is overwhelming. Existence of a "dual" Hamiltonian picture, with a similar advantage, and the use of relations between these two approaches, form a mathematical picture of unparalleled beauty and elegance.

From the very first steps of Variational Calculus it was known that there are important dynamical systems that can not be described by Lagrangian formalism. Linear oscillators with friction, amplifiers and filters in radio-techniques, Burgers equation, systems of balance equations describing irreversible thermodynamical processes and zilions of other **dissipative systems** cannot be studied by the direct use of Lagrangian formalism.

At the end of XIXth and the beginning of XX century it was found that the Lagrangian formalism can be applied to some classes of nonstandard dynamical systems (nonconservative and nonholonomic systems are two examples), provided this formalism is properly modified.

The most well known example of such modification is the use of a Rayleigh dissipation function $\mathcal{R}$ in Mechanics, introduced by Lord Rayleigh in "Theory of Sound, Vol.1" , see also [140], Ch.VIII, Sec.93. It is assumed that R is the function of generalized coordinates q^i, $i = 1, \ldots . m$ and velocities $\dot{q}^i$. Let $L(q, \dot{q}) = T - U$ be a Lagrangian of a mechanical system which is the difference of the kinetic energy $T(q, \dot{q})$ and potential energy $U(q^i)$. Then the Euler-Lagrange Equations of the mechanical system, with the dissipative Rayleigh function, are

$$\frac{d}{dt}\left(\frac{\partial T}{\partial \dot{q}^j}\right) - \frac{\partial U}{\partial q^j} = \frac{\partial R}{\partial q^j}, \ j = 1, \ldots, m. \tag{4.27}$$

The second term in the left side is the potential force with the potential U, the term in the right side is the dissipative force defined by the **Rayleigh function**. If R is homogeneous of second order by the velocities $\dot{y}^i$, dissipative forces are linear functions of velocities,a situation that is quite common in Mechanics.

In 1958, Rayleigh dissipative functions were introduced to the Continuum Mechanics by H.Ziegler (see [141, 142] and references therein) under the name of "*dissipative functions.*" The goal of H.Ziegler's work was to construct the constitutive relations of media with irreversible, dissipative behavior: plasticity, viscous materials, hardening in metals, creep, etc., see [142] for the construction of dissipative functions in a variety of situations.

Later on, G.Maugin ([93, 94]) introduced dissipative functions into the Continuum Thermodynamics under the name of "dissipative potential". In his scheme, dissipative potential of a thermodynamical system with internal variables was used to satisfy the II law of thermodynamics (in the form of Clausius-Duhem inequality) and to construct the dynamical equations for internal variables (appearing in the system in one of two roles: "internal degrees of freedom" and the proper "internal variables". We refer to the Sec. where examples of Dissipative potentials are presented and to the sources [93, 94] for more information.

Another situation that eventually led to the modification of Lagrangian Variational principles was introducing "non-commuting variations" (we use this name for the longer title - "variations non-commuting with the derivatives" or "interchanging operators d and δ") emerged after the works of V. Volterra in 1897,[130],P.Voronets, 1901, L.Boltzman, [8], in 1902 and G.Hamel, 1904 on the variational formulation of equation of motion in **non-holonomic mechanical systems**.

We recall that the variations of dynamical fields y^i used in Lagrangian Mechanics, Lagrangian Field Theory ([10, 44]), as well as in other domains of Variational Calculus have the property that, in a case of one independent variables (Mechanics), is expressed by the relation: $\delta \dot{y}^i = \frac{d}{dt}\delta y^i$, i.e. operations of taking derivative of a dynamical field y^i by time and the operation of taking variation of y^i **commute**. This rule of "interchanging operators d and δ" or, as they were also called by Yu. Neimark, "transpositional relations" was the cause of the fact that the canonical Lagrangian scheme is unable to accomodate the dynamical equations of such systems **in the presence of non-integrable non-holonomic kinematic constraints** (see analysis of this situation given by Yu.Neimark, [104]).

It is these "commutativity rules that were questioned in the works cited above.

It was shown in the works of L.Boltzman, D.Hamel, and P.Voronets, that a proper modification of the commutativity rule, adopted to the non-holonomic kinematical relations allows us to get the dynamical equations of non-holonomic systems in different variational forms: Boltzman equations ([8]), Hamel equations, [57], Voronets equations, [132].

For about 50 years (until approx. 1957) there were two points of view on the character of rules of commutativity of variations and derivatives of dynamical variables $y^i(x^\mu = \{t = x^0, x^1, x^2, x^3\})$. One point of view (Volterra, Hamel) stated that the commutativity rule $\delta y^i_{,x^\mu} = \frac{\partial}{\partial x^\mu}\delta y^i$ has to be valid in all coordinate systems.

Another point of view (Levi-Civita, Amaldi, [83], and Suslov,[123]) postulated that the commutativity of the operators d and δ are valid *only in holonomic coordinate systems*. For non-holonomic systems commutativity rule applies only for the generalized coordinates whose variations (subject to the compatibility with the non-holonomic constraints) are independent. For the remaining coordinates, the commutativity rule, i.e. expressions for $d\delta - \delta d$ have to be derived from the non-holonomic constraints.

A second point of view was predominant between the specialists until, in 1957, Ju.Neimark ([104]) clarified the question about *commutativity relations*. He has shown that the contradiction between two points of view on the commutativity rules is caused by an unsustainable definition of variations of velocities $\dot{q}^i$ **outside of the trajectories of motion**: velocities $\dot{q}^i$ are defined only on the trajectories of motion while variations δy^i in general are different from the direction of actual motion. As a result, composition $d\delta$ is defined on arbitrary (actual or kinematically admissible) trajectory while composition δd has to be defined on the real trajectories. In other words, one has to define the variation δ on the velocities $\dot{y}^i$ in such a way that the compositions $d\delta$ and δd *were compatible, in a proper way, with the non-holonomic kinematical constraints* in the system.

In principle, variations of velocities can be defined arbitrarily outside of actual trajectories (in the tangent space $T(Q)$ of configurational space Q of mechanical system).

This point of view leaves, to a researcher, a vast freedom of choice of variations of derivatives. In this work we study and exploit this freedom and its geometrical meaning in order to extend conventional Lagrangian formalism to a class of dissipative dynamical systems.

The approach suggested by Yu.Neimark achieves, simultaneously, two goals. Firstly, this approach determines that the early controversies between the viewpoints of Hamel and T.Levi-Civita (see [105]) were due to the absence of correct definition of variations of the velocities *outside of real trajectories*.

Secondly, it shows that the early authors (T.Levi-Civita, U.Amaldi (see [83] and the references in [133], Sec. 6.1) implicitly introduced the "**non-commutativity rules**" (as Ju.Neimark and A.Lurie named them) for the variations for generalized velocities of a *mechanical systems with non-holonomic constraints* as a way to adopt conventional Lagrangian formalism to such systems. The monograph of A.Lurie [88] and especially the one of Ju.Neimark and N.Fufaev,[105], deliver the most comprehensive description of classical variational methods in non-holonomic mechanics, including the appropriate non-commutativity rules.

Further development of the use of non-commuting variations belongs to the 1970s, when, in a series of papers, B.Vujanovic, T.Atanackovic and their coauthors have suggested nontivial rules of commutation of time derivative and variations of unknowns y^i as a tool to present some mechanical systems with a non-potential forces and the equations of the heat propagation in solids as Euler-Lagrange equations of a variational principle, see [133], Ch.6 and references therein.

B.Vujanovich have argued ([134]) that while the variations δy^i of dynamical variables and the time derivatives of these variations $\frac{d}{dt}\delta y^i$ ($d\delta y^i$ in the case of Field Theory, where $n > 1$) have *purely kinematic meaning*, variations of velocities $\delta \dot{y}^i$ have a *dynamical character* being related to the non-conservative forces acting on the mechanical system by the Second Newton's law.

Another argument, supporting the use of non-commuting variations, was suggested by A.Lurie in [88], Section 12.9. He argued that it is not necessary to make a comparison of the mechanical system configuration along the true path and along the variated path **at the same time moment**. In other word, if $q^i(t)$ is the true motion of a (mechanical) system in terms of generalized coordinates, one can define an infinitesimally close (and allowed by the constraints, if those are present) neighboring motion by an expression $q^i_*(t + \Delta t)$. Taking into account only terms of the first order, we get $q^i_*(t + \Delta t) \sim q^i_*(t) + \dot{q}^i_*(t)\Delta t$. This defines **asynchronous** variations $\Delta q^i(t)$ of dynamical fields

$$\Delta q^i(t) = q^i_*(t + \Delta t) - q^i(t) = \delta q^i(t) + \dot{q}^i_*(t)\Delta t. \tag{4.28}$$

For any function of time $f(t)$ this gives $\Delta f = \delta f + \dot{f}\Delta t$ and, in particular,

$$\Delta \dot{q}^i = \delta \dot{q}^i + \ddot{q}^i \Delta t = \frac{d}{dt}\delta q^i + \ddot{q}^i \Delta t.$$

In the second equality, here, the usual rule of commuting d and δ was applied. The value of Δt above is an arbitrary function of time, therefore, for the Δ-variation of any function of time $f(t)$, we have $\frac{d}{dt}(\Delta f) = \frac{d}{dt}\delta f + \dot{f}\Delta t + \dot{f}\frac{d}{dt}\Delta t$. As a result,

$$\frac{d}{dt}\Delta q^i = \frac{d}{dt}\delta q^i + \ddot{q}^i \Delta t + \dot{q}^i \frac{d}{dt}\Delta t = \Delta \dot{q}^i + \dot{q}^i \frac{d}{dt}\Delta t. \tag{4.29}$$

The second term in this equation contains the time function that can be defined by a dissipative processes, or by other factors of non-Lagrangian nature. It shows that in contrast to the conventional variation δ, variation Δ does not commute with the time derivative.

Later on, in the works of H.Kleinert, P.Fiziev and A.Pelster [35, 65, 67], on the motion of a spinless particle in space-times with curvature and torsion, a new variational principle was suggested. This principle was also based on the modified rule for the commuting of time derivative and the variations of generalized coordinates. Their considerations were geometrical in their nature, being based on a non-holonomic transformations of the (flat) configurational space. After such a transformation, equations of motion of a particle gain a *torsion* force defined by the (nonmetric) connection in the Cartan space-time.

Non-commuting variations used also in the works of L.Truskinovsky and his collaborators in the work [49] on the weak variations in the problems having Lipschitz (non-smooth) extremals. In other works of the same author, see [87, 77] and[78] non-commuting variations appears in different forms and in very different situations. Finally, we note the works of B.Dimitrov ([19]) on the use of non-commuting variations in the relativistic hydrodynamics and an analysis by Manov in his monograph [89].

Chapter II. Lagrangian Field Theory with non-commuting variations.

5. Introduction.

In this chapter we define and study the **non-commuting variations** for a Lagrangian action $\mathcal{A}_D(s(x))$ defined by a "tensor field" $K_{i\beta}^\alpha$ in the 1-jet space $J^1\pi$. We will call tensor $K_{i\beta}^\alpha$ the **NC-tensor** and the variations defined by tensor K - the K-twisted variations.

Using the modified first variations we obtain the Euler-Lagrange equations modified by the sources (forces in Mechanics):

$$E_\alpha(L) = \frac{\delta L}{\delta y^\alpha} = f_\alpha, \; j = 1,\ldots,m, \tag{5.1}$$

f_α being determined by the NC-tensor K. When $K = K_{i\beta}^\alpha = 0$, system 5.1 reduces to the conventional Euler-Lagrange system.

We will develop the formalism of non-commuting variations for the **first order Lagrangians** $L \in C^\infty(J^1\pi)$ - infinitely differential functions of the variables x^i, y^μ, y_i^μ . At the end of this Chapter we extend our approach to the Lagrangian Field Theory of higher order.

Remark 5. Notice that the "NC-tensor" $K_{i\beta}^\alpha$ behaves tensorially with respect to the **transformations of the fibred charts** (U, x^i, y^α).

Remark 6. In their series of works, C.Muriel and J.Romero introduced the "λ-twisted symmetries" for a scalar ODE based on the modified prolongation procedure defined by a scalar function λ - λ-prolongation. Later on, G.Gaeta, P.Morando and G.Cicogna extended the method of "λ-twisted symmetries" to the "systems of ODEs and PDEs", see [39, 40]. The method of μ-symmetries "NC-tensor" K used by these authors is introduced as the $gl(n)$-valued horizontal 1-form $\mu = K_{i\beta}^\alpha dx^i$. This point of view was very convenient for modifying the geometrical structures on the jet bundles adopting them to the K-modified prolongation procedures. In Chapter 4 we present some basic results of these authors related to our study.

6. Euler-Lagrange equations with K-twisted variations.

In this section we introduce and begin to study the modified construction of variations of jet variables (derivatives of dynamical fields y^α), corresponding to the **modified lift of the vertical vector fields** $\xi \in V(\pi)$ to the vertical vector fields $Pr_K^1(\xi)$ in the 1-jet bundle $\pi_1 : J^1(\pi) \to X$:

$$\xi \to Pr_K^1(\xi) = \xi + (d_i\xi^\alpha + K_{i\beta}^\alpha\xi^\beta)\partial_{y_i^\alpha}, \tag{6.1}$$

where

$$K_{i\beta}^\alpha \in C^\infty(J^1(\pi)) \tag{6.2}$$

are smooth functions of variables $x^i, y^\alpha, y_i^\alpha$ (i.e. functions in the 1-jet space $J^1(\pi)$).

These functions form the NC-tensor $K_{i\beta}^\alpha$ (tensor relative to the group of automorphisms of the configurational bundle π, or, equivalently, with respect to the group of local changes of fibred coordinates (x^i, y^μ, y_i^μ)).

Form the variation of the action $\mathcal{A}_D(s)$ using the prolongation (6.1) of variational vector field $\xi = \xi^\alpha\partial_\alpha$ to the 1-jet bundle $J^1\pi$, we obtain infinitesimal variations

© Springer International Publishing Switzerland 2016
S. Preston, *Non-commuting Variations in Mathematics and Physics*,
Interaction of Mechanics and Mathematics,
DOI 10.1007/978-3-319-28323-4_2

of fields and their derivatives. Notice that base (independent) variables x^i are not variated,

$$\begin{cases} y^\alpha \to y^\alpha + \epsilon\xi^\alpha(x,y), \\ y_i^\alpha \to y_i^\alpha + \epsilon(d_i\xi^\alpha + K_{i\beta}^\alpha\xi^\beta). \end{cases}$$

Now we form the variation of the action

$$\Delta A_D(s)(\epsilon\xi) = A_D(s_\epsilon) - A_D(s) =$$

$$= \int_D [L(x, s^\alpha(x) + \epsilon\xi^\alpha(x, s(x)), s_{,i}^\alpha(x) + \epsilon(d_i\xi^\alpha(x, s(x)) + K_{i\beta}^\alpha(x, j^1 s(x))\xi^\beta(x, s(x))))$$

$$- L(x, s(x), s_{,i}^\alpha(x))]dv =$$

$$= \epsilon \int_D [\frac{\partial L}{\partial y^\beta}\xi^\beta(x, s(x)) + \frac{\partial L}{\partial y_i^\alpha}(d_i\xi^\alpha(x, s(x)) + K_{i\beta}^\alpha(x, \beta^1 s(x))\xi^\beta(x, s(x))))]dv + O(\epsilon^2) =$$

$$= \epsilon \int_D [\frac{\partial L}{\partial y^\beta} - d_i\left(\frac{\partial L}{\partial y_i^\beta}\right) + K_{i\beta}^\alpha(x, j^1 s(x))\frac{\partial L}{\partial y_i^\alpha}]\xi^\beta(x, s(x))dv + O(\epsilon^2).$$

As a result the system of Euler-Lagrange equations takes the form

$$\frac{\partial L}{\partial y^\beta} - d_i\left(\frac{\partial L}{\partial y_i^\beta}\right) = f_\beta = -K_{i\beta}^\alpha(x, j^1 s(x))\frac{\partial L}{\partial y_i^\alpha} = -K_{i\beta}^\alpha\pi_\alpha^i, \ \beta = 1, \ldots, m. \quad (6.3)$$

The right side of this equation represents the generalized (non-potential) sources

$$f_\beta = -K_{i\beta}^\alpha(x, j^1 s(x))\frac{\partial L}{\partial y_i^\alpha} = -K_{i\beta}^\alpha\pi_\alpha^i \quad (6.4)$$

acting in the system. Here $\pi_\alpha^i = L_{,y_i^\alpha}$ are the *momenta* corresponding to the Lagrangian L.

Thus, using the non-commuting variations in the conventional variation formalism we get the systems of equations of the form

$$\frac{\delta L}{\delta y^\beta} = f_\beta, \ f_\beta = -K_{i\beta}^\alpha\pi_\alpha^i. \quad (6.5)$$

Example 1. Look at the simplest case of one ordinary differential equation of second order where $n = 1, m = 1$ (i.e. where there is one independent variable x and one dynamical variable y). Let Lagrangian be a function $L = L(x, y, y')$ of x, y and the first derivative $y' = \frac{dy}{dx}$.

Let $K = K(x, y, y')$ be a NC-tensor (scalar in this case). The Euler-lagrange Equation $\frac{\delta L}{\delta y} = -\frac{\partial L}{\partial y'}K$, has the form

$$\frac{\partial L}{\partial y} - d_x\left(\frac{\partial L}{\partial y'}\right) = -\frac{\partial L}{\partial y'}K,$$

where $d_x = \frac{\partial}{\partial x} + y'\frac{\partial L}{\partial y} + y''\frac{\partial L}{\partial y'}$ is the **total derivative** of lagrangian L by x. In coordinates this equation has the form

$$\frac{\partial^2 L}{\partial x\partial y'} + y'\frac{\partial L}{\partial Y} + y''\frac{\partial L}{\partial Y'} - \frac{\partial L}{\partial y} = -K\frac{\partial L}{\partial y'}.$$

that differs from the conventional EL-equation by the force/source in the right side.

Example 2. In a case of Mechanics (n=1), there is one independent variable t and m dynamical variables y^μ. A NC-tensor K^α_β defines K-twisted variations of variables y^μ and their derivatives:

$$\delta y^\mu = \xi^\mu(t,y),\ \delta \dot{y}^\mu = d_t\xi^\mu + K^\mu_\beta\xi^\beta. \tag{6.6}$$

The full prolongation formula contains the second (horizontal) component K^ν (see bellow, Section 16 for the general prolongation procedure). As a result, the K-twisted 1-prolongation of a vector field $v = \xi\partial_t + \xi^\mu\partial_{y^\mu}$ is

$$Pr^1_K v = (\xi\partial_t + \xi^\mu\partial_\mu) + [(d_t(\xi^\mu - \dot{y}^\mu\xi) + (K^\mu_\nu\xi^\nu - K^\mu\xi)]\partial_{\dot{y}^\mu}. \tag{6.7}$$

Corresponding Euler-Lagrange equations (6.3) takes, in the case of "Mechanics", the form:

$$\frac{\partial L}{\partial y^\beta} - d_t\left(\frac{\partial L}{\partial \dot{y}^\beta}\right) = -K^\alpha_\beta(t,y^\alpha,\dot{y}^\alpha)\frac{\partial L}{\partial \dot{y}^\alpha} = -K^\alpha_\beta\pi_\alpha,\ \beta = 1,\ldots,m, \tag{6.8}$$

where $\pi_\alpha = \frac{\partial L}{\partial \dot{y}^\alpha}$ are momenta corresponding to the variables y^α.

Remark 7. Bellow, in Chapter V, we will study the relation between the representation of the forces in the form $f_\beta = K^\alpha_\beta\pi_\alpha$ and the Rayleigh representation, in terms of "dissipative potentials" ([116]).

In Section 10 below we present a variety of examples of Euler Lagrange equations (and systems of Euler-lagrange-equations) with non-commuting variations. NC tensors K defining the twisted properties of variations in these systems reflects specific dynamical and symmetry properties of these equations.

Remark 8. Prolongation procedures of the form (6.1,6.8) of vector fields from the configurational space Y to the jet bundles of arbitrary order $k \geq 1$ were introduced by C.Muriel and J.L.Romero, [101, 102] in the case where $n = 1$ (ordinary differential equations). Their prolongation procedure was defined by a C^∞-function $\lambda(x,y,\frac{dy}{dx})$ defined on the 1-jet space: $\lambda \in C^\infty(J^1\pi)$. In our notations, for $m = 1$, tensor K^μ_ν in (6.8) reduces to one function K. The procedure of prolongation (6.8) coincides with the λ-prolongation introduced by C.Murial and J.Romero. These authors introduced a new class of symmetries (λ-**symmetries**) for ordinary differential equations. Usage of λ-symmetries allowed the authors to define **new general procedure for the reduction of ODE**. In particulary, they found that some nonlinear ODE having nontrivial Lie derivatives could be integrated using λ-symmetries. Their method of reduction contains, as the special cases, many known procedures of reduction.

Chapter 4 of these notes is a short presentation of the theory of λ-symmetries and its generalization to the case of PDE and systems of PDE by G.Gaeta, G.Gicogna, P.Morando and their collaborators.

Remark 9. Introduction of non-commutative variations modifies the Euler-Lagrange Equations by the sources/forces $f_\beta = -K^\alpha_{i\beta}\pi^i_\alpha$. At the same time, sections $s(x^i) = \{s^\alpha(x)\}$ delivering local extremal values (local or global,l minima or maxima) to the action functional $A_D(s)$ stays (local or global) minimum or maximum although now they satisfy a different system of differential equations.

7. Noether Theorem, Energy-Momentum balance law.

Infinitesimal condition for a Lie group G of (diffeomorphic) transformations of the manifold Y to be a **group of variational symmetries of Lagrangian L** is the relation $Pr^{\{k\}} \cdot L + LDiv(\xi) = 0$ (see (79.5). At the same time,condition for a Lie group G of (diffeomorphic) transformations of the manifold Y to be a **group of divergent symmetries of Lagrangian** L is that for some n-tuple $B = \{B_i\}$ of functions in Y (i.e. functions of variables x^i, y^μ) has the invariant form

$$Pr^k \xi + Ldiv(\xi) = Div(B),$$

(see [106], Sec.4.4). In a case of balance laws this relation is *appropriately modified by the NC-tensor* $K^\nu_{\mu i}$ (see Section 81) or, equivalently, by the source form $f_\mu dy^\mu$. Notice that the defining property of a one-parameter group ϕ^t of symmetries of Lagrangian is that the action of transformations ϕ^t, $t \in R$ transform solutions of Euler-Lagrange Equations to the solutions.

Therefore, phase fields of such one-parameter groups act as the infinitesimal variations of the action $\mathcal{A}_D(s)$ corresponding to L.

Let $\zeta \in \mathfrak{g}$ be an arbitrary element of Lie algebra $\mathfrak{g}$. Let $Pr^1_K(\zeta)$ be the prolongation of vector field ζ to the 1-jet bundle $J^1(\pi)$ modified by a NC-tensor K (see Appendix II, (79.6)). Then,a vector field ζ is the infinitesimal divergent symmetry of L, and generates the (local) one-parameter group of symmetries of L, if and only if there exists a horizontal 1-form $B = B_k(x^i, y^\mu, y^\mu_i)dx^i$ in $J^1(\pi)$ such that

$$Pr^1_K(\zeta)L + L(div(\overline{\zeta}) = Div(B). \tag{7.1}$$

Divergence is taken using the volume form $d_g v$ defined by the metric g in he base X.

Remark 10. Notice that the symmetry condition contains the term $K^\mu_{\nu i}\zeta^\mu \partial_{y^\mu_i} L$ depending on the NC-tensor K.

Theorem 2. *Let L be a Lagrangian of order k and let*

$$E_\mu(L) = f_\mu = -K^\beta_{i\mu}\pi^i_\beta, \quad \mu = 1, \ldots, m \tag{7.2}$$

be an Euler-Lagrange system with the Lagrangian L of order k and the sources $f_\mu, \mu = 1, \ldots, m$. Let ξ be a vector field in Y - infinitesimal generator of variational symmetry of the Lagrangian L (see Sec.78) and let $Pr^k(\xi)$ be its prolongation of order k.Then there exist an n-tuple of the smooth functions A^i such that for all solutions of the system (80.6) the following equality is fulfilled

$$Div(A + L\xi) = -Q^\mu \cdot f_\mu \tag{7.3}$$

Here $Q = \{Q^\mu = \xi^\mu - y^\mu_i \xi^i\}$ is the characteristic of the vector field ξ (see Appendix III, Sec.79).

In the case of **divergent symmetry** corrresponding balance law is modified by a introducing the form B (see Sec.79). As a result, we get, for solution of the Euler-Lagrangian system with NC-variations the **Noether balance law in the presence of non-commuting variations**

$$\begin{cases} \mu Div(P) = -Q^\nu K^i_{i\nu}\pi^i_\mu, \text{ where} \\ P^i = \zeta^\mu L_{,y^\mu_i} + \zeta^i L - \zeta^j y^\nu_j L_{,y^\mu_i} - B. \end{cases} \tag{7.4}$$

Here, Q^ν are components of the characteristic $Q = \{Q^\nu \partial_\nu = (\zeta^\nu - y_i^\nu \zeta^i)\partial_\nu\}$ of the vector field ζ. Balance laws obtained in Theorem 2 will be called K-**twisted conservation or balance laws** for the Lagrangian L.

As an example, illustrating the modified Noether conservation law, we present the canonical Stress-Energy-Momentum balance law.

7.1. **Stress-Energy-Momentum balance law.** Here we write down the canonical stress-energy-momentum (CEM) balance law corresponding to a Lagrangian $L(x^i, y^\alpha, y_i^\alpha)$ - function of independent variables x^i, dynamical fields y^α and their first derivatives $y_{,x^i}^\alpha$. We will be using the approach of [106], modified to produce the balance law, see Appendix III, Sec.81.

Consider a case where vector field $\xi = \xi_k = \partial_{x^k}$ is lifted to the space Y by a connection Γ in the bundle $\pi : Y \to X$: $\partial_{x^k} \to \hat{\xi}_k = \partial_{x^k} + \Gamma_k^\mu \partial_\mu$. Introduce the *characteristic* Q of the vector field $\hat{\xi}_k$ - the quantity playing the principal role in the prolongation of vector fields and in studying symmetries and conservation laws associated with the differential equations (see [106] and sections 72,75 of Appendix. Characteristic of the vector field $\hat{\xi}_k$ has the components $Q^\mu = \Gamma_k^\mu - y_k^\mu$ and we define the Energy-Momentum tensor as

$$T_k^i = L\delta_k^i - (\Gamma_k^\mu - y_k^\mu)L_{,y_i^\mu}. \tag{7.5}$$

As a result the balance equation (see Appendix III, Sec.81, equation 81.4), corresponding to this vector field (*stress-energy-momentum balance law*) has the form

$$d_i T_k^i = d_i \left(L\delta_k^i - y_k^\mu L_{,y_i^\mu} \right) = -\frac{\partial L}{\partial x^k}\bigg|_{expl} - (\Gamma_k^\mu - y_k^\mu)K_{i\mu}^\nu(x, \jmath^1 s(x))\frac{\partial L}{\partial y_i^\nu} =$$

$$= -\frac{\partial L}{\partial x^k}\bigg|_{expl} + (\Gamma_k^\mu - y_k^\mu)f_\mu, \ k = 0, 1, 2, 3. \tag{7.6}$$

where $f_\nu = -K_{i\nu}^\mu \pi_\mu^i$.

The CEM-tensor $T_k^i = \left(L\delta_k^i - y_k^\mu L_{,z_i^\mu} \right)$ has the standard form ([80]).

In particular, in a case where connection Γ is trivial ($\Gamma_k^\mu = 0$), for $k = 0$ we get the energy balance law in the form

$$d_i T_0^i = d_i \left(L\delta_0^i - y_0^\sigma L_{,y_i^\sigma} \right) = -\frac{\partial L}{\partial x^0}\bigg|_{expl} - y_0^\nu K_{i\nu}^\mu(x, j^1 s(x))\frac{\partial L}{\partial y_i^\mu} = -\frac{\partial L}{\partial x^0}\bigg|_{expl} - y_0^\nu \cdot K_{i\nu}^\mu \pi_\mu^i. \tag{7.7}$$

Notice the linear dependence of the second force term on the velocities y_0^μ and on the momenta π_μ^i. This terms allows the system to have positive or negative dissipation.

Introduce the following notion:

Definition 2. *A source $f_\mu dy^\mu$ is called conservative if the are no energy dissipation in the process described by the Euler-Lagrange equations with the source $f_\mu dy^\mu$.*

From the energy-momentum balance law 7.6 it follows that, if through any point $(x^i, y^\mu, y_i^\mu) \in J^1(\pi)$ there passes a solution of the system (6.9), a source $f_\mu dy^\mu$ is conservative if and only if

$$K_{\nu i}^\mu : y_0^\nu \pi_\mu^i = y_0^\nu f_\nu = 0. \tag{7.8}$$

Remark 11. Notice that the energy dissipation $K^\mu_{\nu i} : y^\nu_0 \pi^i_\mu = y^\nu_0 f_\nu$ has the bilinear form where terms y^ν_0 have the meaning of velocities (rate of change of the field y^ν) while the components f_ν are related to the "forces" acting on the system. As a result, expression for the energy dissipation is similar to the expression for the entropy production in the Continuum Thermodynamics, see "Onzager Principle" in [55, 94].

Example 3. Let the bundle π be the tensor-like bundle and tensor K have the canonical form (9.8), see Sec.9). Then, the energy balance laws has the form

$$d_\mu T^\mu_0 = -\frac{\partial L}{\partial x^0}\Big|_{expl} - \dot{y}^i f_i$$

.

Exercise. Consider an inner variation $x^i \rightarrow x^i + \xi^i(x)$, its prolongation to the space Y defined by a **connection** Γ **(see Sec.17, Ch. 3)** and lift ζ of this vector field to the 1-jet bundle $J^1\pi$ induced by an Ehresmann connection K (more specifically, by its component K^μ_{ki}), see Definition 8, Sec.17. Work out the variation equation $\frac{\delta A_D}{\delta y^\mu}(\zeta) = 0$.

Show that if the connection Γ is trivial and if vertical connection K is such that $K^\mu_{ik} = y^\nu_k K^\mu_{\nu i}$, then the obtained equation coincide with the Energy-Momentum balance law with the K-induced dissipation (7.6).

Remark 12. Compare this case of the vertical connection K with the first λ, μ -prolongation in Chapter 4.

8. On the non-unicity of NC-representation (6.5).

By introducing the presentation (6.4) of the source (force) terms in the systems 6.3, it is natural to inquire, for which sources $\{f_\beta\}$ does a such representation exists, under which conditions "such" representation is unique. Finally, if such a representation is not unique, it is interesting to determine the degree of such non-unicity. This information may be useful for the choice of the most convenient representation (6.6), for instance, the one leading to the minimal dissipation.

More then this, it is interesting to study these questions by qualifying the sources f_β and the NC-tensors $K^\alpha_{i\beta}$ in terms of the type and order of derivatives of the fields y^α they may depend on. Answers to these questions depend on the order of derivatives entering f_β and NC-tensor $K^\alpha_{i\beta}$ as well as on the form of this dependence.

Here we discuss the case where sources f_β and NC-tensor K depend on the points of the space Y and on the first derivatives of dynamical fields y^α, i.e when $f_\beta, K^\alpha_{i\beta} \in C^\infty(J^1(\pi))$.

Let there be two different representations of the form (6.4) for the sources f_β:

$$f_\beta = K^\alpha_{i\beta} \pi^i_\alpha = P^\alpha_{i\beta} \pi^i_\alpha.$$

Subtracting these equations we have, for the difference $Q^\alpha_{i\beta} = K^\alpha_{i\beta} - P^\alpha_{i\beta}$, the equality

$$Q^\alpha_{i\beta} \pi^i_\alpha = 0. \tag{8.1}$$

Denote by $\mathcal{R}$ the vector space of all solutions $Q^\alpha_{i\beta}$ of this equation with the components in the space $C^\infty(J^1(\pi))$. If, for a given sources $f_\beta \in C^\infty(\pi)$ there exist a representation (6.4) with the NC-tensor $K^\alpha_{i\beta} \pi^i_\alpha$ $f_\beta = K^\alpha_{i\beta} \pi^i_\alpha$, then the set of **all possible** NC-tensors representing sources f_β as $f_\beta = K^\alpha_{i\beta} \pi^i_\alpha$ has the form

$\{K^{\alpha}{}_{i\beta} + Q^{\alpha}_{i\beta}\}$ where $Q^{\alpha}_{i\beta} \in \mathcal{R}$. Notice that the space $\mathcal{R}$ does not depend on the choice of the sources $f_{j\beta}$ but only on the Lagrangian L. In addition to this, variables x^i, y^α appear in the equation 8.1 as parameters only.

We describe the space $\mathcal{R}$ in the case where $L(x, y, y_i^\alpha)$ is a regular Lagrangian of the first order, i.e. let the matrix $\dfrac{\partial^2 L}{\partial y_i^\alpha \partial y_j^\beta}$ be non-degenerate. In such a case, Legendre transformation $(x, y, y_i^\alpha) \rightarrow (x, y, \pi^i_\alpha)$ is (local) diffeomorphism and the momenta π^i_α define the coordinates in the fibres of the bundle $\pi_{10} : J^1(\pi) \rightarrow Y$.

Consider now a formal analog of the system 8.1. Let R^k be the k-dim vector space endowed with the conventional Euclidian metric. Let u^i be orthogonal Cartesian coordinates in R^k. Consider the radial vector field $\partial_r = \sum_i u^i \partial_{,u^i}$. A vector field $F = f^i(u^j)\partial_{,u^i}$ is orthogonal to the radial vector field ∂_r if and only if

$$\sum_i f_i(u)u^i = 0. \tag{8.2}$$

This condition is equivalent to the condition that for any point $x \in R^k$, vector field $F(x)$ is tangent to the sphere with center at the origin $\mathbf{0}$ and radius $|r|$.

Since the tangent bundle to the unit sphere (and, therefore, to all spheres centered at the origin) are generated by the vector fields $F^{ij} = u^i \partial_{u^j} - u^j \partial_{u^i}$, all vector fields F satisfying to the equation (8.2) have the form

$$F(u) = f_{ij}(u^k)(u^i \partial_{u^j} - u^j \partial_{u^i}).$$

Notice that the components of vector fields $F^{ij} = F^{ijs}\partial_{u^s}$ are defined by the condition

$$F^{ijs} = \begin{cases} -u^j \text{ if } s = i, \\ = u^i \text{ if } s = j, \\ 0 \text{ if } s \neq i, j. \end{cases} \tag{8.3}$$

Using Legendre transformation and applying these arguments to the fibers of the 1-jet bundle $J^1(\pi)$ with the coordinates $u^j = \pi^i_\alpha$ and the metric defined by the condition that the vector fields $\partial_{\pi^i_\alpha}$ form the orthonormal (Cartesian) basis we prove the following

Lemma 1. *Let L be a **regular Lagrangian**, i.e. let the matrix $\dfrac{\partial^2 L}{\partial y_i^\alpha \partial y_j^\beta}$ be non-degenerate. Then all the components Q^β of the solutions of the system of equations 8.1 have the form*

$$K = \sum q_{ij}()F^{ijs}, \tag{8.4}$$

where functions F^{ijs} are defined in 8.3 and q_{ij} are arbitrary functions from $C^\infty(J^1(\pi))$.

This result gives the description of all possible representations of the source form $f = f_\alpha dy^\alpha$ in the form (6.6) provided one such representation exists. Below we will show that such a representation exists on all the natural bundles $\pi : Y \rightarrow X$.

On the other hand, in Section (22) below, we will show that if there exists a representation (6.4) with a NC-tensor $K \in C^\infty(Y)$, such representation **is unique**.

Example 4. Case: Components $K^{\alpha}_{i\beta}$ are linear by y_i^α.
Tensors $K^{\alpha}_{i\beta}$ linear by first order derivatives

$$K^{\alpha}_{i\beta} = L^{\alpha j}_{i\beta\kappa}(x, y)y_j^\kappa$$

(*linear functions of 1-jet variables*) were used by different authors, see papers by H.Kleinert, P.Fiziev and A.Pelster,[35, 65], works of Dj.Djukich and B.Vujanovic and T.Atanaskovich, see (VJ,Vu, At,At2). In the notations of the papers of H.Kleinert and his coworkers on the variational form of Mechanics in the Cartan space-time, $K^\alpha_{t\beta} = 2S^i_{\kappa\beta}y^\kappa_t$ where $S^i_{\kappa\beta}$ is the torsion tensor of an affine connection in the Cartan space-time (see Ch.5, below). When the torsion of this connection vanish, variations (4.10) reduce to the conventional variation (3.1).

Consider a case where $n = 1$ and let a Lagrangian have the form $L(x, y^\alpha, \dot{y}^\alpha) = a_\alpha(x, y^\beta)\dot{y}^\alpha$. In this case the equality (8.1) takes the form

$$Q^\alpha_\kappa(x, y)a_\alpha(x, y)\dot{y}^\kappa = 0.$$

This equation is equivalent to the system of linear equations $Q^\alpha_\kappa(x, y)a_\alpha(x, y) = 0$, $\kappa = 1, \ldots, m$. The association $(x, y) \to a^\alpha(x, y)$ defines the vertical vector field $a = a^\alpha\partial_\alpha$ in Y and any 1-form $\nu = p_\alpha dy^\alpha$ annulating the vector field a delivers the (1.1) tensor $Q = \partial_{y^\alpha} \otimes \nu$ satisfying to the previous equality.

Turning to the question of existence of a representation (6.4) we notice that the tensor field $K^\alpha_{i\beta}(x^k, y^\alpha, y^\alpha_k)$ can be considered as the mapping of vector bundles with the finite-dimensional fibers $K : \pi^*_{10}V(\pi) \to V(\pi_{10})$ over $J^1(\pi)$ sending the vertical over X vector field in Y $\xi^\alpha\partial_\alpha$ to the vertical over Y vector field in $J^1(\pi)$ $K^\alpha_{i\beta}\xi^\beta\partial_{y^\alpha_i}$. Conjugate mapping of bundles

$$K^* : V(\pi_{10})^* \to \pi^*_{10}V(\pi)^* \tag{8.5}$$

is such that

$$K^*d_{v_{10}}L = f_\beta dy^\beta. \tag{8.6}$$

$d_{v_{10}}L$ here is the vertical differential of the function L.

Vice versa, if K^* is **any** bundle mapping 8.5 satisfying the condition 8.6, then its conjugate K^{**} of this bundle mapping $\pi^*_{10}V(\pi) \to V(\pi_{10})$ is such that taking $K = K^{**}$ in defining the variation of jet variables (4.1) we get the Euler-Lagrange system (6.4).

Locally, over a domain in $J^1(\pi)$ where both bundles in 8.5 are trivial, this can be done simply, by defining K^* to satisfy (8.6) and to be an arbitrary mapping in a subspace of $V(\pi_{10})^*$ complemental to the linear span of the vertical covector $d_{v_{10}}L$ (taking, for instance, an orthogonal complement in the fibers with respect to some natural metric).

Thus, locally, near the points where $d_{v_{10}}L \neq 0$, tensor K can be defined in such a way that the corresponding mapping (8.5) has the "rank one". In the next section we formalize these arguments and construct the canonical NC-tensor $K^\alpha_{i\beta}$ on the bundles $\pi : Y \to X$ having the property of "metric prolongation" or on the bundles.

9. Case of natural (tensor-like) bundles: Canonical NC-tensor $K^i_{\mu j}$

. In this section we consider the case where the configurational bundle $\pi : Y \to X$ has some properties similar to the property of tensor and tensor density bundles of "metric prolongation" (see Appendix I). More specifically, we assume that the configurational space Y is endowed with the metric q such that the projection $\pi : Y \to X$ is the isometry. Tangent bundle $T(Y)$ splits as the direct q-orthogonal sum

$$T(Y) = T(V)(\pi) \oplus H, \quad H_y = V_y(\pi)^\perp. \tag{9.1}$$

In a case of "normal bundles", including bundles of tensors and tensor densities, the metric g in X defines the metric q_x on the fibers Y_x of the bundle π. In local fibred chart (x^i, y^α), where the basis of the tangent bundle to the fiber Y_x at a point $(x, y) \in Y_x$ can be taken to be $\partial_\alpha, \alpha = 1, \ldots, m,$

$$< \partial_\alpha, \partial_\beta >_{(x,y)} = q_{(x,y)\ \alpha\beta}.$$

In particular, the vertical subbundle $\nu : V(\pi) \to Y$ of the tangent bundle $T(Y) \to Y$ is endowed with the induced metric.

At the same time, metric g in X defines the covariant metric $g^{\sigma\lambda}$ on the fibres of the cotangent bundle $\tau^* : T^*(X) \to X$.

Let Γ be a connection on the bundle π defined by the decomposition (9.1). If π is a tensor or tensor density bundle, we can take Γ to be prolongation of Levi-Civita connection Γ^g to $\pi : Y \to X$ using products and operation of conjugation, see ([33]).

Connection Γ defines (and is defined by) the section $j_\Gamma : Y \to J^1(Y)$. This section defines an original point in affine fibers $J^1(\pi)_y$ over points $y \in Y$. As a result, it defines the isomorphism **of the bundles over** Y

$$J^1(\pi) \simeq \pi^* T^*(X) \otimes V(\pi). \tag{9.2}$$

Metrics given above in the fibers of bundles $T^*(X) \to X$ and in $V(\pi) \to Y$, define the metric G in the fibers of the bundle (9.2) over Y. For instance, one can choose local frames $\{e_j, j = 1, \ldots, n\}; \{f_\mu, \mu = 1, \ldots, m\}$ orthonormal in the fibers of respective bundles and define the metric G **requesting the frame** $\{e_\mu \otimes f_k\}$ **formed by sections of 9.2 to be** G**-orthonormal**.

In terms of local frames dx^i, ∂_μ and identifying vector spaces - fibers of the bundle (4.8) over Y with the tangent space to these fibers this metric has the form

$$G(dx^i \otimes \partial_\mu, dx^j \otimes \partial_\nu) = g^{ij} q_{Y\ \mu\nu}. \tag{9.3}$$

Coming back to the $J^1(\pi)$ and using the fact that the π_{10}-vertical tangent bundle of $J^1(\pi) \to Y$ is the vector bundle with the basis $\partial_{y^i_\mu}$, we get the metric in the fibres of this bundle defined by

$$G(\partial_{y^{mu}_i}, \partial_{y^\nu_j}) = g^{ij} q_{Y\ \mu\nu}. \tag{9.4}$$

The dual metric in the fibers of the bundle $V(\pi_{10})^*$ have the form

$$G^*(dy^\mu_i, dy^\nu_j) = G^{\mu\nu}_{ij} = g_{ij} q^{\mu\nu}_Y. \tag{9.5}$$

Lagrangian L defines the section $d_{v_{10}} L = L_{,y^\mu_i} dy^\mu_i$ of the bundle $V(\pi_{10})^* \to J^1(\pi)$ (where the "*vertical differential*" with respect to the variables in the fiber of the bundle $\pi_{10} : J^1(\pi) \to Y$ was used) To the section $d_{v_{10}} L$ there corresponds the orthogonal complement $dL^\perp$ in the metric G^* .

Define the morphism $K^* : V(\pi_{10})^* \to V(\pi)^*$ between the dual bundles of vertical bundles ν_{10} and ν_π by the conditions

$$\begin{cases} K^*(d_{v_{10}} L) = f_\mu dy^\mu, \\ K^*|_{dL^\perp} = 0. \end{cases} \tag{9.6}$$

This defines the tensor K^* and, therefore, the dual tensor $K = \{K^\alpha_{i\beta}\}$ that can be considered as the unique mapping $K : V(\pi) \to V(\pi_{10})$. More concretely, the

G^*-orthogonal projector P to the element $\pi = d_{v_{10}}L = \pi^i_\mu dy^\mu_i$ has the form

$$Pk = \frac{G^*(k, d_{v_{10}}L)}{\|d_{v_{10}}L\|^2_{G^*}} d_{v_{10}}L = \frac{G^{\mu\nu}_{ik} k^k_\nu \pi^i_\mu}{\|\pi\|^2_{G^*}} d_{v_{10}}L.$$

The corresponding bundle mapping $K^* : V(\pi_{10})^* \to V(\pi)^*$ has the form

$$\mathcal{K}^*k = \frac{G^{\mu\nu}_{ik} k^\nu_k \pi^i_\mu}{\|\pi\|^2_{G^*}} f_\alpha dy^\alpha.$$

the dual mapping $\mathcal{K} : V(\pi) \to V(\pi_{10})$, defined by the condition

$$< K^*k, \xi^\mu \partial_\mu >=< k, K(\xi^\mu \partial_\mu) >,$$

is

$$K(\xi^\mu \partial_\mu) = (\xi^\mu f_\nu) \frac{\pi^i_\mu}{\|\pi\|} G^{\nu\kappa}_{ik} \partial_{y^\kappa_k}. \tag{9.7}$$

As a result, **the "canonical" tensor** $\mathcal{K}$ has the form

$$K^\mu_{\nu i} = \left(\frac{\pi^j_\alpha}{\|\pi\|^2} G^{\mu\alpha}_{ij} \right) f_\nu. \tag{9.7'}$$

This tensor is defined at all points $(x^i, y^\mu, y^\mu_j) \in J^1(\pi)$ except those where $\pi = d_{v_{10}}L = 0$.

Theorem 3. *Let π be a tensor type bundle and let Γ be a connection in the bundle π. Let G be the metric in the fibers of the vertical bundle ν_{10} in the bundle $J^1(\pi) \to Y$. Then, the tensor*

$$K^\mu_{\nu i} = \left(\frac{\pi^j_\alpha}{\|\pi\|^2} G^{\mu\alpha}_{ij} \right) f_\nu. = \frac{\partial ln(\|\pi\|)}{\partial \pi^i_\mu} f_\nu = \chi^\mu_i f_\nu, \tag{9.8}$$

where

$$\chi^\mu_i = \frac{\partial ln(\|\pi\|)}{\partial \pi^i_\mu} \tag{9.9}$$

was introduced, is such that the Euler-Lagrange system with the variations of derivatives, modified by the tensor K has the form (6.4) of the Euler-Lagrange system with Lagrangian L and the source covector (force if $n = 1$) $f = f_\beta dy^\beta$.

Bundle mapping $K^* : V(\pi_{10})^* \to V(\pi)^*$ of the form

$$\mathcal{K}^*k = \frac{G^{\mu\nu}_{ik} k^\nu_k \pi^i_\mu}{\|\pi\|^2_{G^*}} f_\alpha dy^\alpha.$$

has minimal norm between all the mappings of these bundles such that $K^*(d_{v_{10}}L) = f_\beta dy^\beta$.

Remark 13. If the source functions f_β do not depend on the derivatives, i.e. if $f_\beta \in C^\infty(Y)$, then the tensor K has the "quasi-potential form" form

$$K^\mu_{j\nu i} = \frac{\partial}{\partial \pi^i_\mu}(ln(\|\pi\|) \cdot f_\nu). \tag{9.10}$$

Example 5. (n=1). If we take

$$K^\mu_{i\nu} = \frac{\partial ln(\|\pi\|)}{\partial \pi^i_\mu} Q_{\beta\nu} \dot{y}^\beta, \tag{9.11}$$

with a (0,2)-tensor $Q_{\beta\nu}$ in $J^1(\pi)$, the Euler-Lagrange Equations for a Lagrangian $L \in C^\infty(J^1(\pi))$ takes the form

$$\dot{y}^\beta = Q^{\beta\mu}\frac{\delta L}{\delta y^\mu}. \tag{9.12}$$

Following this scheme one can realize a large group of reaction-diffusion equations () as Euler-Lagrange equations of an appropriate Variational Principle.

Remark 14. Being universal, canonical form (9.7-9.7') of the NC-tensor K, defining non-commuting variations, is rarely the most convenient. As the examples presented in this text shows, most natural expressions for the non-commutativity tensor K are constructed using the specific quantities in the systems to study. Yet, in some examples, this form appears naturally in the construction of forces (source terms). For instance, linear connections participating in the "geometrization" of a mechanical system with non-potential forces (see Ch.V., Sec. 5.1) are directly related with the canonical form of the NC-tensor K.

10. Examples.

In this Section we present several examples of Lagrangian systems with non-commuting variations and classes of such systems. These examples illustrate the range and flexibility of method of NC-variations.

Example 1. Harmonic oscillator.

Here $n = 1$, t-time is the only independent variable, $x = x(t)$ is the only dynamical variable. There is only one 1-jet variable $\dot{x}$. Lagrangian L has the form

$$L = \frac{1}{2}\dot{x}^2 - \frac{\omega^2}{2}x^2.$$

Vertical variations have the form $\xi = \xi^1(t,x)\partial_x$. Being modified by a "NC-tensor" (here - scalar) $K(t,x,\dot{x})$ lift to 1-jet space has the form

$$\xi_K^{(1)} = \xi^1\partial_x + [d_t\xi^1 + K^t,x,\dot{x})\xi^1]\partial_{\dot{x}}.$$

Euler-Lagrange equation of harmonic oscillator with a non-commutative variation(s) has the form

$$L_{,x} - d_t(L_{,\dot{x}}) = -KL_{,\dot{x}} \Leftrightarrow \ddot{x} + (-K)\dot{x} + \omega^2 x = 0. \tag{10.1}$$

This is the standard harmonic oscillator with the (positive) dissipation for $K < 0$ and negative dissipation for $K > 0$. the energy balance law here has the form

$$d_t E = d_t(\frac{1}{2}\dot{x}^2 + \frac{\omega^2}{2}x^2) = K\dot{y}^2.$$

Example 2. Linear system with dissipation of rate type. In a linear system for a vector function $u : T \to R^m$,

$$\ddot{u} = Au + B\dot{u}, \tag{10.2}$$

with a symmetric $m \times m$-matrix A and negative definite symmetrical matrix B energy dissipates provided matrix B is negative definite : $B < 0$ (see [139]).

This system has the form (6.4) with the Lagrangian $L = \frac{1}{2}\|\dot{u}\|^2 - \frac{1}{2}(Au, u)$ (where standard Euclidian metric in the fibers R^m of the configurational bundle is used) and the tensor-potential $K^\mu_{i\nu} = -\delta^0_i B^\mu_\nu$.

Example 3. Heat equation.

Let

$$\theta_{,t} - D\Delta\theta = 0 \tag{10.3}$$

be a standard heat equation in the 4-dim Galilean space-time with the Euclidian metric and Cartesian coordinates $(x^0 = t, x^A, \ A = 1, 2, 3.$ Let $\theta(t, x)$ be the absolute temperature in the 4-dim domain $\Omega = [t_0, t_1] \times V \subset R^4$. Temperature is the only dynamical field.

Consider an action

$$A_\Omega(\theta) = \int_{t_0}^{t_1} \int_V \left[\frac{\partial\theta}{\partial t} + \frac{D}{2}\|\nabla\theta\|^2\right] dvdt. \tag{10.4}$$

Introduce the tensor K of the type $K^\mu_{i\nu}$, see Section 6 above, with the components $K^\theta_{i\theta} = \theta_{,t}\delta^0_i, \mu = 0, 1, 2, 3$. Variations of the derivatives defined as in (6.1) have the form

$$\begin{cases} \delta\frac{\partial\theta}{\partial x^A} = \frac{\partial}{\partial x^A}\delta\theta, \\ \delta\frac{\partial\theta}{\partial t} = \frac{\partial}{\partial t}\delta\theta + \frac{\partial\theta}{\partial t}\delta\theta. \end{cases}$$

It is easy to see that the Euler-Lagrange equation (6.9) takes the form (10.3).

Example 4. Laplace Equation.

Consider the case where $m = 1, n > 1$, base X is the Euclidian vector space E^n and the function $u(x^1, \ldots, x^m)$ satisfies to a EL-equation with the Lagrangian $L(u) = \frac{1}{2}\sum_{i=1}^{i=n}(u_{,x^i})^2$. NC-variation is defined by a "NC-tensor"" K^i . EL-equation (6.9) with NC variations has the form

$$\Delta u + K^i\partial_{x^i}u = 0. \tag{10.5}$$

Example 5. Burgers Equation.

Burgers equation

$$u_{,t} + uu_{,x} = \alpha u_{,xx},$$

can be realized, modifying the construction of Example 2, as the Euler-Lagrange Equations with the Lagrangian $L = u_{,t} + \frac{\alpha}{2}(u_{,x})^2$ and the action

$$A(u) = \int_{t_0}^{t_1} \int_{V^3} [u_{,t} + \frac{\alpha}{2}(u_{,x})^2]dvdt$$

and the commuting relations for variations with

$$K^u_{iu} = \begin{cases} u_{,t}, \ i = 0, \\ u, \ i = 1 \end{cases}$$

Moments π^i have the form: $\pi^0 = 1$, $\pi^1 = \alpha u_{,x}$. Energy density has the form $E = \frac{\alpha}{2}(u_{,x})^2$ and the energy balance law takes the form

$$d_t(\frac{\alpha}{2}(u_{,x})^2) + d_x(-\alpha(u_x)^2) = -(u^2_{,t} + uu_{,t}u_{,x}). \tag{10.6}$$

Similarly, modifying the construction of Example 2, one can realize the Burgers-KdV equation, Kuramoto-Sivashinsky equation and the Ginzburg-Landau equation (see [81]).

Example 6. Maxwell equations.

In this example we are using notations usual in classical electrodynamics, [80]. In particular, indices of fields are the same as indices of space-time coordinates since the fields here are tensor fields in the space-time.

Let $\xi \to M^4$ be the complex line bundle over the pseudo-riemannian manifold (M, g) with the space-time coordinates x^μ, $\mu = 0.1.2.3$. $A = A_\mu dx^\mu$ is a connection form in the bundle ξ, $F = dA = F_{\mu\nu} dx^\mu \wedge dx^\nu$, $F_{\mu\nu} = A_{\nu,\mu} - A_{\mu,\nu}$ - corresponding curvature form. Operator $*$ is the "star operator" corresponding to the metric g, $d_g v$ - corresponding volume form. Action for the connection A is

$$\mathcal{A}_{D^4}(A) = \int_D \|F\|_g^2 d_g v = \int_D F \wedge *F.$$

Maxwell equations in empty space are

$$\begin{cases} dF = 0, \\ d*F = 0. \end{cases}$$

Let $A_\sigma \to A_\sigma + \epsilon \xi^\sigma$ be a a variation of A corresponding to the vertical vector field $\xi = \xi^\sigma \partial_{A_\sigma}$. Let $K^\nu_{\mu\sigma}$ be tensor (corresponding to a vertical connection). Variation of the derivatives $z^\sigma_\mu = A_{\sigma,\mu}$ corresponding to the above variation of A_σ is $z^\sigma_\mu \to z^\sigma_\mu + (d_\mu \xi^\sigma + K^\sigma_{\lambda\mu} \xi^\lambda)$. Since only antisymmetric combinations of derivatives $A_{\sigma,\mu}$ enter Lagrangian, we may assume that tensor K is antisymmetric by corresponding variables.

Then, Euler-Lagrange equations for action $\mathcal{A}_{D^4}(A)$ with the variation rule (4.1) have the form

$$d*F = K*F \Leftrightarrow, \text{ in components } - \quad d_\alpha(*F)^{\alpha\beta} = -K^\beta_{\gamma\delta}(*F)^{\gamma\delta}. \tag{10.7}$$

The energy-momentum balance law takes the form

$$d_\mu(\|F\|_g^2 \delta^\mu_\sigma - 2F^\mu_\sigma) = -K^{\nu\lambda}_\mu F_{\nu\sigma}(*F)^\mu_\lambda. \tag{10.8}$$

Maxwell equations in a media with the dissipation of energy (generated heat) quadratic by the field force are well known in the Electrodynamics of Continuum, see, for example, [48], Ch.13.

If we introduce complex coefficients of dielectric (ϵ) and magnetic (μ) permutability into the tensor K in such a way that it is possible to separate electrical and magnetic terms we can describe a situation where the energy dissipation of a way of frequency ω (in the right side of 10.8) will have the form

$$\frac{\omega}{4\pi}(Im(\epsilon)\|E\|^2 + Im(\mu)\|H\|^2)$$

coinciding with the expression of energy dissipated **in a wave of frequency** ω in dispersive media, [79], Sec.61. NC-tensor K constructed in such way carries information about the dispersive properties of media.

Example 7. Rate type dissipation.

Let $n = 4, i = 0, 1, 2, 3$. Take $K^\mu_{i\nu} = k\delta^0_i \delta^\mu_{\mu_0} \delta^{\nu_0}_\nu$. Then, the lift of a vertical variation $\xi = \xi^\mu \partial_\mu$ to $J^1(\pi)$ has the form

$$\tilde{\xi} = \xi + (d_i \xi^\mu + k\delta^0_i \delta^\mu_{\mu_0} \xi^{\nu_0}) \partial_{z^\mu_i} = \xi^1 + k\xi^{\nu_0} \partial_{z^{\mu_0}_0}.$$

As a result the only *noncommutativity* here is the one of the time derivative ∂_t and the variation of μ_0-th dynamical field y^{μ_0}. The noncommutativity term is proportional to the variation of y^{ν_0}:

$$\delta \partial_t y^{\mu_0} - \partial_t \delta y^{\mu_0} = k \delta y^{\nu_0}.$$

The Euler-Lagrange system has the form

$$L_{y^\nu} - d_i(L_{,z_i^\nu}) = -k\delta_\nu^{\nu_0} L_{z_0^{\mu_0}} = -k\delta_\nu^{\nu_0} \pi_{\mu_0}^0. \tag{10.9}$$

Thus, the force appears only in the ν_0-th equation and is proportional to the μ_0**-th component of linear momentum** $\pi_{\mu_0}^0$. The simplest case is, of course, when $\mu_0 = \nu_0$ and the force that appears in the equation for y^{μ_0} is proportional to π_{μ_0}. Energy-momentum law takes the following form

$$d_i T_0^i = -\frac{\partial L}{\partial x^0}\bigg|_{expl} - k\dot{y}^{\mu_0}\pi_{\nu_0}^0 \stackrel{\mu_0 = \nu_0}{=} -\frac{\partial L}{\partial x^0}\bigg|_{expl} - k\dot{y}^{\mu_0}\pi_{\mu_0}^0. \tag{10.10}$$

In a classical case when Lagrangian L is the sum of kinetic energy $K = \frac{1}{2}m_{\mu\nu}\dot{y}^\mu\dot{y}^\nu$ and potential energy (often - a function of fields y^μ only) and where we take $\mu_0 = \nu_0$, the dissipation term $k\dot{y}^{\mu_0}L_{,\dot{y}^{\mu_0}}$ is proportional to the square of the rate of change of the field y^{μ_0}.

Example 8. Gradient type dissipation. If we take $K_{0j}^i = 0$, the dissipative force in the right side of Euler-Lagrange equations (6.6-6.9) takes the form (6.9), $-\sum_{A=1,2,3} K_{A\nu}^\mu \pi_\mu^A$ and the second term in the energy balance (7.6) becomes equal to $-\dot{y}_\nu K_{A\nu}^\mu \pi_\mu^A$. As a result, this dissipation term is proportional to the spacial gradient of the fields y^i, more specifically, their momenta π_μ^A.

Example 9. Controlled oscillator. In the next example, evolution of one dynamical variable is influenced by the rate of change of the other variable. Consider a system of two oscillators with the variables $y^\mu, \mu = 1, 2$ and the Lagrangian $L = \frac{m_1}{2}(\dot{y}^1)^2 + \frac{m_2}{2}\dot{y}^2)^2 - U(y^1, y^2)$. Let K_ν^μ be of the type $K_\nu^\mu = \partial_\nu P^\mu$ with some functions $P^\mu(y^\mu, y^2)$.

Then, the EL-equations with the force induced by "tensor" K takes the form (see Sec.)

$$\begin{cases} m_1\ddot{y}^1 + U_{,y^1} = \partial_{y^1}(m_1 P^1 \dot{y}^1 + m_2 P^2 \dot{y}^2), \\ m_2\ddot{y}^2 + U_{,y^2} = \partial_{y^2}(m_1 P^1 \dot{y}^1 + m_2 P^2 \dot{y}^2), \end{cases} \tag{10.11}$$

Consider the case where $P^1 = 0$, $m_2 P^2 = \phi(y^\mu)$. Then the last system takes the form

$$\begin{cases} m_1\ddot{y}^1 + U_{,y^1} = \phi_{,y^1}\dot{y}^2, \\ m_2\ddot{y}^2 + U_{,y^2} = 0. \end{cases} \tag{10.12}$$

Thus, the second oscillator influences the dynamical behavior of the first one through the rate of change $\dot{y}^2$ of the variable y^2.

Remark 15. In Sec. (17) it will be shown that such a tensor K corresponds to a *vertical connection* in the bundle $J^1(\pi) \to Y$ with zero curvature.

11. Weak and strong minimizers for the systems with NC-variations.

In this section we compare week and strong minimizers for Lagrangian systems with and without NC-variations. We show that in both these cases strong minimizers are always the same and weak minimizers are the same provided the NC-tensor K is bounded from above and below. We refer reader to the fundamental monograph [47], Vol.2 Chapter 5 for more details and examples.

Local minimizers are sections $s(x)$ of the configurational bundle that delivers minimum to the action functional in some neighborhood of section $s(x)$. In order to specify a type of local minimizers one has to specify the *type of neighborhoods U* of "potential local minimizer" s so that from such neighborhoods of s we take the sections $p(x)$ for comparison of value of action $\mathcal{A}_D(p)$ with the value $\mathcal{A}_D(s)$.

This is done by the choice of metric (or norm) in the space of sections $s(x)$ used in the problem.

In this section we assume that the configurational bundle $\pi : Y \to X$ is a vector bundle with the standard fiber $F \equiv R^m$ (see Appendix I,Sec.64). In particular, in the domain U of a fibred chart (U, x^i, y^μ) (over the domain $\bar{U} = \pi(U) \subset X$) where bundle π is trivial: $\pi_U \equiv (\bar{U} \times F \to \bar{U}$, sections $s : \bar{U} \to Y$ defines and are defined by the mappings $u : \bar{U} \to F$.

Sections $s(x)$ whose components $y^\alpha(x)$ have continuous derivatives $D_I y^\alpha$. up to the order k, $|I| \leqq k$ form the Banach space $C^k(\bar{U})$ with the norm

$$\|s\|_k = \underset{x \in U}{Sup}\{\|D_I u(x)\|, |0 \leqq \alpha \leqq k|.$$

Here $I = (i_1, i_2, \ldots, i_n)$ is a multiindex (i_k are natural numbers) and $D^I y^\alpha = \partial_1^{i_1} \cdot \partial_n^{i_n}$ - corresponding partial dervative of order $|I| = i_1 + i_2 + \ldots + i_n$ of the function $y^\alpha(x)$.

Let $L(x^i, y^\mu, y_i^\mu)$ be a first order Lagrangian and let $D \subset X$ be an open subset with the closure $\bar{D}$. Action $A_D(s)$ is defined for the sections $s \in C^1(\bar{D}, Y)$ $(C^1(\bar{D}, F)$ in the domain of a fibred chart) of the class C^1 *up to the boundary*. This means that for any chart $(U \subset D, x^i, y^\mu)$ components $y^\mu(x))$ of a section $s(x)$ are functions of the class C^1 up to the border ∂D.

Now we define

Definition 3. *For a section $s \in C^1(\bar{D}, Y)$ and for some $\epsilon > 0$, the set*

$$N_\epsilon(s) = \{p \in C^1(\bar{D}, Y) : \|s - p\|_{C^0} < \epsilon\} \tag{11.1}$$

is called a (standard) **strong neighborhood of section s.**

On the other hand,

Definition 4. *For a section $s \in C^1(\bar{D}, Y)$ and for some $\epsilon > 0$, the set*

$$N_\epsilon^1(s) = \{p \in C^1(\bar{D}, Y) : \|p - s\|_{C^1} < \epsilon\} \tag{11.2}$$

is called a (standart) **weak neighborhood of section s.**

It is clear that $N_\epsilon^1(s)$ is the subset of $N_\epsilon(s)$ and that $N_\epsilon^1(s) \subset N_\epsilon(s)$ is the (strict) inclusion.

Now we define the notions of **weak and strong minimizers**.

Definition 5.   (1) *A section $s : D \to Y$ of the class $C^1(\bar{D}, \pi)$ is called a **weak local minimizer** of action $\mathcal{A}_D(s)$ if there exists a weak ϵ-neighborhood $N^1_\epsilon(s)$ of section s such that*

$$\mathcal{A}_D(s) \leqq \mathcal{A}_D(p), \text{ for all sections } p \in N^1_\epsilon(s). \qquad (11.3)$$

(2) *A section $s : D \to Y$ of the class $C^1(\bar{D}, \pi)$ is called a **strong local minimizer** of action $\mathcal{A}_D(s)$ if there exists a strong ϵ-neighborhood $N_\epsilon(s)$ of s such that*

$$\mathcal{A}_D(s) \leqq \mathcal{A}_D(p), \text{ for all sections } p \in N_\epsilon(s). \qquad (11.4)$$

It is obvious that any strong (local) minimizer is, at the same time, the weak (local) minimizer. Next example shows that the opposite statement would be wrong.

Example 6. -(Exercise) ([118],Sec.2.1.)). Prove that zero function $u(x) = 0$ is a weak minimizer but is not **the strong** minimizer of the functional

$$\mathcal{A}(u) = \int_0^\pi (y^2(x))(1 - y'^2(x))dx,$$

with the initial condition $y(0) = y'(0) = 0$.
Hint: consider $y(x) = \frac{1}{\sqrt{n}} sin(nx)$.

If the boundary ∂D is piecewise smooth, conditions for a section s to be a strong or weak minimizer (see Definition 5) are equivalent to the following **strong and weak minimum properties**.

Definition 6.   (1) *A section $s : D \to Y$ of the class $C^1(\bar{D}, \pi)$ satisfies to the **strong minimum property** if for some $\epsilon > 0$, such that inequality $\mathcal{A}_D(s) \leqq \mathcal{A}_D(s + \phi)$ holds for all $\phi \in C^1(\bar{D})$ such that $\|\phi\|_{C^0(\bar{D})} < \epsilon$.*
(2) *A section $s : D \to Y$ of the class $C^1(\bar{D}, \pi)$ satisfies to the **weak minimum property** if for some $\epsilon > 0$, such that inequality $\mathcal{A}_D(s) \leqq \mathcal{A}_D(s^* + \phi)$ holds for all $\phi \in C^1(\bar{D}, \pi)$ such that $\|\phi\|_{C^1(\bar{D})} < \epsilon$.*

11.1. Minimizers for systems with NC-variations.

Let the configurational bundle $\pi : Y \to X$ be a vector bundle (See Appendix I).

Properties of a section $s : D \to Y$ to satisfy the strong or weak minimum property is formulated in terms of variations $s \to s + \phi$ of sections s. For the strong minimum property variation ϕ is estimated in the norm $\|\phi\|_{C^0}$ and, as a result, is calculated using the metric (norm in the case of a vector bundle) in fibers Y_x of the configurational bundle π and the Sup norm for the continuous sections of π.

A strong ϵ-neighborhood of a section s is defined by the condition: $p(D) \subset O_\epsilon(s)$ if

$$\|p - s\|_{C^0(D)} = Sup_{x \in D}|p(x) - s(x)| \leqq \epsilon$$

This neighborhood does not depend on a prolongation of section s to the 1-jet bundle. As a result, **strong ϵ-neighborhoods are the same for conventional and NC-modified variations.** Therefore, **strong minimizers are the same whether one is using conventional or NC-modified prolongations.**

For the weak minimum property, the norm of variation ϕ is calculated using the prolongation of sections s and $s + \phi$ to the 1-jet bundle $\pi_1 : \; J^1(\pi) \to Y \to X$ and estimating the C^1-norm of variation $\|\phi\|$.

If we are using a non-commutative variation approach, prolongation of variation ϕ differs from the conventional one. As a result, weak ϵ-neighborhoods for the system with NC-variations could be different from those in the case of conventional variations. To study this difference we need to look more attentively at the form of C^1-neighborhoods of 1-jet prolongations of sections of the configurational bundle.

More specifically, in the definition of weak ϵ-neighborhood of a section s, one has to use (in the case of a vector bundle π) variations $s(x) \to s(x) + \phi(x)$ such that

$$\|\phi\|_{C^1}(D) = \|Pr_K^1(\phi)\|_{C^0(D)} < \epsilon. \tag{11.5}$$

This condition contains the Sup norms of the components $d_\mu \phi^i + K_{\mu j}^i \phi^j$ of vector field $Pr_K^1 \phi$ in $J^1(\pi)$ (weak variation components).

In the case of a vector bundle, ϵ-neighborhood $O_\epsilon(s)$ (in sense of any of our norms) of a section s has the form $s + O_\epsilon(0)$ where $O_\epsilon(0)$ is the ϵ-neighborhood of zero section of the bundle π. Thus, to compare minimality properties of sections s, it is sufficient to consider the case where s is zero section.

Considering a more general setting, assume that the bundle π is endowed with a Riemannian metric (in the fibers Y_x of the bundle π) and introduce a **tubular neighborhood** $O_\epsilon(s_*)$ of a section $s_* : D \to Y$ (see [6], Sec.8.1 or [86]) and the description of tubular neighborhood below) of the image $s_*(D) \subset Y$ of section s (notice that $s(D)$ is diffeomorphic to D and is the regular submanifold of Y) **generated by geodesics of metric in the fibers Y_x.**

A tubular neighborhood is obtained by using the exponential mappings $exp : T_{s_*(x)}(Y_x) \to Y \colon \gamma_v(t) : exp(tv)$ for unit vectors $v \ \|v\| = 1$. Then, for small ϵ the subsets

$$\{(x, \gamma_v(t)) | v \in T_{s_*(x)}(Y_{s_*(x)}), \ 0 \leqq t \leqq \epsilon\}$$

scan the ϵ-neighborhood of $s_*(D)$. To prove this, it is sufficient to notice that the geodesic are shortest curves for small ϵ, and, if a section $p : D \to Y$ is such that $\|s_* - p\| \leqq \epsilon$, then $p(D) \subset \mathcal{O}_\epsilon(s)$.

To describe weak neighborhoods of a section s we will extend the description of strong tubular ϵ-neighborhoods of the image $s(X)$ of a section $s : D \to Y$ above to the variations of 1-jets $j^1 s$ of sections s.

To do this we have to employ a metric in the fibers of double bundle $J^1(\pi) \to Y \to X$. To introduce a metric, we recall that the bundle $\pi_{10} : J^1(\pi) \to Y$ is the vector bundle provided the configurational bundle $\pi : Y \to X$ is. Namely, a connection ν in the vector bundle π defines (and is defined by) the section $q_\nu : Y \to J^1(\pi)$ (See Appendix I, Sec.75). This section defines the origin $0_y = q_\nu(y)$ at any fiber $J_y^1(\pi)$ specifying the structure of vector space in this fiber and the splitting

$$T(J^1(\pi)) = V(\pi_{10}) \oplus q_{\nu*}(T(Y))$$

at the points of $q(Y)$. The action of a linear group acting in the fibers of $\pi_{10} : J^1(\pi) \to Y$ extends this decomposition to the whole 1-jet bundle.

Choose a norm in the fibers of the vector bundle $J^1(\pi) \to Y$ smoothly depending on (x, y). Together with the norm in the fibers Y_x of the bundle π this delivers the norm in the fibers of the vector bundle $J^1(\pi) \to Y \to X$ over X. This norm (smoothly depending on $x \in X$) defines the Riemannian metric in the fibers $J^1(\pi)_x$ of $\pi_1 : J^1(\pi) \to X$.

Now we repeat arguments used for the previous construction to the case of 1-jets $j^1(p)$ of sections p in the neighborhoods of the section $s_*(x)$.

Take an infinitesimal variation presented by a π-vertical vector field $\xi = \xi^i \partial_i$ and let ϕ^t be the flow of this vector field. This vector field induced the variation $\phi^t(x) s_*(x)$ of the section $s_*(x)$.

Let $Pr^1(\xi) = \xi^i \partial_{y^i} + d_\mu^i \xi^i \partial_{y_\mu^i}$ be the flow prolongation of $\xi^i \partial_{y^i}$- vector field in the 1-jet space $J^1(\pi)$ whose phase flow is acting on the 1-jet $j^1 s_*$ of section s_*

$$j^1 \phi^t s = \widehat{\phi}^t j^1 s_* = e^{t(\xi^i \partial_i + d_\mu^i \xi^i \partial_{y_\mu^i})} j^1 s_*(x) \tag{11.6}$$

Such prolongations for all vector fields ξ (normal to the surface $s_*(D) \subset Y$) and for small t, say for $t \leq \epsilon$, scan the neighborhood of 1-jet section $j^1 s_*$.

Using the *Sup*-norm for sections of 1-jet bundle $\pi_1 : J^1(\pi) \to X$ we obtain the (equivalent of) C^1-norm for the sections $s_*(x)$ of the bundle π.

For a chosen $\epsilon > 0$, there exists $\delta > 0$ such that if $t \leq \delta$, the distance between $\phi_t s_*$ and s_* is less then ϵ in the C^1-norm. Taking ϵ small enough we prove that to check the condition of weak minimum for section s_*, it is sufficient to use weak neighborhoods obtained this way - by using flows of variational vector fields.

In the case of a non-commuting variations formalism defined by a tensor $K_{i\mu}^\nu$, expression (11.6) for the flows scanning the weak neighborhood of a potential weak minimizer s_* has the form

$$Pr^1 \phi_K^t j^1 s_* = e^{t(\xi^i \partial_i + (d_\mu^i \xi^i + K_{\mu j}^i \xi^j) \partial_{y_\mu^i})} j^1 s_*(x). \tag{11.7}$$

Calculate now the decline of 1-jets of variated sections from the 1-jet of the section s_* itself:

$$Pr^1 \phi_K^t j^1 s_* - j^1 s_* \approx (t(\xi^i(s(x)) + (d_\mu^i \xi^i + K_{\mu j}^i \xi^j)(j^1 s_*(x))) + O(t^2). \tag{11.8}$$

Estimating the terms in the right side we get

$$\|Pr^1 \phi_K^t j^1 s - j^1 s\|_{C^1(D)} \approx t \sum_i Sup_{x \in D} |\xi^i(s(x))| + t Sup_{x \in D} \sum_i |(d_\mu \xi^i + K_{j\mu}^i \xi^j)| + o(t).$$

$$\tag{11.9}$$

Now we notice that in a case where $|K_{j\mu}^i| \leq C$, the right hand side is less then or equal to

$$t \sum_i Sup_{x \in D} |\xi^i(s(x))| + t Sup_{x \in D} \sum_i |(d_\mu \xi^i))| + Ct \|\xi\|_{C^0} \leq$$

$$\leq (1 + C) t \|\xi\|_{C^0} + t \sum_i \|(d_\mu \xi^i))\|_{C^0} \leq (1 + C) t \|Pr^1(\xi)(j^1 s(x))\|. \tag{11.10}$$

and, therefore, it is less or equal to $(1 + C) \|Pr \phi^t j^1 s - j^1 s\|_{C^0}$. Thus, any K-weak C^1-ϵ-neighborhood of a section s is contained in the conventional weak C^1-$\frac{\epsilon}{1+C}$=neighborhood of a section s. As a result, if the weak minimum property is fulfilled for the section s in the conventional sense, it is, at the same time, fulfilled in the $K_{i\mu}^\nu$-modified sense.

Along the phase curves of variational vector fields, the tensor K can behave as $t^{-\alpha}$ with $\alpha > 1$. In such a case, no diminishing of delta can prevent the difference of the norms of jets of section s_* and the variated one to leave the weak ϵ-neighborhood of section s_* - weak minimizer for the case of commuting variations.

Thus, we get to the conclusion:

Proposition 1. *For a Lagrangian variational system with a Lagrangian L of the first order and the non-commuting variations defined by a NC-tensor $K^{\nu}_{i\mu}$,*

(1) *Strong minimizers of the action $\mathcal{A}_D(s)$ with Lagrangian L and with commuting variations and only them are strong minimizers of action with the same Lagrangian and a non-commutativity tensor $K^{\nu}_{i\mu}$.*

(2) *If tensor $K^{i}_{\mu j}$ of the non-commutativity of variations is bounded: $|K^{i}_{j\mu}(s)| \leqq C$, then the weak minimizers of action with a first order Lagrangian L and the conventional variations is, at the same time, local weak minimizer in the K-modified formalism.*

Conjecture:

(1) If tensor $K^{\nu}_{i\mu}$ has singularities, weak minimizers in the conventional sense may cease to be weak minimizers provided the NC-variations are used,

(2) If $c \leqq \|K^{\nu}_{i\mu}\| \leqq C$ for positive constants $c < C$, then the weak variations are the same for the conventional variations and for NC-variations defined by the tensor $K^{\nu}_{i\mu}$.

12. Second variation.

The Euler-Lagrange Equations for a system with a Lagrangian L and a non-commutativity tensor K state that the *first variation of action functional* $\delta\mathcal{A}(s)$ is zero along a solution $s(x) = \{y^\mu(x)\}$. *The second variation* along a solution $s(x)$ - $\delta^2\mathcal{A}(s)$ may then be used to recognize a solution $s(x)$ as a local minimum or local maximum of the action functional (Legendre and Jacobi necessary conditions, see [44], Ch.5). In addition, study of the second variation leads to the sufficient conditions of weak extremum ([44], Ch.5). See also [47], V.I,Ch.3 for a more modern exposition of these results.

Thus, it is natural to investigate what form the results mentioned above take in the case of non-commuting variations.

In the case for a Lagrangian L of order one, we have for an action $\mathcal{A}_D(s + \delta s)$ calculated for a a solution $s(x) = \{y^\mu(x)\}$ of the Euler-Lagrange system subject to a variation $\delta y^\mu + \delta y_i^\mu$ of the form $\epsilon[\xi^\mu \partial_{y^\mu} + (d_i\xi^\mu + K^\mu_\sigma \xi^\sigma)\partial_{y_i^\mu}]$,

$$\mathcal{A}_D(s + \delta s) = \int_D L(x^i, y^\mu + \delta y^\mu, y_i^\mu + \delta y_i^\mu)d^n x =$$

$$= \int_D [L(j^1 s(x)) + \epsilon L_{,y^\mu}(j^1 s(x))\xi^\mu(s(x)) + \epsilon L_{,y_i^\mu}(j^1 s(x))(d_i\xi^\mu + K^\mu_{i\kappa}\xi^\kappa)(j^1 s(x))+$$

$$+ \frac{1}{2}\epsilon^2[L_{y^\mu y^\nu}\xi^\mu\xi^\nu + L_{y^\mu y_i^\nu}\xi^\mu(d_i\xi^\nu + K^\nu_{i\kappa}\xi^\kappa) + L_{,y_i^\mu y_j^\nu}(d_i\xi^\mu + K^\mu_{i\kappa}\xi^\kappa)(d_j\xi^\nu + K^\nu_{j\kappa}\xi^\kappa)]$$

$$+ HOT(\epsilon)]d^n x = \mathcal{A}_D(s) + \epsilon \int_D \left[L_{,y^\mu} - d_i L_{,y_i^\mu} + K^\kappa_{i\mu}L_{,y_i^\kappa}\right](j^1 s(x))\xi^\mu(s(x))d^n x+$$

$$+ \frac{\epsilon^2}{2}\int_D [(L_{,y^\mu y^\nu} + L_{,y^\mu y_i^\kappa}K^\kappa_{\nu i} + L_{y_i^\kappa y_j^\lambda}K^\kappa_{i\mu}K^\lambda_{j\nu})\xi^\mu\xi^\nu +$$

$$+ (L_{y^\mu y_i^\nu}\xi^\mu d_i\xi^\nu + L_{,y_i^\kappa y_j^\nu}K^\kappa_{i\mu}\xi^\mu d_j\xi^\nu + L_{,y_i^\mu y_j^\kappa}K^\kappa_{j\nu}\xi^\nu d_i\xi^\mu)+$$

$$+ L_{,y_i^\mu y_j^\nu}d_i\xi^\mu d_j\xi^\nu]d^n x + HOT(\epsilon). \quad (12.1)$$

As a result, the **second variation** of the action functional $A_D(s)$ is the quadratic form of the arguments $(\xi^\mu, d_i\xi^\mu)$ the form

$$\delta^2 A(\epsilon\xi^\mu, \epsilon d_i\xi^\mu) = \frac{\epsilon^2}{2}\int_D [(L_{,y^\mu y^\nu} + L_{,y^\mu y_i^\sigma}K^\sigma_{\nu i} + L_{y_i^\sigma y_\nu^\varsigma}K^\sigma_{i\mu}K^\varsigma_{j\nu})\xi^\mu\xi^\nu +$$

$$+ (L_{y^\mu y_i^\nu}\xi^\mu d_i\xi^\nu + L_{,y_i^\sigma y_j^\nu}K^\sigma_{i\mu}\xi^\mu d_j\xi^\nu + L_{,y_i^\mu y_j^\nu}K^\sigma_{j\nu}\xi^\nu d_i\xi^\mu)+$$

$$+ L_{,y_i^\mu y_j^\nu}d_i\xi^\mu d_j\xi^\nu]d^n x. \quad (12.2)$$

In a case where $n = 1$ (Mechanics), the second variation has the form

$$\delta^2 A(\xi^\mu, d_t\xi^\mu)(\epsilon\xi, \epsilon\xi) = \frac{\epsilon^2}{2}\int_a^b [(L_{,y^\mu y^\nu} + L_{,y^\mu \dot{y}^\sigma}K^\sigma_\nu + L_{,\dot{y}^\sigma \dot{y}^\varsigma}K^\sigma_\mu K^\varsigma_\nu)\xi^\mu\xi^\eta +$$

$$+ (L_{,y^\mu \dot{y}^\nu}\xi^\mu d_t\xi^\nu + L_{,\dot{y}^\sigma \dot{y}^\nu}K^\sigma_\mu\xi^\mu d_t\xi^\nu + L_{,\dot{y}^\mu \dot{y}^\sigma}K^\sigma_\nu\xi^\nu d_t\xi^\mu)+$$

$$+ L_{,\dot{y}^\mu \dot{y}^\nu}d_t\xi^\mu d_t\xi^\nu]dt. \quad (12.3)$$

If, additionally, $m = 1$, the second variation has the following form (comp [44], Ch.V)

$$\delta^2 A(\xi, d_t\xi) =$$

$$= \frac{\epsilon^2}{2} \int_a^b [(L_{,yy} + L_{,y\dot{y}}K + L_{,\dot{y}\dot{y}}K^2)\xi\xi + 2(L_{,y\dot{y}} + L_{,\dot{y}\dot{y}}K)\xi d_t\xi + L_{,\dot{y}\dot{y}}d_t\xi d_t\xi]dt.$$

$$(12.4)$$

Now we integrate by parts the expression for the second variation (12.3) in the case where $n = 1$ (Mechanics). We assume that variations ξ^i vanish on the boundary ∂D of the domain of integration D. We also assume, as is done in [44], Sec. 29, that $L_{y^\mu \dot{y}^\nu} = L_{y^\nu \dot{y}^\mu}$. Finally, to simplify the obtained expression we add the condition that $L_{\dot{\sigma}\mu}K_\nu^\sigma = L_{\dot{\sigma}\nu}K_\mu^\sigma$ has to be valid for all μ, ν. As a result, we get the following expression for the second variation

$$\delta^2 A(\xi^\mu, d_t\xi^\mu)(\epsilon\xi, \epsilon\xi) =$$

$$= \frac{\epsilon^2}{2} \int_D \{[L_{,\dot{y}^\mu \dot{y}^\nu}]d_t\xi^\mu d_t\xi^\nu +$$

$$+[(L_{,y^\mu y^\nu} + L_{,y^\mu \dot{y}^\sigma}K_\nu^\sigma + L_{,\dot{y}^\sigma \dot{y}^\varsigma}K_\mu^\sigma K_\nu^\varsigma) - \frac{1}{2}d_t(L_{,y^\mu \dot{y}^\nu} + L_{,\dot{y}^\sigma \dot{y}^\nu}K_\mu^\sigma + L_{,\dot{y}^\mu \dot{y}^\sigma}K_\nu^\sigma)]\xi^\mu \xi^\nu)]\}dt =$$

$$= \epsilon^2 \int_a^b [P_{\mu\nu}\dot{\xi}^\mu \dot{\xi}^\nu + Q_{\mu\nu}\xi^\mu \xi^\nu]dt. \quad (12.5)$$

Here,

$$\begin{cases} P_{\mu\nu} = \frac{1}{2}L_{,\dot{y}^\mu \dot{y}^\nu}, \\ Q_{\mu\nu} = \frac{1}{2}[(L_{,y^\mu y^\nu} - d_t L_{,y^\mu \dot{y}^\nu}) + (L_{,y^\mu \dot{y}^\sigma}K_\nu^\sigma + L_{,\dot{y}^\sigma \dot{y}^\varsigma}K_\mu^\sigma K_\nu^\varsigma - d_t(L_{,y^\mu \dot{y}^\nu} + L_{,\dot{y}^\sigma \dot{y}^\nu}K_\mu^\sigma + L_{,\dot{y}^\mu \dot{y}^\sigma}K_\nu^\sigma)]. \end{cases}$$

$$(12.6)$$

Notice that the (0,2)-form P does not depend on the non-commutativity tensor K. If we assume, as it is suggested in [44], Sec.29, that the matrix $L_{y^\mu \dot{y}^\nu}$ **is symmetric**, then both matrices $P_{\mu\nu}, Q_{\mu\nu}$ are symmetric and the classical theory of sufficient conditions of Legendre-Jacobi is applied immediately with the corresponding modification of the Jacobi's equation.

In particular, necessary Legendre condition has the usual form:

Theorem 4. *A necessary condition for the quadratic functional*

$$\xi \to \int_a^b [(P\dot{\xi}, \dot{\xi}) + (Q\xi, \xi)]dt$$

to be non-negative for all $\xi(t)$ such that $\xi(t)_{t=a,b} = 0$, is that the matrix P_{ij} is non-negative definite.

12.1. **Jacobi equation.** Expression (12.5) defines the quadratic functional

$$J(h, h) = \epsilon^2 \int_a^b [P_{\mu\nu}\dot{h}^\mu \dot{h}^\nu + Q_{\mu\nu}h^\mu h^\nu]dt$$

for a vector function $h : R \to R^m$. Euler-Lagrange system of this functional is called the **Jacobi equation:**

$$-\frac{d}{dt}P_{\mu\nu}\dot{h}^\nu + Q_{\mu\nu}h^\nu = 0, \quad \mu = 1, \ldots, m. \quad (12.7)$$

Remark 16. The principal part of Jacobi equations do not depend on the tensor $K^i_{\mu j}$ and is the same as in classical theory. At the same time, zero order terms in Jacobi equations depend on the tensor K linearly and quadratically.

The classical theory of quadratic functionals, the Jacobi equation and the conjugate points ([44]) are applied here without any restrictions. In particular, the following two classical result are valid here:

Theorem 5. *(Legendre necessary condition for $\mathcal{A}(s)$, case $n = 1$). If the vector function (solution of Euler-Lagrange equations) $s(t) = \{y^\mu(t)\}$ yields a **local minimum** of the functional $\mathcal{A}_D(s)$, then the matrix function*

$$P_{\mu\nu}(s(t)) = \frac{1}{2}L_{,\dot{y}^\mu \dot{y}^{j\nu}}(s(t))$$

is nonnegative definite $(P_{ij}(s(t)) \geqq 0)$ along the solution $s(x)$.

Theorem 6. *Jacobi necessary condition for $\mathcal{A}(s)$).*

If the regular (i.e.such that the condition $det(\frac{\partial^2 L}{\partial \dot{y}^\mu \partial \dot{y}^\nu})(s(t)) \neq 0$ is fulfilled along all the curve $s(t)$) extremal $s(t) \in C^2([a,b] \to R^m)$ of variational problem yields a local minimum for the action functional $\mathcal{A}(s)$ with $L(t, y^\mu, \dot{y}^\nu) \in C^3(J^1(\pi))$, then the interval (a, b) does not contain points conjugate to the point a.

Example 7. Consider the harmonic oscillator with dissipation (Example 1, Sec. 10) with the NC-tensor (being a scalar function $K(t, x, \dot{x})$) K. Jacobi equation for a function h is

$$\ddot{h} + \omega^2 - 2d_t K - K^2 = 0. \tag{12.8}$$

If K is constant and $K^2 - \omega^2 > 0$, then, there are no no points $t > 0$ conjugate to the point $t = 0$ and the lower position of equilibrium is stable.

Remark 17. Necessary and sufficient second variation criteria for weak minimizers were studied in terms of spectral properties (minimal eigenvalue) of Jacobi operator - linearization of Euler-Lagrange operator, starting with the works of H.A.Schwartz (1885), see [47], Part II, Ch.5. In the paper [5], T.Atanascovich obtained a sufficient second variation condition for local extremum in a mechanical system with non-conservative forces for the variational principle with the non-commuting variations. This condition has the form $\lambda_{min} + \gamma > 0$ where λ_{min} is the minimal eigenvalue of the appropriate Jacobi operator while γ is the constant participation in the positive definiteness condition of the quadratical form of second variation with coefficients depending on the non-conservative forces.

13. Hamiltonian systems and the NC-variations.

In this section we present the Hamiltonian systems corresponding to the Euler-Lagrange equations with non-commuting variations. We will follow the classical approach of H.Rund ([117]).

Let L be a regular Lagrangian of the first order (i.e. such that determinant $Det(\frac{\partial^2 L}{\partial y^\mu_i y^\nu_j})$ of Hessian of the function L is nonzero) and let $K^\mu_{i\nu}$ be a NC-tensor as in Section 6. Regularity condition allows to apply Legendre transformation

$$J^1(\pi) \xrightarrow{\quad \mathcal{L} \quad} J^{1*}(\pi)$$
$$\downarrow \qquad\qquad \downarrow$$
$$Y \xrightarrow{\quad = \quad} Y$$

from the 1-jet space to the dual space $J^{1*}(\pi)$ of variables (xi, y^μ, π^i_μ):

$$(x^i, y^\mu, y^\mu_i) \to (x^i, y^\mu, \pi^i_\mu = L_{,y^\mu_i}).$$

One can consider this transformation as defining the new local chart in $J^1(\pi)$ (Legendre coordinates, see [44]). Due to the regularity condition, this transformation is invertible, i.e., in the domain of original fibred coordinates one has

$$y^\mu_i = \phi^\mu_i(x, y, \pi) \tag{13.1}$$

with the smooth functions $\phi \in C^\infty(J^{1*}(\pi))$.

Introduce the Hamiltonian function

$$H(x, y, \pi) = \pi^i_\mu \phi^\mu_i(x^i, y^\mu; \pi^j_k) - L(x^i, y^\mu; \phi^\mu_i(x, y, \pi)). \tag{13.2}$$

Using 13.1 we calculate (using, at the last equality, the definition of π)

$$\frac{\partial H}{\partial \pi^j_\nu} = -\frac{\partial L}{\partial \phi^\mu_i}\frac{\partial \phi^\mu_i}{\partial \pi^j_\nu} + \pi^i_\mu \frac{\partial \phi^\mu_i}{\partial \pi^j_\nu} + \phi^\nu_j = \phi^\nu_j = y^\nu_j. \tag{13.3}$$

Similarly, one proves that

$$\frac{\partial H}{\partial y^\mu} = -\frac{\partial L}{\partial y^\mu}, \quad \mu = 1, \ldots, m, \tag{13.4}$$

and

$$\frac{\partial H}{\partial x^i} = -\frac{\partial L}{\partial x^i}, \quad i = 1, \ldots, n.$$

Consider now the Euler-Lagrange Equations with the sources f_ν:

$$d_\mu\left(\frac{\partial L}{\partial y^\nu_i}\right) - \frac{\partial L}{\partial y^\nu} = f_\nu, \quad \nu = 1, \ldots, m,$$

substitute $\pi^i_\nu = L_{,y^\mu_i}$ here and use (13.4). This equation takes the form

$$d_i \pi^i_\nu = -\frac{\partial H}{\partial y^\nu} + f_\nu. \tag{13.5}$$

Combining this equation with (13.3) we will get Hamiltonian (canonical) system of equations equivalent to the Euler-Lagrange system with the sources f_ν

$$\begin{cases} y^\mu_{,i} = \frac{\partial H}{\partial \pi^i_\mu}, \\ d_i \pi^i_\nu = -\frac{\partial H}{\partial y^\nu} + f_\nu. \end{cases} \tag{13.6}$$

If the sources f_ν in (13.5) are generated by the tensor K, Hamiltonian system (13.6) takes the form

$$\begin{cases} y^\mu_{,i} = \frac{\partial H}{\partial \pi^i_\mu}, \\ d_i \pi^i_\nu = -\frac{\partial H}{\partial y^\nu} + K^\mu_{i\nu} \pi^i_\mu. \end{cases} \tag{13.7}$$

Consider the case where $n = 1$, i.e. the case of finite-dimensional dynamical systems, and use the notation t for the independent variable. In this case we have (using the notation π_ν for momenta instead of the traditional p_ν)

$$E_\nu(L) = \frac{d\pi_\nu}{dt} + \frac{\partial H}{\partial y^j{}_\nu}. \tag{13.8}$$

Thus, the Euler-Lagrange equations $E_\nu(L) = f_\nu$ are equivalent to the Hamiltonian dynamical system

$$\begin{cases} y^\nu_{,t} = \phi^\nu_{,t} = \frac{\partial H}{\partial \pi_\nu} \\ \frac{d\pi_\nu}{dt} = -\frac{\partial H}{\partial y^\nu} + f_\nu =^{NC-case} = -\frac{\partial H}{\partial y^\nu} + K^\mu_\nu \pi_\mu. \end{cases} \tag{13.9}$$

To get the energy balance for Hamiltonian system (13.9) we calculate $\frac{d}{dt}H$ and get the **energy balance law**

$$\frac{d}{dt}H = \frac{\partial H}{\partial t} - \dot{y}^\sigma f_\sigma =^{NC-case} \frac{\partial H}{\partial t} - \dot{y}^\sigma K^\mu_\sigma \pi_\mu = \frac{\partial H}{\partial t} - K^\mu_\sigma \dot{y}^\sigma L_{,\dot{y}^\mu}, \tag{13.10}$$

with the energy dissipation term

$$-K^\mu_\sigma \dot{y}^\sigma L_{,\dot{y}^\mu} = -K^\mu_\sigma \dot{y}^\sigma \pi_\mu. \tag{13.11}$$

Let $J \in C^\infty(J^*(\pi))$ be a function. This function is the **first integral of the dynamical system 13.9** if (and only if) $d_t J = 0$ along the solution of system (13.9).

Calculate this total derivative of a function J we get

$$d_t J = \frac{\partial J}{\partial t} + \frac{\partial J}{\partial y^\mu}\frac{\partial H}{\partial \pi_\mu} + \frac{\partial J}{\partial \pi_\nu}\left(-\frac{\partial H}{\partial y^\nu} + f_\nu\right) = \left(\frac{\partial J}{\partial t} + f_\nu\frac{\partial J}{\partial \pi_\nu}\right) + \left(\frac{\partial J}{\partial y^\mu}\frac{\partial H}{\partial \pi_\mu} - \frac{\partial J}{\partial \pi_\nu}\frac{\partial H}{\partial y^\nu}\right). \tag{13.12}$$

Using the Poisson bracket of the functions on the (symplectic) fibers of the Hamiltonian bundles $J^*(\pi) \to R_t$: $\{F, G\} = \frac{\partial F}{\partial y^\mu}\frac{\partial G}{\partial \pi_\mu} - \frac{\partial G}{\partial y^\mu}\frac{\partial F}{\partial \pi_\mu}$ we write (13.12) for any function $F \in C^\infty(J^*(\pi))$ in the form

$$d_t F = \left(\frac{\partial F}{\partial t} + f_\nu\frac{\partial F}{\partial \pi_\nu}\right) + \{F, H\} \tag{13.13}$$

In particular,

Lemma 2. *A function $J \in C^\infty(J^*(\pi))$ is the first integral of the Hamiltonian system (13.9) with a Hamiltonian H and the "force" f_ν (13.9) if and only if*

$$\left(\frac{\partial J}{\partial t} + f_\nu\frac{\partial J}{\partial \pi_\nu}\right) + \{J, H\} = 0. \tag{13.14}$$

13.1. **Comparison with the metriplectic model.** Here we would like to compare, in the case where $n = 1$, the NC-variations Lagrangian model of introduction of forces into a Hamiltonian dynamical system with the other geometrical model having similar purpose - "Metriplectic systems" (known also under the names: double bracket systems (in Control Theory), GENERIC systems (in Europe) and, in more general setting as the dynamics on "Leibniz manifolds",[108].

Rewrite the system (13.7) in the form

$$\frac{d}{dx^i}\begin{pmatrix} y^\mu \\ \pi^i_\nu \end{pmatrix} = \begin{pmatrix} 0 & 1 \\ -1 & 0 \end{pmatrix} \nabla_{\begin{pmatrix} y \\ \pi \end{pmatrix}} H + \begin{pmatrix} 0 \\ K^\sigma_{i\nu}\pi^i_\sigma \end{pmatrix}. \tag{13.15}$$

Introduce the kinetic energy written in terms of the momenta π^i_σ: $T = \frac{1}{2}G^{ls}_{ij}(x,y)\pi^i_l\pi^j_s$ and present the previous system in the form

$$\frac{d}{dt}\begin{pmatrix} y^\mu \\ \pi^i_\nu \end{pmatrix} = \begin{pmatrix} 0 & 1 \\ -1 & 0 \end{pmatrix} \nabla_{\begin{pmatrix} y \\ \pi \end{pmatrix}} H + \begin{pmatrix} 0 & 0 \\ 0 & M \end{pmatrix} \begin{pmatrix} T_{,y^l} \\ T_{,\pi^\nu_s} \end{pmatrix} \tag{13.16}$$

with some symmetrical matrix function M_{ls}. Calculating derivatives in the second term of the right side we see that non-Hamiltonian term in the last system has the form

$$\begin{pmatrix} 0 \\ M_{js}(x^i, y^\mu, y^\mu_i)G^{sl}(x,y)\pi_l. \end{pmatrix}$$

Introduce now the tensor

$$K^l_j = M_{js}(x^i, y^\mu, y^\mu_i)G^{sl}(x,y)\pi_l. \tag{13.17}$$

Then, the metriplectic dynamical system (13.12) with the degenerate metric $\tilde{M} = \begin{pmatrix} 0 & 0 \\ 0 & M \end{pmatrix}$ in the space $T^*(Y)$ is the representation of the Legendre transformation of the Lagrangian system with the non-commutative variations defined by the tenor K. Formula (13.17) establishes a relationship between the tensor K and the degenerate (covariant) metric M characterizing the dissipative conditions in the phase space of metriplectic system (10.12).

14. HAMILTON-JACOBI EQUATION.

The first order nonlinear Hamilton-Jacobi PDE ([44, 117] and [47], vol II,Ch.9) plays a prominent role in the explicit determination of solutions of variational problems. Combined with the method of separating of variables, the Hamilton-Jacobi equation represents, by the opinion of many classical mathematicians, the best tool for construction of explicit solutions in Classical and Quantum Mechanics, Control theory and other domains.

That is why it is interesting to see, if this method works in the presence of non-commutative variations and if not, if it is possible to modify it to be able to use Hamilton-Jacobi method for K-twisted Euler-Lagrange or/and equivalent Hamiltonian systems (see previous Section).

Recall that the interest in the development of Hamilton-Jacobi Theory in a non-classical settings led to a series of works extending, in different forms, the Hamilton-Jacobi Theory to the Classical Field Theory (classical works of Caratheodory, Weyl, Rund, etc., see [117]), to the singular Lagrangian systems and to the non-holonimic systems (see, for example, [85, 84]).

Refering the reader to the sources cited here and below, we restrict our presentation to two simple aspects. First we show that the classical Hamilton-Jacobi Theory is not applicable to a Lagrangian mechanical system with NC-variations or, more generally, to the Euler-Lagrange equations with the generic sources. More specifically, while the first Jacobi Theorem holds with some modifications, the proof of the second Jacobi Theorem (see [44], Sec.23, Theorem 2 fails in the case of Hamiltonian system with a non-conservative forces.

We suggest to our reader an Exercise to construct an example of such Hamiltonian system with failing Jacoby Theorem 2.

At the next step we show, that, following the idea of "embedding surfaces of solutions" similar to the use of "geodesic fields" in H.Weyl's Theory of multiple integrals ([117, 47]) some classes of solutions of Hamiltonian systems with sources can be constructed starting with the general solutions of appropriately constructed partial differential equations. We follow the method suggested by B.Vujanovich([133], Chapter 6) for dynamical systems (with one independent variable), a method that, in a variety of cases, replaces the Hamilton-Jacobi method in a case of non-conservative, dissipative, dynamical systems. We present his method below, refering reader to the monograph ([133]) and the cited papers for more details.

14.1. **Case $n = 1$. Formula for variations.**

Consider an action functional for the curves $[t_0, t_1] \to Q$ in the space Q of variables $y^i, i = 1, \ldots, m$

$$A(s = \{y^i(t)\}) = \int_{t_0}^{t_1} L(t, y^i(t), \dot{y}^i(t))dt. \tag{14.1}$$

We start with the case $n = 1$ using notation t for an independent variable and consider the interval $[t_0, t_1]$ as the domain D of integration. We introduce the Hamilton-Jacobi equation using the variational formula for action. Here we follow the [44], Sec.13, 23.

We begin with the formula for variation of action including the variations of endpoints ([44], Sec.13, (7))

$$\delta\mathcal{A} = \int_{t_0}^{t_1} \sum_{i=1}^{m} \left(L_{,y^i} - \frac{d}{dt}L_{,\dot{y}^i} + K_i^k(j^1 s(t))L_{,\dot{y}^k} \right) \xi^i dt + \sum_{i=1}^{m} F_{,\dot{y}^i}\delta y^i|_{t=t_0}^{t=t_1}$$
$$+ \left(L - \sum_{i=1}^{m} \dot{y}^i L_{,\dot{y}^i} \right) \delta t|_{t=t_0}^{t=t_1}, \tag{14.2}$$

The term with the tensor K_i^k appeared when we calculate the slope of variations on the ends of the interval - domain of the variated curve, see [44],Sec.13, p.55.

The notation $\delta t|_{t=t^j} = \delta t^j$; $\delta y^i|_{t=t_j} = \delta y_i^j, j = 0,1$ are used here.

Introduce the dual (momentum) variables $p_i = L_{,\dot{y}^i}$ and assume that the Jacobian $Det(\frac{\partial p_i}{\partial \dot{y}^j}) \neq 0$.

Then we can solve for $\dot{y}^i$ and get

$$\dot{y}^i = \psi^i(y^k, p_l)$$

.

Introduce the Hamiltonian function

$$H = \sum_{i=1}^{i=m} \dot{y}^i p_i = L. \tag{14.3}$$

Here $\dot{y}^i$ are considered as the functions of y^k, p_l.

Performing the Legendre transformation (i.e. using the canonical variables (t, y^i, p_j)) in the expression for Hamiltonian variations (Sec.13) we get

$$\delta\mathcal{A} = \int_{t_0}^{t_1} \sum_{i=1}^{m} \left(L_{,y^i} - \frac{d}{dt}L_{,\dot{y}^i} + K_i^k L_{,\dot{y}^k} \right)_{(t,y^s(y),p_k(t))} \xi^i(t, y^s(t))dt + \left(\sum_{i=1}^{m} p_i\delta y^i - H\delta t \right) |_{t=t_0}^{t=t_1}. \tag{14.4}$$

Now let the curve $s(t) = \{y^i(t)\}$ be a solution of Euler-Lagrange Equations with the NC-variations defined by the tensor K and connecting points $A = (t_0, y_0^i)$ and $B = (t_1, y_1^i)$. Then the first term in (14.3) vanishes and the variation $\delta\mathcal{A}$ takes (in canonical variables) the standard form (see [44], Sec.13)

$$\delta\mathcal{A} = \left(\sum_{i=1}^{m} p_i\delta y^i - H\delta t \right) |_{t=t_0}^{t=t_1} \tag{14.5}$$

14.2. Case n=1, Hamilton-Jacoby equation and failure of Second Jacoby Theorem. Here we follow (up to a point) [44], Sec. 23.

Consider a domain W in the space Q of variables y^i such that for any two points of W there exists a unique extremal in the domain W connecting these two points. Consider the functional $\mathcal{A}(s(t) = \{y^i(t)\}$ and fix the initial point $A = s(t_0)$. As a result, we get the function of the endpoint B

$$S = \int_{t_0}^{t_1} L(t, y^i, \dot{y}^j)dt \tag{14.6}$$

evaluated along the extremal γ joining the points $A = (t_0, y_i^0)$ and $B = (t_1, y_i^1)$ - a *geodetic distance* ([44],Sec.23) between A and B. Fix the initial point and consider S as the function of final point B.

Calculating partial derivatives $\frac{\partial S}{\partial t}, \frac{\partial S}{\partial y^\alpha}$ of function $S(t, y^k(t))$ we get the relation $dS = \delta \mathcal{A}$ where variation of action is calculated at the fixed extremal γ. Using formula (14.4) for the general variation (including variations at endpoints) of an extremal we get

$$\frac{\partial S}{\partial t} = H, \ \frac{\partial S}{\partial y^i} = p_i = L_{,\dot{y}^i}, \tag{14.7}$$

where $H = H[t, y^i; p_k(t, y^k)]$ is the Hamilton function introduced in Sec.13 (see (13.2)). From these two equations it follows that the function $S(t, y^i)$ as the function of endpoint B satisfies the (conventional) Hamilton-Jacoby Equation (HJ-equation)

$$\frac{\partial S}{\partial t} + H(t, y^1, \ldots, y^m, \frac{\partial S}{\partial y^1}, \ldots, \frac{\partial S}{\partial y^m}) = 0. \tag{14.8}$$

Basic relations between solutions of the Hamilton-Jacoby equation and the integrals of the Euler-Lagrange system to be modified in the case of non-commuting variations. Namely,

Theorem 7. *(Jacoby) Let $S = S(t, y^i, \alpha_1, \ldots, \alpha_k)$ be a solution of HJ-equation depending on $k \leq m$ parameters α_j. Then, each derivative $\frac{\partial S}{\partial \alpha_s}$ is the first integral for the solutions $(t, y^i(t), p_l(t))$ of the (K-modified) Hamiltonian system:*

$$\frac{d}{dt}\frac{\partial S}{\partial \alpha_s} = \frac{\partial^2 S}{\partial y^j \partial \alpha_s} f_j = 0. \tag{14.9}$$

Proof. Proof of this theorem repeats literally the proof in the non-twisted case ([44], Sec.23, Thm.1) due to the fact that the first part of Hamilton equations in K-twisted case coincide with the classical one. $\qquad\square$

Consider now the second (basic) Theorem of Jacoby.

Theorem 8. *(Jacoby) Let $S = S(x, y^\mu, \alpha_i, \ i = 1, \ldots, m)$ be a **complete integral of the Hamilton-Jacoby Equation** depending on m parameters α_i. Let the determinant of $m \times m$ matrix*

$$\left(\frac{\partial^2 S}{\partial \alpha_i \partial y^\mu} \right)$$

be nonzero and let $\beta_j, \ j = 1, \ldots, m$ be arbitrary constants.

Then the functions $y^\mu = y^\mu(x, \alpha_i, \beta_j)$ defined from the relations

$$\frac{\partial}{\partial \alpha_i} S(x, y, \alpha) = \beta_i, 1 = 1, \ldots, m, \tag{14.10}$$

together with the functions

$$p_k = \frac{\partial}{\partial y^k} S(x, y, \alpha) \tag{14.11}$$

where y are defined at (), constitute the general solution of Hamitonian Equations

$$y^\mu_{,x} = \frac{\partial H}{\partial p_\mu}, \ p_{k,x} = -\frac{\partial H}{\partial y^k}.$$

It is easy to see that the conventional proof ([44]) does not work in the present situation where $K \neq 0$. More specifically, starting with a general solution of the Hamilton-Jacoby Equation (14.8) and following the proof we get the conventional Hamiltonian system rather then the K-modified one.

As a result, in its classical form, Hamiton-Jacoby Theory can not be applied to the K-twisted Euler-Lagrange systems.

Remark 18. It is interesting to see if some modification of the HJ-equation and/or procedure of construction of functions y^μ and p_ν could lead to the K-modified hamiltonian system.

14.3. Basic Field Equation for non-conservative dynamical systems.
In the works (), B. Vujanovich suggested a method that in a variety of cases, replaces for the dynamical systems (one independent variable) the Hamilton-Jacoby method. He called the PDE replacing the Hamitlon-Jacoby Equation the "Basic Field Equation" and applied it to a variety of dynamical systems: modified Hamilton equations with non-conservative forces, equations of vibrations, non-conservative coupled oscillators, diffusion in a Tubular reactor and others, see [133].

We will present here a short description of this method and two examples of its application.

We start with the general idea of choosing an "embedding" -surface in the configurational space where solutions of a dynamical system take values.

Let

$$\dot{x}^i = X_i(t, x^k), \; i = 1, \ldots, m \qquad (14.12)$$

be a dynamical system for a vector function $x(t) = (x^1(t), \ldots x^m(t)$.

Look for solutions of this system satisfying the condition

$$\Phi(t, x^1, \ldots, x^m) = 0, \qquad (14.13)$$

i.e. belonging to the hypersurface $\Phi \subset Y$.

Assume now that $\frac{\partial \Phi}{\partial x^1} \neq 0$. Then we can (at least locally) rewrite condition (14.12) in the form

$$x^1 = \Psi(t, x^2, \ldots, x^m). \qquad (14.14)$$

Taking the total derivative by time and using the equations (14.12) for $i > 1$ we get the first order PDE for solutions of system (14.12) "embedded" into the surface Φ

$$\frac{\partial \Psi}{\partial t} + \frac{\partial \Psi}{\partial x^i} X_i(t, \Psi, x^k, k > 1) = X_1(t, \Psi, x^k, k > 1) = 0. \qquad (14.15)$$

We will call this equation the **basic field equation** (BFE) defined by the surface Φ.

This equation is a quasi-linear PDE and as such, is simpler then the Hamiton-Jacoby Equation.

14.4. Basic Field Equation for a Hamiltonian system with nonconservative forces.
Let

$$\begin{cases} \dot{x}^i = \frac{\partial H}{\partial p_l}, \; i = 1, \ldots, m \\ \dot{p}_j = -\frac{\partial H}{\partial x^j} + f_j(t, x^i.p_l), j = 1, \ldots, m. \end{cases} \qquad (14.16)$$

be a Hamiltonian system with arbitrary forces f_j.

This system can be obtained from (and is actually equivalent to) the modification of D'Alambert variational principle

$$\frac{d}{dt}(p_l \delta x^i) - \delta(p_i \dot{x}^i - H) - f_i \delta x^i = 0, \; i = 1, \ldots, m. \qquad (14.17)$$

Let $U(t, x^i, p_l, \; l = 2, \ldots, m)$ be a function of the listed variables. Adding and subtracting the variation $U\delta x^1$ in this equation, we write it in the form

$$\frac{d}{dt}(p_l \delta x^i - U\delta x^1) - \delta(p_i \dot{x}^i - H) - f_i \delta x^i + \frac{d}{dt}(U\delta x^1) = 0, \; i = 1, \ldots, m. \qquad (14.18)$$

Now we **define** the first component of momentum

$$p_1 = U(t, x^i, p_l, \ l = 2, \ldots, m). \tag{14.19}$$

This is the "embedding condition" mentioned in the beginning to this Section. From now on it is convenient to use the notation z_a for p_a with $a = 2, \ldots, m$.

Using this in the previous variational equation we write it in the form

$$\frac{d}{dt}(z_a \delta x^a) - \delta(U \dot{x}^1 - z_a \dot{x}^a - H) - f_i \delta x^1 - f_a \delta x^a + \frac{d}{dt}(U \delta x^1) = 0, \ i = 1, \ldots, m. \tag{14.20}$$

Separating the coefficients of independent variations $\delta x^1, \delta x^a, \delta z_a$ we get the system of equations

$$\begin{cases} \delta x^1: & \frac{dU}{dt} + \frac{\partial H}{\partial x^1} + \frac{\partial H}{\partial U}\frac{\partial U}{\partial x^a} + \frac{\partial U}{\partial x^a}\dot{x}^a + \frac{\partial U}{\partial z_a}\dot{z}_a - f_i = 0. \\ \delta x^a: & \dot{z}_a - \dot{x}^1 \frac{\partial U}{\partial x^a} + \frac{\partial H}{\partial x^a} + \frac{\partial H}{\partial U}\frac{\partial U}{\partial x^a} - f_a = 0, \\ \delta z_a: & \dot{x}^a + \dot{x}^1 \frac{\partial U}{\partial z_a} - \frac{\partial H}{\partial U}\frac{\partial U}{\partial z_a} - \frac{\partial H}{\partial z_a} = 0. \end{cases} \tag{14.21}$$

Calculating derivatives $\dot{x}^a$ and $\dot{z}_a$ from the last two equations and substituting them into the first equation we get the field equation (BFE) for the scalar function U:

$$\frac{dU}{dt} + \frac{\partial H}{\partial x^1} + \frac{\partial H}{\partial U}\frac{\partial U}{\partial x^1} + \frac{\partial H}{\partial z_a}\frac{\partial U}{\partial x^a} + \left(f_a - \frac{\partial H}{\partial x^a}\right) - f_1 = 0. \tag{14.22}$$

Remark 19. This way to work out "basic field equation", being based on the D"Alambert equation, proves the invariant character of obtained equation.

Remark 20. Notice that any solution of a Hamiltonian dynamical system with non-conservative sources (14.16) satisfies to the condition (14.19) with some function U. Therefore, if starting with a general solution of (14.22) we construct corresponding solutions of (14.16). As a result, we find ALL solutions of system (14.16).

14.5. **Complete solution of BFE and related conservation laws.**

Definition 7. *A general solution of the BFE is the relation*

$$p_1 = U(t, x^1, \ldots, x^n; z_2, \ldots, z_n; C_1, \ldots, C_{2m}), \tag{14.23}$$

where C_i are 2m arbitrary constants such that if we substitute it to the BFE-equation 14.22 it is satisfied identically.

Every complete solution satisfies the property: if we eliminate all 2mn constants C_i from 2m+1 relations: $14.23, \frac{\partial U}{\partial t}, \frac{\partial U}{\partial x^i}, \frac{\partial U}{\partial z_a}, i = 1, \ldots, m; a = 2, \ldots, m$, we obtain the basic field equation 14.22.

Next we notice that the general solution 14.23 contains in itself 2m conservation laws for the dynamical system (14.16). We can recover these equations by fixing values of 2m-1 constants C_i and leaving values of the left constant arbitrary. For instance if we fix $C_i = C_2 = \ldots = C_{k-1} = C_{k+1} = \ldots, C_{2m} = 0$ for $k = 1, \ldots, 2m$ and obtain 2n relations

$$p_1 = U_k(t, x^1, \ldots, x^m; z_2, \ldots, z_m; C_k),$$

and solve these relations for C_k then, assuming that these relations are mutually independent (i.e. under the condition that $\frac{\partial U_k}{\partial C_k} \neq 0$), 2m conservation laws

$$\Phi_k(t, x^1, \ldots, x^m; z_1, \ldots, z_m) = C_k \tag{14.24}$$

represent a complete set of conservation laws for the dynamical system (14.16).

14.6. General solution of modified Hamiltonian system (14.16) from the general solution of BFE.

Let (14.23) be a complete solution of the BFE equation. Choose now one of the constants in this solution, say, consider C_1 as an arbitrary function of other constants $C_2, \ldots, C_{2m}$: $C_1 = C_1(C_2, \ldots, C_{2m})$. Taking the derivative of the BFE by C_A where $A = 2, \ldots, 2m$, we get $2m - 1$ functional equations

$$\frac{\partial U}{\partial C_A} + \frac{\partial U}{\partial C_1} D_A = 0, \tag{14.25}$$

where

$$D_A = \frac{\partial C_1}{\partial C_A}, \quad A = 2, \ldots, 2n \tag{14.26}$$

are new constants.

Now, having a general solution of BFE in the form 14.23, solve it for C_1 to get

$$C_1 = \Psi_1(t, x^1, \ldots, x^m, U, z_2, \ldots, z_m; C_2, \ldots C_{2m}). \tag{14.27}$$

Find constants D_A from () and then substitute the function C_1 just obtained in these expressions. We get the resulting equations

$$\begin{cases} \Psi_2(t, x^1, \ldots, x^m, U, z_2, \ldots, z_m; C_2, \ldots C_{2m}) = D_2, \\ \ldots \\ \Psi_{2m}(t, x^1, \ldots, x^m, U, z_2, \ldots, z_m; C_2, \ldots C_{2m}) = D_{2m}. \end{cases} \tag{14.28}$$

Relations (14.27) and (14.28) together **form the (implicitly written) general solutions of the dynamical system** (14.16)

Parameters $C_A, A = 2, \ldots, 2m$ can take arbitrary values in their relevant domain. by substituting some values and returning to the variables $U = p_1$, $z_2 = p_2, \ldots, z_m = p_m$ we write system of relations (14.27,14.28) delivering (although implicitly) the general solution of the system (14.16) in the final form

$$\begin{cases} \Psi_1(t, x^1, \ldots, x^m, p_1, p_2, \ldots, p_m; C_2, \ldots C_{2m}) = C_1, \\ \Psi_2(t, x^1, \ldots, x^m, U, p_2, \ldots, p_m; C_2, \ldots C_{2m}) = D_2, \\ \ldots \\ \Psi_{2m}(t, x^1, \ldots, x^m, p_2, \ldots, p_m; C_2, \ldots C_{2m}) = D_{2m}. \end{cases} \tag{14.29}$$

To illustrate described method we apply it to the following example

Example 8. Consider the 2-dim dynamical system on the time interval $[1, \infty)$

$$\begin{cases} \dot{x}^1 = x^2, \\ \dot{x}^2 = \frac{x^{2\,2}}{x^1} - \frac{x^2}{t}, \end{cases} \tag{14.30}$$

with the initial data

$$\begin{cases} x^1(1) = a, \\ x^2(1) = b. \end{cases}$$

Take $x^2 = U(t, x^1)$ as the **basic field** we find the basic field equation (14.22) to be

$$\frac{\partial U}{\partial t} + U \frac{\partial U}{\partial x^1} - \frac{U^2}{x^1} + \frac{U}{t} = 0. \tag{14.31}$$

Separate variables x^1 and t by the assumption $U(t,x^1) = \frac{F(x^1)}{f(t)}$ from which we find that

$$\frac{\dot{f}}{f} = (F' - \frac{F}{x^1})\frac{1}{f} - \frac{a}{t}. \tag{14.32}$$

The overdot here is the time derivative while prime is the derivative with respect to x^1. This relation can be written in the form

$$\dot{f} - \frac{f}{t} = (F' - \frac{F}{x^1}).$$

Equating both expressions to constant 1 and integrating both obtained expressions by corresponding variables we find the general solution of the BFE equation

$$x_2 = \frac{x^1}{t}\frac{ln(x^1) + A}{ln(t) + B}, \tag{14.33}$$

where A, B are constants.

Substituting (14.33) to the dynamical system (14.30) with given initial conditions we get $A = B\frac{b}{a} - ln(a)$. Putting this back into (14.33) we obtain the conditional form solution of BFE

$$x^2 = \bar{U}(t,x^1,a,b,B) = \frac{x^1}{t}\frac{ln(x^1/a) + B(b/a)}{ln(t) + B}, \tag{14.34}$$

where B is arbitrary.

Using the relation $\frac{\partial \bar{U}}{\partial B} = 0$ we see that under the condition $(ln(t) + M) \neq 0$, the last equation gives

$$x^1 = at^{b/a}.$$

Substituting this into the expression for x^2 we get the second component of the solution

$$x^2 = bt^{(b/a)-1}.$$

To find the general solution of the starting dynamical system using the above described method, we choose the relation between arbitrary constants $A = A(B)$ to be true. Equation $\frac{\partial x^1}{\partial B} = 0$ is now equivalent to the relation

$$ln(x^1) = \frac{dA}{dB}(ln(t) + B) - A. \tag{14.35}$$

Using this in () we find

$$x^2 = \frac{x^1}{t}\frac{dA}{dB} = \frac{x^1}{t}D_1, \ D_1 = \frac{dA}{dB} \tag{14.36}$$

and equation (0 now gives

$$\ln(x^1) = D_1 ln(t), \ D_2 = B\frac{dA}{dB} - A. \tag{14.37}$$

Last two equations represent the general solution of dynamical system (14.30).

Remark 21. (Conjecture:Case of Field Theory.) Now, let

$$\partial_t y^\mu + \sum_{i=1}^{i=3} \partial_{x^i} F_\mu^i = \Pi_\mu, \ \mu = 1,\ldots,m \tag{14.38}$$

be a system of balance equations, where flux components F_μ^i and sources Π_μ are functions of $t = x^0, x^i, i = 1,2,3$ and of the fields y^μ.

Place the embedding condition

$$w_1 = \Phi(t, x^i, w_\nu \nu = 2, \ldots, m) \tag{14.39}$$

Acting as in the case of dynamical systems we get for a function Φ the BFE

$$\frac{\partial \Phi}{\partial t} + \sum_{\mu > 1} \frac{\partial \Phi}{\partial w^\mu} (\Pi_\mu - \partial_{x^i} F_\mu^i) - (\Pi_1 - \partial_{x^i} F_1^i) = 0. \tag{14.40}$$

This quasilinear BFE for Field Theory can probably be used in the same way as the equation 14.22 in the case of Mechanics.

Part II. Vertical connections and the twisted prolongations.

In this, geometrical, part of this work, we relate the procedures of K-twisted prolongations of vector fields to the "vertical connections" at the 1-jet bundle $J^1(\pi) \to Y$ over the the configurational space Y. Later on, we extend K-twisted prolongation and the geometrical interpretation of such prolongations to the k-jet bundles $J^k(\pi)$ with $1 \leqq k \leqq \infty$).

In all cases it is (Ehresmann) connections in the bundle $\pi_{10} : J^1(\pi) \to Y$ (respectively, in the k-jet bundle $J^k(\pi) \to Y$) (see Appendix II) that plays the main role. This opens a way to recognize and study existing relations of the procedures of non-commutative prolongations of variations and the related properties of the geometrical structures on the jet bundles $J^k(\pi)$ (e.g., connections, curvature and torsion of connections, contact structures on the jet bundles, etc.).

To follow the presentation of this theme, a knowledge of basic geometrical notions and constructions of the Variational Calculus is necessary. That is why in the Appendix I of this text we provide a short description of basic geometrical structures necessary in Chapters 3,4 - fiber bundles, Ehresmann connections, curvature, linear and affine connections, their curvature and torsion, automorphisms of bundles, absolute parallelism. In the second part of Appendix - Appendix II we present basic information about of the jet bundles including their contact structures, Lie fields, flow prolongations of vector fields and Ehresmann connections in the jet bundles. Below we will be using introduced geometrical notions and results refereing to the Appendices I,II and the references therein.

Chapter 3. Vertical connections, K-twisted prolongations and the NC-variations.

15. Introduction

In this Chapter we introduce the general "K-twisted prolongations" $\xi \to Pr_K^1(\xi)$ (non-commutative, or NC-prolongations) of vector fields $\xi \in X(Y)$ to the 1-jet bundle $J^1(\pi)$. In the last section of this Chapter we extend this presentation to the higher order bundles (and, correspondingly, to the higher order differential equations and the systems of such equations).

Here, $\mathbf{K} = (K_{i\nu}^\mu, \ K_{i\nu}^k)$ is a (1,2)-tensor field with the components of two types. Such tensor fields define the **Ehresmann connections** in the bundle $\pi_{10} : J^1(\pi) \to Y$, (see Appendix II, Section 77). We study the properties of these, K-**twisted** procedures of prolongation. In particular, we determine the obstructions to the fulfillment of two basic properties of prolongations - invariance of Cartan distribution on $J^1\pi$ under the flow of K-twisted prolongations of vector fields $\xi \in \mathcal{X}(Y)$ and the conservation of Lie bracket of vector fields under this K-twisted prolongation procedure.

More specifically, conventional (flow) prolongations

$$\xi = \xi^i \partial_i + \xi^\mu \partial_\mu \to Pr^1 \xi = \xi + ()\partial_{y_i^\mu}$$

of vector fields from Y to $J^1(\pi)$ have the following two basic properties:

(1) Flow prolongation $Pr^1\xi$ of a vector field $\xi \in \mathcal{X}(Y)$ is the **Lie vector field** in the 1-jet space $J^1(\pi)$ - i.e the phase flow ϕ_t of vector field $Pr^1\xi$ **preserves the Cartan distribution** Ca in $J^1(\pi)$ (or, what is equivalent to this property, preserves the ideal of contact forms, see Appendix II, Section 75.

(2) The mapping $\xi \to Pr^1\xi$ is the **Lie algebra monomorphism** $\mathcal{X}(Y) \to \mathcal{X}(J^1(\pi))$, in other words , for all vector fields $\xi, \chi \in \mathcal{X}(Y)$,

$$Pr^1[\xi, \chi] = [Pr^1\xi, Pr^1\chi]. \tag{15.1}$$

We analyze the obstruction preventing the K-twisted prolongations of vector fields from Y to $J^1(\pi)$ from having these properties. We will show that the obstruction to the fulfillment of the first property is the tensor K itself and that it is impossible to preserve the invariance of the contact structure in $J^1(\pi)$ with respect to the flow of K-prolongation $Pr_K^1(\xi)$ of the vector fields $\xi \in X(Y)$ unless $K = 0$.

Yet, one can modify the conventional condition of invariance of contact structure on the k-jet bundles in such a way that for a large class of K-twisted prolongations such modified condition is fulfilled. This modification was suggested by E.Pucci and G.Saccomandi in the paper [115]. In the Section 17 below, we formulate the condition on the NC-tensor K for which the "invariance of contact structure" as the defining condition for vector fields on the jet space to be prolongation of some vector field on Y holds (see Remark 23).

The obstruction for the K-twisted prolongation $\xi \to \xi_K^{(1)}$ to be the Lie algebra monomorphism consists of two parts: curvature of the vertical connection defined by the tensor K and the non-trivial deformation of the Lie bracket of vector fields defined by tensor K.

In the next Chapter we present the results of series of works of Spanish and

© Springer International Publishing Switzerland 2016
S. Preston, *Non-commuting Variations in Mathematics and Physics*,
Interaction of Mechanics and Mathematics,
DOI 10.1007/978-3-319-28323-4_3

Italian mathematicians, see [39, 40, 101, 115, 98] etc. who introduced and studied the class of K-twisted prolongations ($\lambda-$ and $\mu-$-prolongations). This class of $\lambda-$ and $\mu-$ prolongations has several useful properties: they satidfy to the **modified condition of invariance of contact structure in** $J^1\pi$, they generate a class of λ and μ-symmetries leading to the new type of "conservation laws" (similar to the Noether balance laws defined by a symmetry group of the Lagrangian and the NC-tensor K, see Appendix III, Section 78-80). It is interesting that the λ- and μ-prolongations are related to the conventional flow prolongation **by the gauge transformations**.

The most important fact (a starting point of all their program) that appeared in the paper [101] of C.Muriel and J.Romero) is that the λ **and** μ-**symmetries** can be used for performing the reduction of dimension of dynamical systems and PDE in the same way as in the case of usual symmetries, (see [106], Chapter 3).

16. K-twisted prolongation $\xi \to Pr^1_K(\xi)$.

We start with stating the general definition of K-twisted prolongation of vector fields basic for our scheme. We borrow this name from the works of sited above authors.

Definition 8. *Let*

$$K = dx^i \otimes (\partial_i + K^j_{i\nu}\partial_{y^\nu_j}) + dy^\mu \otimes (\partial_\mu + K^\nu_{j\mu}\partial_{y^\nu_j}) \tag{16.1}$$

be an Ehresmann connection in the bundle $J^1(\pi) \to Y$ (see Appendix III, Sec.77) and let

$$\xi = \xi^i \partial_{x^i} + \xi^\mu \partial_{y^\mu} \in X(Y)$$

*be a vector field in the manifold Y with the characteristic (see Section 79.7 for the definition and use of the"characteristic" of a vector field on the jet bundles) $Q = Q^\mu \partial_{y^\mu}$, $Q^\mu = \xi^\mu - y^\mu_i \xi^i$. The K-**twisted prolongation of vector field** ξ to the 1-jet space $J^1(\pi)$ is the vector field*

$$Pr^{(1)}_K = \xi+[(d^t_i\delta^\alpha_\mu+K^\nu_{\mu i})\xi^\mu-(d^t_i(y^\nu_k\xi^k)+K^\nu_{ki}\xi^k)]\partial_{y^\nu_i} = \xi+[d^t_i(\xi^\nu-y^\nu_k\xi^k)+(K^\nu_{\mu i}\xi^\mu-K^\nu_{ki}\xi^k))]\partial_{y^\nu_i} =$$
$$\xi + [d^t_iQ^\nu + (K^\nu_{\mu i}\xi^\mu - K^\nu_{ki}\xi^k))]\partial_{y^\nu_i} =^{K^\nu_{ki}=y^\mu_k K^\nu_{i\mu}} \xi + [d^t_iQ^\nu + K^\nu_{\mu i}Q^\mu]\partial_{y^\nu_i}. \tag{16.2}$$

The last expression *for the K-twisted prolongation of a vector field ξ is valid if the horizontal component $K^\nu_{\mu i}$ of connection K satisfies to the condition $K^\nu_{\mu i} = y^\mu_k K^\nu_{i\mu}$ (we will call them the $\lambda\mu$-**prolongations**). This condition defined the class of $K-$ twisted prolongations introduced and studied in the works of C.Muriel, J.Romero, E.Pucci, G.Saccomandi, G.Gaeta, P.Morando and their coworkers, (see next Chapter).*

Remark 22. Notice that only the vertical part of connection K participates in the modification (twisting) of the Euler-Lagrange equations. Prolongation of general vector fields in Y is important for the study of symmetries (and twisted symmetries, see below) as well as for applications including the inner variations, see [49].

16.1. K-**twisted total derivative.** After introduction of a K-twisted prolongation it is natural to compare obtained expression with the one for the flow prolongation. In order to do this we introduce the corresponding modification of the notion of total derivatives - K-covariant total derivatives. Then, the K-twisted prolongation (16.2) will looks like the *naturally modified flow prolongation* of vector fields.

Definition 9. *Fix an Ehresmann connection K (16.1) on the bundle $\pi_{10} : J^1(\pi) \to Y$.*

(1) *K-twisted total derivative by x^i is, is defined in matrix form as follows:*

$$d_i^K = d_i \delta_\beta^\alpha + K_{i\beta}^\alpha$$

,

(2) *The **K-twisted total derivative of a vertical vector field** ξ **by** x^i defined by the connection K is*

$$(d_i^K \xi)^\mu = d_i \xi^\mu + K_{ik}^\mu \xi^\sigma = \partial_i \xi^\mu + y_i^\nu \partial_\nu \xi^\mu + K_{i\nu}^\mu \xi^\nu. \qquad (16.3)$$

(3) *The first **K-twisted prolongation** of a vector field $\xi = \xi^i \partial_i + \xi^\mu \partial_\mu \in \mathcal{X}(Y)$ with the characteristic vector field $Q = Q^\alpha \partial_{y^\alpha}$ is defined as*

$$\xi_K^{(1)} = \xi^i d_i + Q^\mu \partial_\mu + (d_l Q^\mu + K_{l\nu}^\mu Q^\nu) \partial_{y_i^\mu} = \xi + [d_i(\xi^\nu - y_j^\nu \xi^j) + K_{i\mu}^\nu(\xi^\mu - y_k^\mu \xi^k)] \partial_{y_i^\nu}.$$
$$(16.4)$$

Comparing 16.2 with the 16.4 we see that the K-twisted total derivative allows us to write the Euler-Lagrange equations with NC-variables defined by a tensor $K_{i\nu}^\mu$ simply by replacing the total derivative in the standard Euler-Lagrange equations (4.7) by the covariant version 16.2. As a result, obtained system of equations

$$L_{,y^\nu} - (d_i \delta_\nu^\mu + K_{i\nu}^\mu) L_{,y_i^\mu} = 0, \nu = 1, \ldots, m$$

is equivalent to the K-twisted Euler-Lagrange equations (6.4).

Example 9. In particular, the **K-twisted flow prolongation of a vertical vector field** $\xi = \xi^\mu(x, y) \partial_\mu$, defined above, has the form:

$$Pr_K^1 \xi = \xi + d_i^K \xi^\mu \partial_{y_i^\mu} = \xi + [d_i \xi^\mu + K_{\sigma 1}^\mu \xi^\sigma] \partial_{y_i^\mu} = Pr^1 \xi + (K_{i\sigma}^\mu \xi^\sigma) \partial_{y_i^\mu}. \qquad (16.5)$$

Remark 23. Flow prolongation $Pr^1 \xi$ of a vector field ξ in Y allows us to define a local one-parametrical group of diffeomorphisms of $J^1(\pi)$ - local phase flow $\phi_t^{(1)}$ of vector field $Pr^1 \xi$. This phase flow is projectable to Y - its projection is the (local) phase flow of vector field ξ.

substitution of the conventional prolongation $Pr^1 \xi$ with $Pr_K^1 \xi$ for a tensor K (actually using both tensors $K_{lj}^\nu, K_{\mu j}^\nu$) of an Ehresmann connection leads to the corresponding deformation of the phase flow $\phi_t^{(1)} \to \phi_{K\ t}^{(1)}$.

Remark 24. Affine shift in the fiber of π_{10} and partial connection.

One may consider the replacement of the flow lifted vector field $Pr^1 \xi$ by the vector field $Pr^1 \xi + (K_{i\nu}^\mu \xi^\nu) \partial_{y_i^\mu}$ as the composition of conventional flow prolongation of a vector field and the **affine shift** in the π_{10}-vertical direction by the quantity linearly depending on the vector field ξ.

17. NON-CONSERVATION OF CARTAN DISTRIBUTION BY A K-TWISTED PROLONGATIONS OF VECTOR FIELDS.

Here we show that the flow prolongation $\xi \to Pr^1 \xi$ is the *only linear by ξ* prolongation of π-vertical vector fields in Y to the π_1-vertical vector fields in $J^1(\pi)$ *preserving the Cartan distribution Ca.*

An arbitrary linear by ξ, prolongation $\mathcal{V}(\pi) \to \mathcal{V}(\pi_1)$ has the form

$$Pr_K^1 \xi = \xi + (d_i \xi^\mu + K_{i\nu}^\mu \xi^\nu)\partial_{y_i^\mu} = \xi + (d_i^K \xi^\mu)\partial_{y_i^\mu} \qquad (17.1)$$

with some functions $K_{i\nu}^\mu \in C^\infty(J^1(\pi))$. Here we use the notation $d_i^K \xi^\mu = d_i \xi^\mu + K_{i\nu}^\mu \xi^\nu$ introduced above.

The condition that the (local) phase flow of the vector field $Pr_K^1 \xi$ preserves the Cartan distribution has the form of the equality

$$\mathcal{L}_{Pr_K^1 \xi}\omega^\alpha = \lambda_\beta^\alpha \omega^\beta, \ \alpha \in \overline{1,m}$$

that should be valid for all generating contact 1-forms $\omega^\alpha = dy^\alpha - y_i^\alpha dx^i$ with some smooth functions λ_β^α, see [71]. Since the flow prolongation $\xi \to Pr^1 \xi$ satisfies this property and since this equality is linear by ξ, to prove the previous relation, it is sufficient to show that

$$\mathcal{L}_{K_{i\beta}^\alpha \xi^\beta \partial_{y_i^\alpha}}\omega^\alpha = \lambda_\gamma^\alpha \omega^\gamma, \ \alpha \in \overline{1,m}$$

with some (other) functions $\lambda_\gamma^\alpha(\xi)$.

Using the Cartan formula $L_\xi = di_\xi + i_\xi d$ and the fact that $i_{\partial_{y_i^\alpha}}\omega^\alpha = 0$ we get this condition in the form

$$\mathcal{L}_{K_{i\beta}^\alpha \xi^\beta \partial_{y_i^\alpha}}\omega^\alpha = (di_{K_{i\beta}^\alpha \xi^\beta \partial_{y_i^\alpha}} + i_{K_{i\beta}^\alpha \xi^\beta \partial_{y_i^\alpha}}d)(dy^\alpha - y_i^\alpha dx^i) = i_{K_{i\beta}^\alpha \xi^\beta \partial_{y_i^\alpha}}d(-y_i^\alpha dx^i) =$$

$$= -K_{i\beta}^\alpha \xi^\beta dx^i = -\lambda_\gamma^\alpha(\xi)\omega^\gamma = -\lambda_\gamma^\alpha(\xi)(dy^\gamma - y_i^\gamma dx^i), \ i \in \overline{1,m}, \quad (17.2)$$

which has to be fulfilled for arbitrary ξ. This is possible only if $\lambda_k^i(\xi) = 0$ and $K_{i\beta}^\alpha \xi^\nu = 0$ for all ξ, i.e., when tensor K vanishes.

Exercise.

Prove that the *deformed* Cartan distribution in $J^1(\pi)$ generated by the forms $\tilde{\omega}^\alpha = dy^\alpha - T_i^\alpha dx^i$ is invariant under the flows of the fields $\hat{\xi}$ given by (17.1) if and only if $K_{i\beta}^\alpha = 0$ and $T_i^\alpha = y_i^\alpha$.

Thus, we have proved the following Proposition (last statement in this Proposition follows from the Frobenius Theorem, [122])

Proposition 2.　　　(1) *Let ξ be a vector field $\xi \in \mathcal{X}(Y)$. Vector field $Pr_K^1(\xi)$ preserves the Cartan distribution in $J^1(\pi)$ (and, therefore, preserve the contact structure in $J^1(\pi)$) if and only if $K_{i\beta}^\alpha \xi^\beta = 0$ for all α, i ,*

(2) *All Vector fields $Pr_K^1(\xi), \xi \in V(\pi)$ obtained by the K-twisted prolongations of vertical vector fields $\xi = \xi^\beta \partial_\beta$ **preserves the Cartan (contact) distribution** Ca^1 if and only if tensor K vanishes: $K = 0$,*

(3) *Introduce the collection of one-forms $\chi_i^\alpha = K_{i\beta}^\alpha dy^\beta$ in $J^1(\pi)$. Let Δ_K is the distribution of subspaces - subdistribution of vertical subbundle $V(\pi) \subset T(Y)$ defined by vertical (relative to the projection $\pi : Y \to X$) 1-formes $\chi_i^\alpha = K_{i\beta}^\alpha dy^\beta \in V^*(\pi)$.*

Then, the K-twisted prolongation $Pr_K^1(\xi)$ of a vertical vector field $\xi = \xi^i \partial_i$ preserves the contact structure of $J^1(\pi)$ (is the Lie field) if and only if

$$\xi \in \Delta_K,$$

(4) *Assume that the distribution Δ_K has constant dimension. Then, Δ_K is completely integrable if and only if the space of vector fields in $\xi \in \mathcal{X}(J^1(\pi))$ whose K-twisted prolongation preserves the Cartan distribution Ca^1* **is the subalgebra of the Lie algebra of vector fields in** Y, *or, equivalently, if for some tensors $c_{i\beta\gamma}^{\alpha jk} \in C^\infty(J^1(\pi))$,*

$$d\chi_i^\alpha = \sum_{\beta\gamma}^{jk} c_{i\beta\gamma}^{\alpha jk} \chi_j^\beta \wedge \chi_k^\gamma, \quad c_{i\beta\gamma}^{\alpha jk} \in C^\infty(J^1(\pi)).$$

Proof. First and second statements follows from (16.3). Third statement follows from the definition of Δ_K. (last statement in this Proposition follows from the Frobenius Theorem, [122], Ch.3. $\qquad\qquad\square$

Example 10. Distribution Δ_K is integrable when the vertical 1-forms χ_i^α are all closed, for instance, when $K_{i\beta}^\alpha = \frac{\partial Q_i^\alpha}{\partial y^\beta}$.

Remark 25. In Chapter 4 we present the generalization of the conventional equivalence of the following properties of a vector field on a k-jet bundle: $\xi \in \mathcal{X}(J^k(\pi))$: vector field ξ **is the k-prolongation** of a vector field in the configurational space Y if and only if the phase flow of vector field ξ preserves the contact structure in $J^k(\pi)$, i.e. preserves the Cartan distribution or, what is equivalently, the differential ideal of contact exterior forms in $J^k(\pi)$. This **generalized condition** was found, for the case of ordinary differential equations by E.Pucci and G.Saccomandi in the work [115]. Later on this condition was extended to the case of partial differential equations and systems of such equations by G.Gaeta and P.Morando in [38].

Being translated from the language of one-forms μ (see Section 30) to the language of NC-tensor $K_{i\beta}^\alpha$ in the 1-jet space $J^1(\pi)$, this condition states that the **(vertical) connection defined by the tensor $K_{i\beta}^\alpha$ is curvature free** (curvature of the vertical connection corresponding to an arbitrary NC-tensor $K_{i\beta}^\alpha$ is calculated in the next Section (formula (18.6)).

18. Obstruction for a K-twisted prolongation $\xi \to \xi_K^{(1)}$ to be the Lie algebra morphism.

Here we will study the obstruction for the K-twisted prolongation procedure $\xi \to Pr_k^1\xi : V(\pi) \to \mathcal{X}(J^1(\pi))$ of the form 6.1 to be the **Lie algebra morphism**.

Let ξ, η be two vector fields in Y and let $Pr_K^1\xi$, $Pr_K^1\xi$ be K-twisted prolongations of these vector fields. Measure of non-satisfying for the K-twisted prolongation to be the Lie algebra homomorphism is the following bilinear form

$$N(\xi, \eta) = [Pr_K^1(\xi), Pr_K^1(\eta)] - Pr_K^1([\xi, \eta]). \tag{18.1}$$

This bilinear form is similar to the curvature form for a connection K on the 1-jet bundle $J^1\pi \to Y$, see Appendix III, Sec.(77). We show here that the curvature of connection K is the part of obstruction $N(\xi, \eta)$ for the preservation of Lie bracket by the K-prolongation procedure.

To calculate the bilinear form $N(\xi, \eta)$ for two arbitrary vector fields ξ, η we start with the calculating of the form N for the vertical vector fields $\xi\partial_\mu$ and $\eta\partial_\nu$ for

$\xi, \eta \in C^\infty(Y)$. In this calculation we will use shorter notations $\tilde{\xi}$ and $\tilde{\eta}$ for K-twisted prolongations of vector fields ξ, η 17.1:

$$\tilde{\xi} = Pr_K^1 \xi = \xi \partial_\mu + (d_k \delta_\alpha^\mu \xi + K_{k\mu}^\alpha \xi) \partial_{y_k^\alpha}, \quad \tilde{\eta} = Pr_K^1 \eta = \xi \partial_\nu + (d_s \delta_\beta^\nu \eta + K_{s\nu}^\beta \eta) \partial_{y_s^\beta},$$

The measure of non-commutativity of the K-prolongation is the bilinear form depending on the components of vector fields as well as the first derivatives of these components (ξ, η in the above setting)

$$\mathcal{N}(\xi \partial_\mu, \eta \partial_\mu) = \widetilde{[\xi \partial_\mu, \eta \partial_\nu]} - [\widetilde{\xi \partial_\mu}, \widetilde{\eta \partial_\nu}].$$

Let us calculate both parts of this expression.
We have

$$I = \widetilde{[\xi \partial_\mu, \eta \partial_\nu]} = \widetilde{(\xi \eta_{,\mu} \partial_\nu - \eta \xi_{,\nu} \partial_\mu)} =$$
$$= (\xi \eta_{,\mu} \partial_\nu + (d_i(\xi \eta_{,\mu}) \delta_\nu^\sigma + K_{i\nu}^\sigma(\xi \eta_{,\mu})) \partial_{y_i^\sigma}) - (\eta \chi_{,\nu} \partial_\mu + (d_i(\eta \xi_{,\nu}) \delta_\mu^\sigma + K_{\mu i}^\sigma(\eta \xi_{,\nu})) \partial_{y_i^\sigma}) =$$
$$= (\xi \eta_{,\mu} \partial_\nu - \eta \xi_{,\nu} \partial_\mu) + [d_i(\xi \eta_{,\mu}) \delta_\nu^\sigma - d_i(\eta \xi_{,\nu}) \delta_\nu^\sigma) + (K_{i\nu}^\sigma(\xi \eta_{,\mu}) - K_{\mu i}^\sigma(\eta \xi_{,\nu}))] \partial_{y_i^\sigma} =$$
$$= [\xi \partial_\mu, \eta \partial_\nu] + \{[\eta_{,\mu}(d_i \xi) \delta_\nu^\sigma - \xi_{,\nu}(d_i \eta) \delta_\nu^\sigma)] + [\xi(d_i \eta_{,\mu}) \delta_\nu^\sigma - \eta(d_i \xi_{,\nu}) \delta_\mu^\sigma] + [\xi \eta_{,\mu} K_{i\nu}^\sigma - \eta \xi_{,\nu} K_{\mu i}^\sigma]\} \partial_{y_i^\sigma}$$
$$\tag{18.2}$$

At the same time

$$II = [\widetilde{\xi \partial_\mu}, \widetilde{\eta \partial_\nu}] = [(\xi \partial_\mu + (d_i \xi \delta_\mu^\sigma + K_{i\mu}^\sigma \xi) \partial_{y_k^{mu}}), (\eta \partial_{nu} + (d_i \eta \delta_\nu^\sigma + K_{i\nu}^\sigma \eta) \partial_{y_i^\sigma})] =$$
$$= [\xi \partial_\mu, \eta \partial_\nu] + \xi \partial_\mu(d_i \eta \delta_\nu^\sigma + K_{i\nu}^\sigma \eta) \partial_{y_i^\sigma} - \eta \partial_\nu(d_i \xi \delta_\mu^\sigma + K_{\mu i}^\sigma \xi) \partial_{y_i^\sigma} + (d_j \xi \delta_\mu^\lambda +$$
$$K_{j\mu}^\lambda \xi) \partial_{y_\nu^\lambda}(d_i \eta \delta_\nu^\sigma + K_{i\nu}^\sigma \eta) \partial_{y_i^\sigma} - (d_j \eta \delta_\nu^\lambda + K_{\nu j}^\lambda \eta) \partial_{y_j^\lambda}(d_i \xi \delta_\mu^\sigma + K_{i\mu}^\sigma \xi) \partial_{y_i^\sigma} = \tag{18.3}$$

Now we use the condition that $\xi, \eta \in \mathcal{X}(Y)$ to calculate

$$\partial_{y_j^\lambda} d_i \xi = \partial_{y_j^\lambda}(\xi_{,i} + y_i^\nu \xi_{,y^\nu}) = \delta_\lambda^\nu \delta_i^j \xi_{,y^\nu} = \delta_i^j \xi_{,y^\lambda}$$

and, similarly, $\partial_{y_j^\lambda} d_i \eta = \delta_i^j \eta_{,y^\lambda}$.
Thus, continuing, we get

$$II = [\xi \partial_\mu, \eta \partial_\nu] + (\xi d_i \eta_{,\mu} \delta_\nu^\sigma - \eta d_i \xi_{,\nu} \delta_\mu^\sigma) \partial_{y_i^\sigma} + (\xi \partial_\mu(K_{i\nu}^\sigma \eta) - \eta \partial_\nu(K_{\mu i}^\sigma \xi)) \partial_{y_i^\sigma} +$$
$$+ (d_\nu \xi \delta_\mu^\lambda + K_{\mu j}^\lambda \xi)(\delta_i^j \eta_{,y^\lambda} \delta_\nu^\sigma + (\partial_{y_j^\lambda} K_{i\nu}^\sigma) \eta) \partial_{y_i^\sigma} -$$
$$- (d_j \eta \delta_\nu^\lambda + K_{\nu j}^\lambda \eta)(\delta_i^j \xi_{,y^\lambda} \delta_\mu^\sigma + (\partial_{y_j^\lambda} K_{i\mu}^\sigma) \xi) \partial_{y_i^\sigma} =$$
$$= [\xi \partial_\mu, \eta \partial_\nu] + (\xi d_i \eta_{,\mu} \delta_\nu^\sigma - \eta d_i \xi_{,\nu} \delta_\mu^\sigma) \partial_{y_i^\sigma} + \eta \xi(\partial_\mu K_{i\nu}^\sigma - \partial_\nu K_{i\mu}^\sigma) \partial_{y_i^\sigma} + (\xi \eta_{,\mu} K_{i\nu}^\sigma - \eta \xi_{,\nu} K_{i\mu}^\sigma) \partial_{y_i^\sigma} +$$
$$+ (d_j \xi \delta_\mu^\sigma \delta_i^j \eta_{,y^\sigma} \delta_\nu^\sigma - d_j \eta \delta_\nu^\sigma \delta_i^j \xi_{,y^{lambda}} \delta_\mu^\sigma) \partial_{y_i^\sigma} + (d_j \xi \delta_\mu^\lambda \partial_{y_j^\lambda} K_{i\nu}^\sigma) \eta - \xi d_j \eta \delta_\nu^\lambda \partial_{y_j^\lambda} K_{\mu i}^\sigma) \partial_{y_i^\sigma} +$$
$$+ (K_{j\mu}^\lambda \xi \delta_i^j \eta_{,y^\lambda} \delta_\nu^\sigma - K_{\nu j}^\lambda \eta \delta_i^j \xi_{,y^{lambda}} \delta_\mu^\sigma) \partial_{y_i^\sigma} + \xi \eta(K_{j\mu}^\lambda \partial_{y_j^\lambda} K_{i\nu}^\sigma - K_{\nu j}^\lambda(\partial_{y_j^\lambda} K_{i\mu}^\sigma)) \partial_{y_i^\sigma} =$$
$$= [\xi \partial_\mu, \eta \partial_\nu] + (\xi d_i \eta_{,\mu} \delta_\nu^\sigma - \eta d_i \xi_{,\nu} \delta_\mu^\sigma) \partial_{y_j^\sigma} + \eta \xi(\partial_\mu K_{i\nu}^\sigma - \partial_\nu K_{i\mu}^\sigma) \partial_{y_i^\sigma} + (\xi \eta_{,\mu} K_{i\nu}^\sigma - \eta \xi_{,\nu} K_{\mu i}^\sigma) \partial_{y_i^\sigma} +$$
$$+ ((d_i \xi) \eta_{,\mu} \delta_\nu^\sigma - (d_i \eta) \xi_{,\nu} \delta_\mu^\sigma) \partial_{y_i^\sigma} + (d_j \xi \partial_{y_j^\mu} K_{i\nu}^\sigma) \eta - \xi d_j \eta \partial_{y_j^\nu} K_{\mu i}^\sigma) \partial_{y_i^\sigma} +$$
$$+ (\xi \eta_{,y^\lambda} K_{\mu i}^\lambda \delta_\nu^\sigma - \eta \xi_{,y^\lambda} K_{i\nu}^\lambda \delta_\mu^\sigma) \partial_{y_i^\sigma} + \xi \eta(K_{j\mu}^\lambda \partial_{y_j^\lambda} K_{i\nu}^\sigma - K_{j\nu}^\lambda(\partial_{y_j^\lambda} K_{i\mu}^\sigma)) \partial_{y_i^\sigma}$$
$$\tag{18.4}$$

The sum of the first,second, forth and fifth terms in the last expression for II coincide with I. Therefore,

$$II - I = \eta\xi(\partial_\mu K^\sigma_{i\nu} - \partial_\nu K^\sigma_{\mu i})\partial_{y^\sigma_i} + (\eta d_j \xi \partial_{y^\mu_j} K^\sigma_{i\nu}) - \xi d_j \eta \partial_{y^\nu_j} K^\sigma_{\mu i})\partial_{y^\sigma_i} +$$

$$+ (\xi\eta_{,y^\lambda} K^\lambda_{\mu i}\delta^\sigma_\nu - \eta\xi_{,y^\lambda} K^\lambda_{i\nu}\delta^\sigma_\mu)\partial_{y^\sigma_i} + \xi\eta(K^\lambda_{j\mu}\partial_{y^\lambda_j} K^\sigma_{i\nu} - K^\lambda_{\nu j}(\partial_{y^\lambda_j} K^\sigma_{i\mu}))\partial_{y^\sigma_i} =$$

$$= \eta\xi[(\partial_\mu K^k_{i\nu} + K^\lambda_{j\mu}\partial_{y^\lambda_j} K^\sigma_{i\nu}) - (\partial_j K^\sigma_{i\mu} + K^\lambda_{\nu j}(\partial_{y^\lambda_j} K^\sigma_{\mu i})]\partial_{y^\sigma_i} +$$

$$+ [\xi(\eta_{,y^\lambda} K^\lambda_{\mu i}\delta^\sigma_\nu - d_j\eta\partial_{y^\nu_j} K^\sigma_{\mu i})\partial_{y^\sigma_i} - \eta(\xi_{,y^\lambda} K^\lambda_{i\nu}\delta^\sigma_\mu - d_j\xi\partial_{y^\mu_j} K^\sigma_{i\nu})]\partial_{y^\sigma_i}. \quad (18.5)$$

Notice that the first part of obtained expression - the bilinear form of vertical vectors ξ, η **coincide with the vertical-vertical component $\tilde{R}^\sigma_{\mu i \nu}$ of the curvature tensor of Ehresmann connection** K (see Appendix II, Sec.77).

$$\tilde{R}^\sigma_{\mu i \nu} = (\partial_\mu K^\sigma_{i\nu} - \partial_\nu K^\sigma_{\mu i}) + (K^\lambda_{j\mu}\partial_{y^\lambda_j} K^\sigma_{i\nu} - K^\lambda_{j\nu}\partial_{y^\lambda_j} K^\sigma_{\mu i}). \quad (18.6)$$

Let $\xi = \xi^\mu\partial_{,\mu}$, $\eta = \eta^\nu\partial_{,\nu}$ be arbitrary *vertical* vector fields. Apply the previous formula (18.4) to each couple $\xi^\mu\partial_{,\mu}, \eta^\nu\partial_{,\nu}$ and take there sum. As a result, we get the value of the bilinear form $N(\xi^\mu\partial_{,\mu}, \eta^\nu\partial_{,\nu})$:

$$N(\xi^\mu\partial_{,\mu}, \eta^\nu\partial_{,\nu}) = [\xi = \widetilde{\xi^\mu\partial_{,\mu}}, \eta = \widetilde{\eta^\nu\partial_{,\nu}}] - \widetilde{[\xi, \eta]} = \{\tilde{R}^\sigma_{i\mu\nu}\xi^\mu\eta^\nu +$$

$$+ \xi^\mu[\eta^\nu_{,y^\lambda} K^\lambda_{\mu i}\delta^\sigma_\nu - d_j\eta^\nu\partial_{y^\nu_j} K^\sigma_{\mu i}] - \eta^\nu[\xi^\mu_{,y^\lambda} K^\lambda_{i\nu}\delta^\sigma_\mu - d_j\xi^\mu\partial_{y^\mu_j} K^\sigma_{i\nu}]\}\partial_{y^\sigma_i} =$$

$$= \{\tilde{R}^\sigma_{i\mu\nu}\xi^\mu\eta^\nu + [\xi^\mu\eta^\nu_{,y^\lambda} K^\lambda_{\mu i} - \eta^\nu\xi^\sigma_{,y^\lambda} K^\lambda_{i\nu}] - [\xi^\mu d_j\eta^\nu\partial_{y^\nu_j} K^\sigma_{\mu i} - \eta^\nu d_j\xi^\mu\partial_{y^\mu_j} K^\sigma_{i\nu}]\}\partial_{y^\sigma_i}.$$

$$(18.7)$$

Introduce now the antisymmetric bracket-like operation on the π-vertical vector fields: $V(\pi) \times V(\pi) \to V(\pi_{10})$ taking values in the bundle of $\pi_{10}-$ vertical vector fields in the 1-jet space $J^1\pi$:

$$\lceil\xi, \eta\rceil = [(\xi^\mu\eta^\sigma_{,y^\lambda} K^\lambda_{\mu i} - \eta^\nu\xi^\sigma_{,y^\lambda} K^\lambda_{i\nu}) - (\xi^\mu d_j\eta^\nu\partial_{y^\nu_j} K^\sigma_{\mu i} - \eta^\nu d_j\xi^\mu\partial_{y^\mu_j} K^\sigma_{i\nu})]\partial_{y^\sigma_i} =$$

$$= [(\xi^\nu\eta^\sigma_{,y^\lambda} - \eta^\nu\xi^\sigma_{,y^\lambda})K^\lambda_{i\nu} - (\xi^\nu d_j\eta^\mu - \eta^\nu d_j\xi^\mu)\partial_{y^\mu_i} K^\sigma_{i\nu}]\partial_{y^\sigma_i}. \quad (18.8)$$

Remark 26. Notice that in a case where tensor $K^\sigma_{i\nu}$ does not depend on the derivatives (jet variables y^α_i), the second term in the bracket defined above vanishes.

As a result we get the following value for the bilinear form N on the couples of vertical vector fields

$$N(\xi^\mu\partial_{,\mu}, \eta^\nu\partial_{,\nu}) = \{\tilde{R}^\sigma_{i\mu\nu}\xi^\mu\eta^\nu + \lceil\xi, \eta\rceil. \quad (18.9)$$

The tensor $\tilde{R}^\sigma_{i\mu\nu}$ in the expression (18.9) for the bilinear form N is the vertical-vertical component of curvature tensor of the Ehresmann connection K. In the next statement we collect the forms taken, in some cases, by the curvature $\tilde{R}^k_{i\mu j}$ of the Ehresmann connection K, by the bracket $\lceil\xi, \eta\rceil$ and by the bilinear form $\mathcal{N}(\xi, \eta)$

Theorem 9. (1) *A K-twisted prolongation $\xi \to Pr^1_K\xi$ preserves brackets for basic vertical vector fields $\partial_\mu, \partial_\nu$ (i.e $[P^1_K(\partial_\mu), P^1_K(\partial_\nu)] = 0$), if and only if the vertical-vertical component of Ehresmann connection K in the bundle $\pi_{10} : J^1\pi \to Y$ -*

$$\tilde{R}^k_{i\mu j} = (\partial_i K^k_{\mu j} - \partial_j K^k_{\mu i}) + (K^l_{\nu i}\partial_{y^l_\nu} K^k_{\mu j} - K^l_{\nu j}\partial_{y^l_\nu} K^k_{\mu i}), \ \forall i, j, k, \mu,$$

$$see(Appendix II, Section 77), \textbf{\textit{vanishes}}. \quad (18.10)$$

(2) *An obstruction for the K-prolongation of vertical vector fields to be a Lie algebra morphism has the form:*

$$v\mathcal{N}(\xi,\eta) = R^{\sigma}_{i\mu\nu}\xi^{\mu}\eta^{\nu} + v\lceil\xi,\eta\rceil, \qquad\qquad (18.11)$$

where $\tilde{R}$ is the vertical component of the curvature tensor of the Hessian connection K on the bundle $\pi_{10}: J^1\pi \to Y$ and $v\lceil\xi,\eta\rceil$ is the bilinear form defined in (18.8).

*In particular, if the vertical part of connection R - $\tilde{R}^{\sigma}_{i\mu\nu}$ vanishes (equals zero), then for two arbitrary vertical vector fields $\xi = \xi^{\mu}\partial_{,\mu}$, $\eta = \eta^{\nu}\partial_{,\nu}$ the difference $[Pr^1_K\xi, Pr^1_K\eta] - Pr^1_K[\xi,\eta]$ **vanishes** if and only if*

$$\lceil\xi,\eta\rceil = [(\xi^{\nu}\eta^{\sigma}_{,y^{\lambda}} - \eta^{\nu}\xi^{\sigma}_{,y^{\lambda}})K^{\lambda}_{iv} - (\xi^{\nu}d_j\eta^{\mu} - \eta^{\nu}d_j\xi^{\mu})\partial_{y^{\mu}_j}K^{\sigma}_{iv}]\partial_{y^{\sigma}_i} = 0. \qquad (18.12)$$

Proof. The first statement follows from the fact that coefficients of basic vector fields are trivially constant and, as a result K-prolongation of such vector fields **coincide with the K-horizontal prolongation.**

The second statement follows from the first one or from the calculations above in the case where $\xi = 1 - \eta$.

The arguments, presented before the formulation of the Theorem, proves the Statement 3) of Theorem for two vertical vector fields.

$$\square$$

Now, introduce the bracket N of the K-twisted prolongation of two basic **horizontal** vector fields. It is easy to see that for two basic horizontal vector fields $\partial_k, \partial_s,$

$$[Pr^1_K(\partial_k), Pr^1_K(\partial_s)] = \{[-\partial_k K^{\alpha}_{is} + \partial_s K^{\alpha}_{ik}] + [K^{\beta}_{jk}\partial_{,y^{\beta}_j}K^{\alpha}_{is} - K^{\beta}_{js}\partial_{,y^{\beta}_j}K^{\alpha}_{ik}]\}\partial_{y^{\alpha}_i}.$$

Comparing the obtained expression with the expression of the horizontal component of curvature tensor of the connection K we see that they differs by sign only.

Repeating, for two horizontal vector fields $\xi = \xi^i\partial_i, \eta = \eta_j\partial_j$, the calculations performed above for two vertical vector fields we find that

$$[Pr^1_K\xi, Pr^1_K\eta] = [\xi,\eta] + \left[(\partial_k K^{\alpha}_{si} - \partial_s K^{\alpha}_{ki}) + (K^{\beta}_{kj}\partial_{y^{\beta}_j}K^{\alpha}_{si} - K^{\beta}_{sj}\partial_{y^{\beta}_j K^{\alpha}_{ki}})\right]\xi^k\eta^s +$$
$$+ K^{\alpha}_{ki}(\xi^s\partial_s\eta^k - \eta^s\partial_s\xi^k) + y^{\alpha}_k(\xi^s\partial_s d_i\eta^k - \eta^s\partial_s d_i\xi^k) +$$
$$+ y^{\beta}_k(\partial_{y^{\beta}_j}K^{\alpha}_{si}(d_j\xi^k \cdot \eta^s - \xi^s \cdot d_j\eta^k)\partial_{y^{\alpha}_i}. \qquad (18.13)$$

The second term in this expression (bilinear by ξ,η) is the horizontal-horizontal component of curvature $\tilde{R}$ of the connection K, while the third part (containing the derivatives of components ξ^i, η^j of vector fields ξ,η, form the bilinear form of derivatives. At the same time,

$$Pr^1_K([\xi,\eta]) = [\xi,\eta] - [y^{\alpha}_k d_i(\xi^j\partial_j\eta^k - \eta^j\partial_j\xi^k)) + (K^{\alpha}_{ik}(\xi^s\partial_s\eta^k - \eta^s\partial_s\xi^k))]\partial_{y^{\alpha}_i}. \qquad (18.14)$$

Subtracting we get

$$N(\xi^i\partial_i, \eta^j\partial_j) = \{R^{\alpha}_{i,ks}\xi^k\eta^s +$$
$$+ 2K^{\alpha}_{ki}(\xi^l\partial_l\eta^k - \eta_l\partial_l\xi^k) + y^{\alpha}_k[2(\xi^s\partial_s d_i\eta^k - \eta^s\partial_s d_i\xi^k) + (d_i\xi^j\partial_j\eta^k - d_i\eta^j\partial_j\xi^k)] +$$
$$+ y^{\beta}_k[(\partial_{y^{\beta}_j}K^{\alpha}_{si})((d_j\xi^k\eta^s - (\partial_l\eta^k(\xi^s)]\}\partial_{y^{\alpha}_i}. \qquad (18.15)$$

As we see, the obstruction for the K-twisted prolongation of vector fields ξ, η in Y has the same structure as for vertical vector fields - corresponding component $R^\alpha_{i,ks}$ of the curvature tensor of connection K plus the "horizontal" component of the bilinear form $[\xi, \eta]$.

Finally, we repeat our calculations for a couple of vector fields $\xi = \xi^\mu \partial_\mu$ and $\eta = \eta^i \partial_i$. We get the following result:

$$\mathcal{N}(\xi, \eta) = [(\partial_\mu K^\alpha_{ki} - \partial_k K^\alpha_{\mu i}) - (K^\beta_{\mu j}\partial_{y^\beta_j} K^\alpha_{ki} - K^\beta_{kj}\partial_{y^\beta_j} K^\alpha_{\mu i})]\xi^\mu \eta^k +$$
$$+ [K^\alpha_{ki}\xi^\mu(\partial_\mu \eta^k) - K^\alpha_{\nu i}(\partial_k \xi^\nu)\eta^k] + [y^\alpha_k \xi^\mu \partial_\mu d_i\eta^k - (\partial_k d_i\xi^\alpha)\eta^k] -$$
$$- [d_j\xi^\alpha d_i\eta^j - y^\beta_k(\partial_{y^\beta_j} d_i\xi^\alpha)d_j\eta^k] - [K^\alpha_{\nu j}\xi^\nu d_i\eta^j - K^\beta_{kj}(\partial_{y^\beta_j} d_i\xi^\alpha)\eta^k)] -$$
$$- [(\partial_{y^\beta_j} K^\alpha_{ki})(d_j\xi^\beta)\eta^k - y^\beta_k(\partial_{y^\beta_j} K^\alpha_{\nu i})\xi^\nu d_j\eta^k] \quad (18.16)$$

The first line of the obtained expression represents the action of mixed curvature operator on the couple of vector fields ξ, η while the next three lines *define the bilinear form* $[\xi, \eta]$ acting on the corresponding couples of vectors.

Combining obtained results, we define the bracket $[\xi, \eta]$ on the tangent bundle $T(Y)$ for all pairs of vectors $\xi, \eta \in \mathcal{X}(Y)$. Repeating the arguments proving the Theorem 8, we get the next result. Independence of the requirement that both parts of $\mathcal{N}(\xi, \eta)$ vanish simultaneously follows from the fact that curvature is the bilinear form of vector fields ξ, η while the bracket $[\xi, \eta]$ is the bilinear form of these vector fields *and their derivatives* that vanish together with the vanishing of these derivatives.

Decomposing two arbitrary vector fields ξ, η into horizontal and vertical components and summating the expressions for curvature bilinear forms and bilinear forms $[\xi, \eta]$, we get the expression for curvature (see Appendix II, Sec.77) and that for bilinear form $[\xi, \eta]$ on the whole tangent bundle $T(Y)$ (for all pairs of vector fields ξ, η).

As a result we get the following description of the obstruction for K-prolongation to be Lie algebra morphism:

Theorem 10. (1) *A K-twisted prolongation $\xi \to Pr^1_K\xi$ preserves brackets for all basic vector fields ∂_i, ∂_μ, if and only if the curvature R of the connection K vanishes.*

(2) *Obstruction for the K-prolongation of vector fields to be Lie algebra morphism $T(Y) \to T(J^1)$ is the bilinear form*

$$\mathcal{N}(\xi, \eta) = R(\xi, \eta) + [\xi, \eta], \quad (18.17)$$

where R is the curvature tensor of the connection K and the bilinear form $[\xi, \eta]$ is defined as the sum of vv, vh, hv, hh terms in (20.12, 20.16, 20.19).

In particular, if the curvature R of the connection K vanish (equals zero), then for two arbitrary vertical vector fields ξ, η the difference $[Pr^1_K\xi, Pr^1_K\eta] - Pr^1_K[\xi, \eta]$ vanishes if and only if $[\xi, \eta] = 0$.

19. CURVATURE AND THE SOURCES f_j.

In this section we study the vertical bilinear form $N(\xi, \eta)$ for some simple situations and try to determine a relation between curvature and the form of the source

f_j in the Euler-Lagrange equations in the case of non-commutative variations (see (6.6)).

Remark 27. At the points where functions ξ, η are positive, equating $II - I$ to zero, dividing by $\eta\xi$ and assuming that $\eta, \xi > 0$ we get the condition (28.12) in the form

$$[\widetilde{\xi\partial_i} - \widetilde{\eta\partial_j}] - [\widetilde{\xi\partial_i, \eta\partial_j}] = 0 \Leftrightarrow$$
$$\Leftrightarrow \tilde{R}^k_{i\mu j} + [ln(\eta)_{,y^l} K^l_{\mu i}\delta^k_j - d_\nu ln(\eta)\partial_{y^j_\nu} K^k_{\mu i}] + [ln(\xi)_{,y^l} K^l_{\mu j}\delta^k_i - d_\nu \ln(\xi)\partial_{y^i} K^k_{\mu j})] = 0 \; \forall k, i, \mu, j. \tag{19.1}$$

Here terms containing different functions are explicitly separated.

Example 11. Let $K^i_{\mu j} = \frac{\partial ln(\|\pi\|)}{\partial \pi^\mu_i} f_j = \lambda^i_\mu f_j$ be the "canonical K-tensor", see Sec.9 . "The "curvature" $\tilde{R}^k_{i\mu j}$ corresponding to this tensor is equal to

$$\tilde{R}^k_{i\mu j} = (f_j\partial_i - f_i\partial_j)\lambda^k_\mu + \lambda^k_\mu(f_{j,i} - f_{i,j}) + \lambda^l_\nu\lambda^k_\mu[f_i\partial_{y^l_\nu} f_j - f_j\partial_{y^l_\nu} f_i]. \tag{19.2}$$

In a case of potential forces: $f_i = U_{,i}(x, y)$, the middle term in the expression for $\tilde{R}^k_{i\mu j}$ vanishes. If f_j do not depend on the derivatives of fields, the third term vanishes.

19.1. Special cases. Here we present the expressions for "curvature" $\tilde{R}$ and the bracket $[\xi, \eta]$ in two special cases: the case of Mechanics ($n = 1$) and the case of one field ($m = 1$).

19.1.1. *Case $m = 1$.* Let $y(x^\mu)$ be the only dynamical field of theory. Then, the "curvature" $\tilde{R}$ vanishes identically while the bracket $[\xi, \eta]$ simplifies due to the fact that vector fields $\xi = \xi\partial_y, \eta = \eta\partial_y$ have one component each. As a result, conditions (18.5) and (18.7) (for couples of vertical vector fields $\xi\partial_y, \eta\partial_y$) take the form 18.9, 18.11.

$$\begin{cases} \tilde{R}^1_{1\mu 1} = 0, \\ [\xi, \eta] = [(\xi \cdot \eta_{,y} - \eta \cdot \xi_{,y})K^1_{.1} - (\xi d_i\eta - \eta d_j\xi)\partial_{\dot{y}} K^1_{\mu 1}]\partial_y = 0. \end{cases} \tag{19.3}$$

19.1.2. *Case $n = 1$ - Mechanics.* Let $y^\mu(t), \mu = 1, \ldots, m$ be m dynamical fields. Then, $\xi = \xi^\mu\partial_\mu, \eta = \eta^\nu\partial_\nu$ and

$$\begin{cases} \tilde{R}^k_{i0j} = (\partial_i K^k_j - \partial_j K^k_i) + (K^l_i\partial_{\dot{y}^l} K^k_j - K^l_j\partial_{\dot{y}^l} K^k_i), \\ [\xi, \eta] = [(\xi^j\eta^k_{,l} - \eta^j\xi^k_{,l})K^l_j - (\xi^j d_t\eta^i - \eta^j d_t\xi^i)\partial_{\dot{y}^i} K^k_j]\partial_{\dot{y}^k}. \end{cases} \tag{19.4}$$

Example 12. Interacting oscillators. Consider a system of two interacting oscillators with $y^1(t), y^2(t)$ as dynamical variables and the Lagrangian

$$L = \frac{m_1}{2}\dot{y}^{1,2} + \frac{m_2}{2}\dot{y}^{2,2} - \left(\frac{\omega_1}{2}y^{1,\,2} + \frac{\omega_2}{2}y^{2,\,2}\right)$$

. Let $K = K^i_{0j}$ be the K-tensor modifying the variations of velocities $\dot{y}^i$.
The Euler-Lagrange equations have the form

$$m_j\ddot{y}^j + \omega^j y^j = -K^i_{0j}m_i\dot{y}^i. \tag{19.5}$$

For the "curvature" $\tilde{R}$ we have (with $i, l, k = 1, 2$):

$$\begin{cases} \tilde{R}^k_{0ii} = 0, \\ \tilde{R}^k_{012} = (\partial_1 K^k_2 - \partial_2 K^k_1) + (K^l_1 \partial_{\dot{y}^l} K^k_2 - K^l_2 \partial_{\dot{y}^l} K^k_1). \end{cases}$$

If $K^k_j \in C^\infty(Y)$, second term in the formula last vanishes and the condition $R^k_{012} = 0$ is equivalent to the local representation

$$K^k_i = \partial_{\dot{y}^i} P^k(t, y) \Rightarrow f_j = -\partial_{\dot{y}^j}(P^k m_k \dot{y}^k).$$

Thus, the "curvature" $\tilde{R}$ for the system of two oscillators with the NC-tensor K, that does not depend on the velocities, vanishes if and only if forces are potential and potential is linear by velocities $\dot{y}^i$.

20. Case: Zero order tensor K: $K^{\alpha}_{\beta i} \in C^{\infty}(Y)$.

In this section we specify the considerations of this Chapter to the case where tensor $K^{\alpha}_{\beta i}$ is defined on Y, i.e. depends on variables x^i, y^{μ} but not on the derivatives of fields y^{μ}: $K^{\alpha}_{\beta i} \in \pi^{*}_{10} C^{\infty}(Y)$. Examples of such situations are presented, for instance, in [133], Ch.6.

This case is special in different aspects starting from the fact that if source terms f_{ν} can be generated by a NC-tensor K of non-commuting variations with $K^{\mu}_{\nu i} \in C^{\infty}(Y)$, tensor K **is defined uniquely**. More specifically, we have

Proposition 3. *Let $L \in C^{\infty}(J^1(\pi))$ be a regular Lagrangian and let*

$$\frac{\delta L}{\delta y^i} = f_i, \qquad (20.1)$$

be the Euler-Lagrange Equations with the sources $f_{\mu} \in C^{\infty}(J^1(\pi))$. Then,

(1) *The following statements are equivalent:*
   (a) *Euler-Lagrange system (22.1) allows the representation (6.4,6.6) with the tensor $K^{\mu}_{\nu i} \in \pi^{*}_{10} C^{\infty}(Y)$,*

   (b) *Being written in Legendre coordinates $x^i, y^{\mu}, \pi^i_{\mu}$, sources f_{μ} are linear homogeneous functions of momenta π^i_{μ}.*
(2) *If conditions of part 1) are fulfilled, tensor $K^{\mu}_{i\nu}$ is defined uniquely in $\pi^{*}_{10} C^{\infty}(Y)$.*
(3) *The dissipation term in the energy balance of the Stress-Energy-Momentum balance law has the form*

$$-\dot{y}^{\mu} K^{\nu}_{\mu i}(x,y) \pi^j_{\mu}. \qquad (20.2)$$

(4) *Tensor K can have the "canonical" form (9.8,9.9) if and only if it is identically zero.*

Proof. Proof of equivalence of statements a),b) is straightforward.

We prove the second statement.

Let sources f_i have two presentations of the form (4.7)

$$f_{\mu} = -K^{\mu}_{i\nu}\pi^i_{\mu} = -S^{\mu}_{i\nu}\pi^i_{\mu}, \ \mu, \nu = 1, \ldots, m; i = 1, \ldots n$$

for some functions $K^{\mu}_{i\nu}, S^{\mu}_{i\nu} \in C^{\infty}(Y)$. Subtracting corresponding Euler-Lagrange equations we get

$$Q^{\mu}_{i\nu}\pi^i_{\nu} = (K^{\mu}_{i\nu} - S^{\mu}_{i\nu})\pi^i_{\mu} = 0, \ j = 1, \ldots, m \qquad (20.3)$$

for all values of their arguments $(x^i, y^{\mu}, y^{\mu}_i)$. Due to the regularity of Lagrangian L, we can (locally) change variables in the jet space $J^1(\pi)$ to $(x^i, y^{\mu}, \pi^i_{\mu})$ and equalities (20.3) will hold. Since momenta π^i_{μ} are functionally independent, conditions (20.3) are valid if and only if $K^{\mu}_{i\nu} = S^{\mu}_{i\nu}, \ \mu, \nu = 1, \ldots, m; i = 1, \ldots n.$

The third statement follows from the form of the right side in the energy balance law (7.6).

To prove the last statement it is sufficient to compare the form of canonical tensor K and generated sources f_{μ} with the conditions on this tensor and sources given in part 1 of this proposition. $\qquad \square$

Consider now the form of curvature $\tilde{R}$ and of the bracket $\lceil , \rceil$ of vertical connection defined by tensor K of order zero (i.e., tensor K is defined on the space Y).

Tensor $\tilde{R}$, in this case, has the form

$$\tilde{R}^{\sigma}_{\mu\nu k} = \partial_{\mu} K^{\sigma}_{\nu k} - \partial_{\nu} K^{k}_{\mu k}. \tag{20.4}$$

At the same time,

$$\lceil \xi, \eta \rceil = (\xi^{\mu} \eta^{\sigma}_{,\lambda} K^{\lambda}_{i\nu} - \eta^{\nu} \xi^{\sigma}_{,\lambda} K^{\lambda}_{i\nu}) \partial_{y^{\sigma}_i} = \xi^{\nu} \eta^{\sigma}_{,\lambda} - \eta^{\nu} \xi^{\sigma}_{,\lambda}) K^{\lambda}_{i\nu} \partial_{y^{\sigma}_i}. \tag{20.5}$$

Example 13. If $K^{\lambda}_{\mu i} = c_i(x, y) \delta^{\lambda}_{\mu}$ where $c_i \in C^{\infty}(Y))$, then the previous formula takes the form

$$\lceil \xi, \eta \rceil = [\xi, \eta] c_{\mu} \partial_{y^{\sigma}_i} \tag{20.6}$$

with the standard bracket $[\xi, \eta]$ of vector fields $\xi, \eta \in \mathcal{X}(Y)$. In such a case,

$$f_{\nu} = c_i \pi^i_{\nu}.$$

Dissipation term () in the energy balance law has the form

$$c_i(x, y) \dot{y}^{\mu} \pi^i_{\nu} =^{c_i = c \delta^0_i} c(x, y) \dot{y}^{\mu} \frac{\partial L}{\partial \dot{y}^{\mu}}.$$

Last equality is valid if $c_i = c \delta^0_i$.

20.1. Curvature $\tilde{R}$ and the potential forces.

Consider a case of Mechanics, where $n = 1$. The independent variable will be denoted by t. Let $L \in C^{\infty}(J^1(\pi))$ be a Lagrangian and let K^{μ}_{ν} be a tensor defined by a vertical connection on the bundle $J^1(\pi) \to Y$. Then, the EL-equations with the NC-variations have the form

$$L_{,y^{\mu}} - d_t (L_{,\dot{y}^{\mu}}) = -K^{\mu}_{\nu} \pi_i, \tag{20.7}$$

where $\pi_{\mu} = L_{,\dot{y}^{\mu}}$ are the the components of linear momentum.

The curvature of the vertical connection K has the form (21.3-4)

$$\tilde{R}^{\sigma}_{\mu\nu} = (\partial_{y^{\mu}} K^{\sigma}_{\nu} - \partial_{y^{\nu}} K^{\sigma}_{\nu}) + (K^{\lambda}_{\mu} \partial_{\dot{y}^{\mu}} K^{\sigma}_{\nu} - \dot{y}^{\nu} K^{\sigma}_{\mu}). \tag{20.8}$$

In a case where $K^{\alpha}_{\beta} \in C^{\infty}(Y)$, the second term in the expression of curvature vanish. As a result, $\tilde{R}^{\sigma}_{\mu\nu} = 0$ if and only if

$$\partial_{y^{\mu}} K^{\sigma}_{\nu} - \partial_{y^{\nu}} K^{\sigma}_{\nu} = 0, \ \forall i, j,$$

or, what is the same, locally

$$K^{\sigma}_{\nu} = \partial_{y^{\mu}} P^{\sigma} \tag{20.9}$$

for some functions $P^{\sigma} \in C^{\infty}(Y)$.

Substituting (21.4) to the dynamical equations (22.7) we get

$$d_t (L_{,\dot{y}^{\mu}}) - L_{,y^{\mu}} = f_{\mu} = (\partial_{y^{\mu}} P^{\sigma}) L_{,\dot{y}^{\sigma}},$$

Functions P^{σ} depend on (t, y) but not on the derivatives $\dot{y}^{\mu}$. **In a special case where** $L_{,\dot{y}^{\mu}\dot{y}^{\nu}} = 0 \ \forall \mu, \nu$ (Lagrangian, linear by derivatives) this equation takes the form of an Euler-Lagrange system **with potential forces** having the potential $U = P^{\sigma} L_{,\dot{y}^{\sigma}} = \partial_{\dot{y}^{\mu}} (P^{\sigma} L)$.

$$L_{,y^{\mu}} d_t (L_{,\dot{y}^{\mu}}) - L_{,y^{\mu}} = \frac{\partial U}{\partial y^{\mu}} = \partial_{y^{\mu}} (P^{\sigma} L_{,\dot{y}^{\sigma}}).$$

It is easy to see that in the general situation, Euler-Lagrange Equations have the form (see bellow)

$$L_{,y^\mu} d_t (L_{,\dot{y}^\mu}) - L_{,y^\mu} = \partial_{\dot{y}^\mu}(L\partial_{y^\mu}P^\sigma).\qquad(20.10)$$

In the case of Field Theory, when the K-tensor does not depend on the jet variables y_i^μ, condition $\tilde{R} = 0$ is equivalent to the existence of a local representation

$$K_{i\nu}^\sigma = \partial_\nu P_i^\sigma,$$

with $P_i^\sigma \in C^\infty(Y)$. As a result, source/force terms can be written in one of the forms

$$f_\nu = (\partial_\nu P_i^\sigma)\pi_\sigma^i = \partial_\nu(P_i^\sigma \pi_\sigma^{iu}) = P_i^\mu \partial_\nu \pi_\mu^i = \partial_{y_i^\mu}(L\partial_\nu P_i^\mu).\qquad(20.11)$$

It seems natural to call a source having the form $f_\nu = \partial_{y_i^\mu}T_{i\nu}^\mu$ - a **tensor-potential source** defined by the tensor $T_{i\nu}^\mu$.

UP

21. NC VARIATIONS AND THE "DYNAMICAL CONNECTIONS" IN HAMILTONIAN SYSTEMS.

Connections, playing a dynamical role as the factor of evolution acting in the dynamical systems, are known from the time of Levi-Civita and Ricci. In the course of the last 20-25 years, connections were used as the basic dynamical driver in the finite-dimensional dynamical systems and in the infinite-dimensional systems of Classical Field Theory. In particular, I would like to refer to the works of Belgian and Spanish groups of mathematicians (M.de Leon, P.Rodrigues, D.Martin de Diego, R.Ibanez, a.Ibort, F.Cantrijn, M.Ctampin,F.Sarlet) , as well as to the works of G.Giachetta , L.Mangiarotti and G.Sardanashvily, [27].

Here we introduce connections playing a similar role in the extended configurational bundles of a mechanical or field systems with the non-commutative variations.

Consider a Lagrangian dynamical system with the NC-variations tensor K_i^μ. Denote by y^μ the dynamical variables of this system and introduce the momenta $\pi_\nu = \frac{\partial L}{\partial \dot y^\nu}$. Let $\pi : Y \to T$ be the configurational bundle of this system over the time axis T.

Euler-Lagrange equations

$$\frac{\partial L}{\partial y^\mu} - \frac{d}{dt}\frac{\partial L}{\partial \dot y^\mu} = -K_\mu^\nu \pi_\nu, \ \mu = 1,\ldots,m \tag{21.1}$$

for this dynamical system can be written in the form

$$\frac{\partial L}{\partial y^\mu} = \frac{d}{dt}\frac{\partial L}{\partial \dot y^\mu} - K_\mu^\nu \pi_\nu = \frac{d}{dt}\pi_\mu - K_\mu^\nu \pi_\nu = (\frac{d}{dt}\delta_\mu^\nu - K_\mu^\nu)\pi_\nu, \tag{21.2}$$

i.e.as

$$\left(\frac{d}{dt}\delta_\mu^\nu - K_\mu^\nu\right)\frac{\partial L}{\partial \dot y^\nu} = \frac{\partial L}{\partial y^\mu}. \tag{21.3}$$

Corresponding Hamiltonian system (see(13.6)) obtained using (explicitly or implicitly) Legendre transformation from variables (x^i, y^μ, y_i^μ) to the variables (x^i, y^μ, pi_μ^i) can be written in the form

$$\begin{cases} \left(\frac{d}{dt} - 0\right) y^\nu = \frac{\partial H}{\partial \pi_\nu} \\ \left(\frac{d}{dt}\delta_\mu^\nu - K_\mu^\nu\right)\pi_\nu = -\frac{\partial H}{\partial y^\mu}. \end{cases} \tag{21.4}$$

Here, coefficients K_μ^ν are functions of Hamiltonian variables.

Consider now the **double bundle** $T^*(Y) \to Y \to T$, where $\pi : Y \to T$ is the configurational bundle and introduce a connection ∇_K on this double bundle such that the K-horizontal lift of the basic vector field ∂_t is

$$Hor_K \partial_t = \partial_t - K_\mu^\lambda p_\lambda \partial_{\pi_\mu}. \tag{21.5}$$

This connection is defined by the 1-form

$$dt \otimes (\partial_t - K_\sigma^\mu \pi_\mu \partial_{\pi_\sigma}). \tag{21.6}$$

Notice that the Legendre transformation defines, in a case of regular Lagrangian K, isomorphism of the bundles

Thus, this isomorphism transforms the connection K_ν^μ in the double bundle $\pi_{10} : J^1\pi \to Y \to X$ into the connection (we will call it by the same name) in the double bundle $J^* \to Y \to X$.

Presentation (21.4) shows how a Hamiltonian system can be written as the first order Δ_K-covariant system of ordinary differential equations

$$\nabla_K \begin{pmatrix} y^\nu \\ \pi_\nu \end{pmatrix} = \begin{pmatrix} \frac{\partial H}{\partial \pi_\nu} \\ -\frac{\partial H}{\partial y^\mu} \end{pmatrix}. \tag{21.7}$$

21.1. **Case of Field Theory.** In the case of Field Theory of order 1 (when Lagrangian L is the function on the 1-jet space $J^1\pi$ of the configurational bundle $\pi : Y \to X$) with n independent variables x^i and m fields y^μ, Hamiltonian system has the form (11.7). Consider the double dual bundle

$$\kappa : J^*(\pi) \to Y \to X,$$

where.$J^*\pi \to Y$ is the dual bundle of the bundle $J^1\pi \to Y$ (see Appendix I, Sec.66). A (local) section s of this double bundle has the form $s : \bar{x} = \{x^i\} \to \{(x^i, i = 1, \ldots, n; y^\mu(\bar{x}); \pi^k_\sigma(\bar{x}))\}$.

Rewrite Hamiltonian system 21.7 in the form

$$\begin{cases} \partial_i y^\mu = \frac{\partial H}{\partial \pi^i_\mu}, \\ (\partial_i \delta^\mu_\nu - K^\mu_{i\nu}) \pi^i_\mu = -\frac{\partial H}{\partial y^\nu}. \end{cases} \tag{21.8}$$

The operator $(\partial_i \delta^\mu_\nu - K^\mu_{i\nu})$ has the form of the covariant derivative of a section of bundle κ. First equation of system (23.8) can be written in a similar form -

$$(\partial_i \delta^\nu_\mu - \mathbf{0}) y^\mu = \frac{\partial H}{\partial y^\nu}.$$

This leads us to introduce a K-connection on the bundle $\kappa : J^*(\pi) \to X$ with the form

$$dx^i \otimes (\partial_{x^i} - K^\mu_{i\sigma} \pi^j_\mu \partial_{\pi^j_\sigma}). \tag{21.9}$$

The horizontal lift of a basic vector field ∂_{x^i} with respect of this connection has the form

$$\partial_{x^i} \to \partial_{x^i} - K^\mu_{i\sigma} \pi^j_\mu \partial_{\pi^j_\sigma}. \tag{21.10}$$

Using the introduced connection, one can write the Hamitlonian system using the covariant derivative ∇_K as follows:

$$\begin{cases} \nabla_i \bar{y}^\mu = \frac{\partial H}{\partial \pi^i_\mu}, \\ \nabla_i \bar{\pi}^i_\mu = -\frac{\partial H}{\partial y^\mu}. \end{cases} \tag{21.11}$$

Remark 28. Notice that the source covector f_σ (π-vertical 1-form) present in the Hamiltonian form of the system (13.6, (13.7)) is the **trace** of the connection coefficients

$$f_\sigma = Tr(K^\mu_{i\sigma} \pi^j_\mu) = K^\mu_{j\sigma} \pi^j_\mu. \tag{21.12}$$

Next, we calculate the curvature of introduced connection defined by the NC-tensor $K^\mu_{i\sigma}$. Taking $K - twisted$ prolongations of the basic vector fields ∂_i, ∂_j and calculating their commutator, we get

$$[\partial_{x^i} - K^\mu_{i\sigma} \pi^j_\mu \partial_{\pi^j_\sigma}, \partial_{x^i} - K^\mu_{i\sigma} \pi^j_\mu \partial_{\pi^j_\sigma}] = [(\partial_i K^\nu_{j\lambda} - \partial_j K^\nu_{i\lambda}) \pi^s_\nu \partial_{\pi^s_\lambda} +$$
$$+ K^\mu_{i\sigma} \pi^s_\mu \partial_{\pi^s_\sigma}(K^\nu_{j\lambda} \pi^s_\nu) - K^\mu_{j\sigma} \pi^s_\mu \partial_{\pi^s_\sigma}(K^\nu_{i\lambda} \pi^s_\nu)] \partial_{\pi^s_\lambda}. \tag{21.13}$$

since $[\partial_i, \partial_j] = 0$, previous formula gives the curvature tensor of connection ∇_K in the form of collection of coefficients of the vector field $* * \frac{\partial}{\partial \pi^s_\mu}$:

$$R(K)_{ij} = (\partial_i K^\nu_{j\lambda} - \partial_j K^\nu_{i\lambda})\pi^s_\nu \partial_{\pi^s_\lambda} + + K^\mu_{i\sigma}\pi^s_\mu \partial_{\pi^s_\sigma}(K^\nu_{j\lambda}\pi^s_\nu) - K^\mu_{j\sigma}\pi^s_\mu \partial_{\pi^s_\sigma}(K^\nu_{i\lambda}\pi^s_\nu)]$$

$$(21.14)$$

Notice some analogy of obtained expression with the formula (18.13) containing the curvature component for the K-twisted lift of two horizontal vector fields to the 1-jet space $J^1\pi$.

22. Kinematical connection (Γ, S), dynamical connection K, and the energy-momentum balance law.

In the considerations of Chapter 2, vertical connections K on the bundle $\pi_{10} : J^1(\pi) \to Y$ have been used to define variations of the action $\mathcal{A}$ in the directions of the jet variables y_i^μ. Thus, a vertical connection K has *dynamical meaning* and represents a machinery by which the non-potential sources enter the dynamical equations.

On the other hand, determination of the balance laws corresponding to the symmetries of a given Lagrangian Field Theory using the E.Noether formalism requires determining the action of geometrical transformations - transformations acting on the base X and space Y of the configurational bundle on the dynamical fields y^μ and their derivatives. In other words, one has to use prolongation of vector fields from X to Y and from Y to the jet bundle $J^k(\pi)$, where k is the order of Lagrangian L.

In generic situations this prolongation is achieved by using a connection Γ on the configurational bundle $\pi : Y \to X$ and a connection S on the bundle $\pi_k : J^k(\pi) \to Y$. Below we consider, what effect the coexistence of general *"kinematic"* connections Γ and S with the dynamical connection K will have for the conservation (or balance) laws obtained by E.Noether method. We consider the case $k = 1$.

Thus, we introduce a connection Γ on the bundle π with the 1-form $\omega_\Gamma = dx^i + \Gamma_i^\mu dy^\mu$ defining the horizontal lift of vector fields from X to Y:

$$\xi = \xi^i \partial_i \to \hat{\xi} = \xi^i \partial_i + \Gamma_\mu^i \xi^\mu \partial_i. \tag{22.1}$$

Now we use a connection S on the bundle $\pi_{10} : J^1(\pi) \to Y$ to lift vector fields of the form $\hat{\xi}$ and all the others vector fields in Y to $J^1(\pi)$. Applying the operation of S-horizontal lift to the vector field $\hat{\xi}$ we get

$$\hat{\xi} \to \hat{\xi}^1 = (\xi^i \partial_i + S_{i\nu}^j \xi^i \partial_{y_j^\nu}) + (\Gamma_i^\mu \xi^i \partial_\mu + S_{\mu j}^\nu \Gamma_i^\mu \xi^i \partial_{y_j^\nu}) =$$
$$= \xi^i \partial_i + \Gamma_i^\mu \xi^i \partial_\mu + ((S_{i\nu}^j + S_{\mu j}^\nu \Gamma_\mu^i)\xi^i)\partial_{y_j^\nu}. \tag{22.2}$$

In particular, having in mind the stress-energy-momentum balance law for a Lagrangian field Theory with the sources, we define the lift of translational vector fields on X to $J^1(\pi)$. We get

$$\widehat{\partial_k}^{(1)} = (\partial_k + S_{k\nu}^j \partial_{y_j^\nu}) + (\Gamma_i^\mu \partial_\mu + S_{\mu j}^\nu \Gamma_k^\mu \partial_{y_j^\nu}) = \partial_k + \Gamma_{ka}^\mu \partial_\mu + (S_{k\nu}^j + S_{\mu j}^\nu \Gamma_k^\mu)\partial_{y_j^\nu}. \tag{22.3}$$

Remark 29. As we show now, connection K *should not be used* as the "kinematical connection" for defining symmetries of the action and participating in the formulation of Noether Theorem in the presence of dissipation, in particular, in the Stress-Energy-Momentum Balance law.

Infinitesimal condition of divergence invariancy of the Action (Lagrangian) with respect to the vector field $\widehat{\partial_k}^{(1)}$ in terms of this prolongation (meaning the **Divergent invariance** with respect to the translation transformations) has the form (see

Appendix III.)

$$Div(B) = \widehat{\partial_k}^1_S L + L Div(\partial_k) = Div(L\partial_k) + (\hat{\xi}_Q)^1_S L = Div(L\partial_k) + (Q^i \partial_\mu)^1 \cdot L$$

$$\Leftrightarrow$$

$$Div(B - L\partial_k) = (Q^\mu \partial_\mu)^1_S \cdot L = Q^\mu \partial_\mu L + S^\mu_{\nu i} Q^j L_{y^\mu_i} + (d_i Q^\mu) L_{,y^\mu_i} =$$

$$= d_i(Q^\mu L_{,y^\mu_i}) + (S^\mu_{\nu i} - K^\mu_{i\nu})Q^\nu L_{y^\mu_i} + Q^\mu(\partial_\mu L - d_i L_{,y^\mu_i} + K^\nu_{i\mu} L_{y^\nu_i})$$

$$\Leftrightarrow Div(B - L\partial_k - Q^\mu L_{,y^\mu_i}) = (S^\mu_{i\nu} - K^\mu_{i\nu})Q^\mu L_{y^\mu_i} + Q^\mu(\partial_\mu L - d_i L_{,y^\mu_i} + K^\nu_{i\mu} L_{y^\nu_i}).$$

$$(22.4)$$

where $Q^\mu = \omega^\mu(\hat{\xi}^\mu) = \hat{\xi}^\mu - \xi^j y^\nu_j = \Gamma^\mu_k - y^\mu_k$ is the characteristic of the vector field $\hat{\xi}^1$. Along a solution of Euler-Lagrange equations (6.4;6.6) the second term on the right side vanishes and we get the Stress-Energy-Momentum balance law

$$Div(B - L\partial_k - (\Gamma^\mu_k - y^\mu_k)L_{,y^\mu_i}) = (S^\mu_{\nu i} - K^\mu_{\nu i})(\Gamma^\nu_k - y^\nu_k)L_{y^\mu_i}, \tag{22.5}$$

or

$$Div(B - L\partial_k - (\Gamma^\mu_k - y^\mu_k)L_{,y^\mu_i}) = (S^\mu_{\nu i} - K^\mu_{\nu i})(\Gamma^\nu_k - y^\nu_k)\pi^i_\mu, \quad k = 0, \ldots, 3. \tag{22.6}$$

This is the form of the stress-energy-momentum conservation (or balance) law delivered by the first Noether Theorem (see [106], Ch.VI)¿ The only difference by comparison with the conventional situation is in the form of characteristic $Q^i = -y^i_\mu$ of the μ-th component of vector field ξ_Q (see Appendix) is replaced by $\Gamma^i_\mu - y^i_\mu$. Here Γ is the connection in the bundle $\pi : Y \to X$ used to lift vector fields ∂_{x^μ} to Y.

This shows that in order to get the balance law taking the dissipative sources into account one should employ a (kinematical) connection S whose vertical part is different from that of the connection K, to lift the vector fields from Y to the space $J^1(\pi)$.

Remark 30. In a case where Γ is the trivial connection (as in Chapter 2), one can take connection S to be trivial as well and get the dissipation term in the energy balance law.

Remark 31. Notice, that two connections S, K differs by the tensor: $T^i_{\mu j} = S^i_{\mu j} - K^i_{\mu j}$. This tensor should play some role in the description of dissipation of energy.

23. Infinitesimal Variational Calculus and the non-commutative variations.

Infinitesimal Variational calculus pioneered in the works of A.Poincare and E.Cartan. Further deep development was done in 1935-46 in the works of H.J.Lepage, see, for instance, [74]. Next, important progress was done in the paper [50]. Finally, in 70th, this region of Variation Calculus assumed its present form in works of P.Dedecker, D.Krupka, O.Krupkova P.L.Garcia, A.Anderson, M. Francaviglia, D.Saunders and other specialists. We refer to the monograph [74] for acquaintance with the present state of this domain.Here we make a short introduction to the

23.1. Poincare-Cartan formalism, case of the first order. As before, we introduce a configurational fiber bundle $\pi : Y \to X$ with an $n - dim$ connected paracompact (see Appendix I for definition of this term) smooth (C^∞)manifold X as the base and a total space Y, $dim(Y) = n + m$. The fiber of the bundle π is an m-dim connected smooth manifold F. The manifold X is endowed with a (pseudo)-Riemannian metric G. Denote by η the volume n-form corresponding to the metric G.

We will be using fibred charts (W, x^i, y^μ), W being an open subset of Y (see Ch.1).

Volume form η permits us to introduce the vectival endomorphism - tensor field on the 1-jet bundle $J^1\pi$ of the type $(1, n)$ (1 - contravariant and n-covariant)

$$S_\eta = (dy^\mu - y_i^\mu dx^i) \wedge \eta_j \otimes \frac{\partial}{\partial y_i^\mu}. \tag{23.1}$$

Let $L\eta$ be a Lagrangian n-form with a smooth function $L \in C^\infty(J^1\pi)$.
Poincare-Cartan n-form is defined as follows:

$$\Theta_L = L\eta + S_\eta^*(dL), \tag{23.2}$$

where S^* is the adjoint operator of the vertical endomorphism S_η. In coordinates, we have

$$\Theta_L = \left(L - y_i^\mu \frac{\partial L}{\partial y_i^\mu}\right)\eta + \frac{\partial L}{\partial y_i^\mu} dy^\mu \wedge \eta_i. \tag{23.3}$$

An extremal of L is a section $s : D \to Y$ if the bundle π (with the domain $D \subset X$) such that **for any vector field** $\hat{\xi}$ on the 1-jet manifold $J^1\pi$,

$$(j^1(s))^*(i_{\hat{\xi}}d\Theta_L) = 0 \tag{23.4}$$

, where $j^1(s)$ is the first jet prolongation of section s.

Next statement can be proved by direct calculation.

Theorem 11. *A section $s : D \to Y$ is is an extremal of the action with the Lagrangian L if and only if it satisfies to the system of Euler-Lagrange equations*

$$\frac{\partial(L \circ j^1(s)\sqrt{|G|})}{\partial y^\mu} - \frac{d}{dx^i}\left(\frac{\partial(L \circ j^1(s)\sqrt{|G|}}{\partial y_i^\mu}\right) = 0, \ \mu = 1, \ldots, m. \tag{23.5}$$

It is known that one can replace in this theorem "all vector fields $\xi \in X(J^1\pi)$" by "all the (conventional) prolongations $Pr^1\xi$ of vector fields ξ in Y".

If we replace here conventional (flow) prolongation with the prolongation defined by a NC-tensor $K_{i\nu}^\mu$ and will be using prolongations $Pr_K^1\xi = Pr^1\xi + K_{i\nu}^\mu\partial_{y_i^\mu}$ of all vector fields in Y, equivalence stated in the last Theorem above holds but the Poincare-Cartan system of equations (23.4) will be replaced with the similar system of equations that has to be fulfilled for the K-twisted prolongation $Pr_K^1\xi$ of any vector field in Y **with the right side defined by the tensor** K:

$$j^{1*}(s)i_{Pr^1\xi}d\Theta_L = -j^{1*}(s)i_{K_{i\nu}^\mu \partial_{y_i^\mu}}d\Theta_L. \tag{23.6}$$

Right side in this system is similar to the source/force terms in Chapter 2. Correspondingly, right side in (23.6) may play the role of forces acting in the system or that of a source of different type.

24. Lifted Poincare-Cartan form of a balance system and its contact source modification.

In the works [113, 114] it was shown that any balance system (for convenience, we will denote this system the $\bigstar$) of order k for m dynamical fields y^μ

$$d_i F_\mu^i(j^k s) + F_\mu^i(j^k s)\lambda_{G,i} = \Pi_\mu(j^k s), \ \mu = 1,\ldots,m, \qquad \bigstar \tag{24.1}$$

can be realized in invariant form - as the requirement that for many enough variational vector fields $\xi \in \mathcal{X}(J^k(\pi))$,

$$j^{k+1}s^* i_\xi \tilde{d}\Theta = 0. \tag{24.2}$$

Here $\tilde{d}$ - is the "con-differential", see Appendix. The form

$$\Theta = F_\mu^i \omega^\mu \wedge \eta_i + \Pi_\mu \omega^\mu \wedge \eta \tag{24.3}$$

is the $n + (n+1)$-form on the k-jet bundle $J^k(\pi)$ - analog of the Poincare-Cartan form for the balance system (3.1) (these forms were called *lifted* in [112, 114]).

Yet, calculation of $i_\xi \tilde{d}\Theta$ shows that unless the vector field ξ satisfies to certain conditions, obtained expression will depend on the derivatives of the vector field ξ and, as a result, can not be used for separation of components of the balance system. It was shown that locally there are always enough variational vector fields $\xi \in \mathcal{X}(Y)$ satisfying this condition and, therefore, ensuring the separation of equations. Yet, by reasons provided in the Introduction it is highly inconvenient to have such restriction to the vector fields used in combination with the Poincare-Cartan form (3.3).

That is why, in the works [113, 114] we suggested to modify the Poincare-Cartan form (24.3) on the jet bundle $J^k(\pi)$ for $k > 1$ and on the 2-jet bundle $J^2(\pi)$ for $k = 1$ by adding an extra source term:

$$\widetilde{\Theta} = F_\mu^i \omega^\mu \wedge \eta_i - \Pi_\mu \omega^\mu \wedge \eta - F_\mu^i \omega_i^\mu \wedge \eta. \tag{24.4}$$

We call such a form - the ML-form - 'modified,lifted" (see [114]) Poincare-Cartan form.

Consider the lift ξ^2 to the bundle $J^2(\pi)$ of a vector field $\xi \in \mathcal{X}(J^1(\pi))$ for $k = 1$ and the lift $\xi^{k+1} \in \mathcal{X}(J^{k+1}(\pi))$ of the vector field ξ^k on $J^k(\pi)$ for $k > 1$. Applying to the form $\widetilde{\Theta}$ the arguments leading to the representation (3.2) of the balance system, we get

$$i_{\xi^{k+1}}\tilde{d}\widetilde{\Theta} = -\omega^\mu(\xi)[d_i F_\mu^i + \lambda_{G,i} F_\mu^i - \Pi_\mu]\eta + Contact form. \tag{24.5}$$

This proves the following

Theorem 12. *For a given balance system (3.1) of order $k \geqq 0$ and a section $s : W \to Y$ of the bundle π the following statements are equivalent:*

(1) *For any vector field $\xi \in \mathcal{X}(J^k(\pi))$ and its arbitrary prolongation to the π_k^{k+1}-projectable vector field ξ^{k+1} on $J^{k+1}(\pi)$,*

$$j^{k+1\,*}(s)i_{\xi^{k+1}}\widetilde{d\widetilde{\Theta}} = 0, \tag{24.6}$$

(2) *Balance equations $\bigstar$:* $\left(F_\mu^i \circ j^k(s)\right)_{;i} = \Pi_\mu \circ j^{k+1}(s), \ \mu = 1, \ldots, m$ *holds.*

Corollary 1. *Any balance system (3.1) of order $k \geqq 0$ can be presented in the infinitesimal variational form (3.2) using lifted Poincare-Cartan form (24.4) and admissible variations of dynamical fields y^μ or using ML-form (3.4) and arbitrary variations of dynamical fields y^μ.*

Remark 32. Using K-twisted prolongation of vector fields we can modify balance equations by adding some terms in the source side as it was done in equation (23.6). This might be useful in studying interactions of components of a physical system one studies.

Remark 33. One of the interesting particularity of the geometrical properties of balance systems is the form of the Noether Theorems corresponding to the symmetries of the balance systems, see [113], Section 7.

Modified Poincare-Cartan form 24.4 (named MPC-form) for the Balance equations and systems of balance equations is the $n + (n + 1)$ form. Correspondingly to this, condition that a Lie group G with the Lie algebra $\mathfrak{g}$, acting on the MPC-form splits into two separate conditions - invariants of both components of MPC-forms. It was shown in [113] that the Noether law corresponding to the invariance of n-part is, **in general, the balance law.** If both components of MOC-form are invariant with respect to a Lie group, second balance law take place for the solutions of balance system. At the same time, MOC-form defines the horizontal cohomology class (that could be called the $\mathfrak{g}$-charge, see [71]). This horizontal cohomology class is the obstraction for the Noether "balance equations" obtained above to be conservation laws.

25. Higher order Lagrangian systems with NC variations.

25.1. Higher order Lagrangian field theory. Let $L \in C^\infty(J^k(\pi))$ be a Lagrangian of order k - a smooth function L defined in a domain of the k-jet space: $L \in C^\infty(J^k(\pi))$ and let

$$\mathcal{A}(s) = \int_D (j^k s)^* L dv$$

be the action for sections $s : D \to Y$, where D is an open subset of X.

Using a fibred coordinate system $(x^i, y^\mu, \ldots y^\mu_I \ldots)$, $|I| \leq k$ in the bundle $\pi_k : J^k(\pi) \to X$ and introducing the higher order total derivatives corresponding to a multiindex $I = (i_1, \ldots i_n)$

$$d_I = d^{i_1}_{x^1} \cdot \ldots d^{i_n}_{x^n},$$

(we recall that the total derivatives $d_i = d_{x^i}$ commute with one another) we can write the higher order Euler-Lagrange system of equations with the Lagrangian L of the order k, $0 < k < \infty$ for the fields y^μ, $\mu = 1, \ldots, m$ in the form

$$\sum_{I| \ k \geq |I| \geq 0} (-1)^{|I|} d_I \frac{\partial L}{\partial y^\mu_I} = 0, \ \mu = 1 \ldots, m, \tag{25.1}$$

[106].

Example 14. For the case of second order Lagrangians (where $k = 2$), Euler-lagrange system has the form

$$\frac{\delta L}{\delta y^\mu} = \frac{\partial L}{\partial y^\mu} - d_i \frac{\partial L}{\partial y^\mu_i} + d_{ij} \frac{\partial L}{\partial y^\mu_{ij}} = 0, \ \mu = 1, \ldots, m. \tag{25.2}$$

Introduce now an Ehresmann connection K of order k, $0 < k \leq \infty$ on the infinite jet bundle $\pi_{\infty 0} : J^\infty(\pi) \to Y$ by defining the K-horizontal lift of the vector fields $\xi = \xi^i \partial_i + \xi^\mu \partial_\mu$ in Y to the space $J^\infty(\pi)$ (see Appendix III, Sec.77) as follows:

$$Hor^\infty_K(\xi) = \xi + \sum_{\nu, I| \ |I| > 0} (K^\nu_{Ii} \xi^i + K^\nu_{I\mu} \xi^\mu) \partial_{y^\nu_I} \tag{25.3}$$

Coefficients $K^\nu_{Ii}, K^\nu_{I\mu}$ are arbitrary smooth functions in $J^k(\pi)$ where $k = |I|$. This condition guarantees that the composition of the projection $\pi_{\infty k} : J^\infty \pi \to J^k \pi$ with the K-horizontal lift of vector fields $\xi \to Hor_K(\xi) : \mathcal{X}(Y) \to \mathcal{X}(J^\infty \pi)$ *defines the connection K_k on the bundle $\pi_{k0} : J^k \pi \to Y$.*

We use such a connection to modify the formula (6.1) for the prolongation of vector fields to the 1-jet bundle $J^1 \pi \to Y$ to the case of higher order prolongation of vector fields. Specifically, **we define K-twisted prolongation of vector fields** $\xi \in \mathcal{X}(Y)$ to $J^\infty \pi$ by the formula

$$\xi^{(\infty)}_K = \xi^i d_i + Q^\mu \partial_\mu + \sum_{|I| > 0} (d_I Q^\mu + K^\mu_{I\nu} Q^\nu) \partial_{y^\mu_I}. \tag{25.4}$$

Here $Q^\mu \partial_\mu = (\xi^\mu - y^\mu_i \xi^i) \partial_\mu$ is the *characteristic* of the vector field ξ (see Appendix III).

Remark 34. To define K-twisted prolongation of vector fields from Y to $J^k(\pi)$, $k < \infty$, we use the fact that the coefficients $K^\mu_{I\nu}$ are defined on $J^k(\pi)$ and simply cut the the terms with $|I| > k$ in the expression (25.3).

Remark 35. Notice that in modifying the prolongation formula we are using coefficients $K^\mu_{I\nu}$ of the "vertical part" of the connection **but not the coefficients** K^μ_{Ii}.

Example 15. As the examples of the first and second order prolongations in the Field Theory we present the μ-**twisted prolongations** (see Sec.30, below). with the $gl(m,R)$-valued horizontal 1-form $\mu = \Lambda_j dx^j = (\Lambda_j)^\alpha_\beta dx^j$. We remind the relation of the 1-forms μ with the vertical connection form $K^\alpha_{i\beta}$: $K^\alpha_{i\beta} = (\Lambda_i)^\alpha_\beta$ (see Chapter 4).

Let $\mu = (\Lambda_i)^\alpha_\beta dx^i$ be a horizontal 1-form satisfying to the conditions of Theorem 11.

Then the formula 30.10 defines the first order μ-twisted prolongation of vector fields from Y to $J^1\pi$: **First order μ-twisted prolongation** of a general vector field $\xi = \xi^i \partial_i + \xi^\alpha \partial_\alpha$ in the space Y to the 1-jet space $J^1\pi$ has the form

$$Pr^1_\mu \xi = \xi + [d_i \phi^\alpha + (\Lambda_i)^\alpha_\beta \phi^\beta - u^\alpha_m[d_j + (\Lambda_j)^\alpha_m]\xi^m]\partial_{y^\alpha_j}. \tag{25.5}$$

The Second order μ-twisted prolongation.

Under the same conditions as in the case of the 1st order prolongation above, the **second order prolongation** of a vector field $\xi = \xi^i \partial_i + \phi^\alpha \partial_\alpha$ to the bundle $J^2\pi$ has the form (030.11):

$$Pr^2_\mu(\xi) = (\xi^i \partial_i + \phi^\alpha \partial_\alpha) + \Psi^\alpha_i \partial_{u^\alpha_i} + \Psi^\alpha_{ij} \partial_{u^\alpha_{ij}} = Pr^1_\mu \xi + \Psi^\alpha_{ij} \partial_{u^\alpha_{ij}}. \tag{25.6}$$

Here $\Psi^\alpha_{ij} = [d_j + \Lambda_j]\Psi^\alpha_J - u^\alpha_{J,m}[d_j + \Lambda_j]\xi^m$.

To compare the prolongation formula (23.4) with the expression (6.1) for the flow prolongation of a vector field $\xi = \xi^i \partial_i + \xi^\mu \partial_\mu$ consider the case where $k = 1$. For a vector field $\xi = \xi^i \partial_i + \xi^\mu \partial_\mu$, expression (25.4) takes the form (we recall that in $J^1(\pi)$ trunkated total derivatives have the form $d_i = \partial_i + y^\mu_i \partial_\mu$):

$$Pr^1_K(\xi) = \xi^i d_i + Q^\mu \partial_\mu + \sum_l (d_l Q^\mu + K^\mu_{l\nu} Q^\nu)\partial_{y^\mu_l} = \xi + [d_k(\xi^\mu - y^\mu_l \xi^\lambda) + K^\mu_{\sigma\nu}(\xi^\nu - y^\nu_l \xi^l)]\partial_{y^\mu_k} =$$

$$= (\xi^i \partial_i + \xi^\mu \partial_\mu) + [d_k(\xi^\mu - y^\mu_l \xi^l) + K^\mu_{k\nu}(\xi^\nu - y^\nu_l \xi^l)]\partial_{y^\mu_k} = \xi + [d_k Q^\mu + K^\mu_{k\nu} Q^\nu]\partial_{y^\mu_k} =$$

$$= \xi + [(d_k \delta^\mu_\nu + K^\mu_{k\nu})Q^\nu]\partial_{y^\mu_k} = \xi + [d_i \xi^\mu - y^\mu_l d_i \xi^l + K^\mu_{k\nu}(\xi^\nu - y^\nu_i \xi^k)]\partial_{y^\mu_i}. \tag{25.7}$$

This expression coincides with the one given by (6.1) in the case of vertical vector field $\xi = \xi^\mu \partial_\mu$ and with the usual (flow) prolongation of a vector field in the case of commuting variations (where $K = 0$).

More then this, for a general vector field $\xi = \xi^i \partial_i + \xi^\mu \partial_\mu$ in Y, this K-twisted prolongation coincides, (for $k = 1$) with the μ-prolongation for $\mu = K^\alpha_{i\beta} dx^i$ developed by the group of Spanish and Italian mathematicians, see next Chapter.

Using the procedure similar to that in Sec.6 we obtain corresponding K-**twisted Euler-Lagrange equations of higher order with the source terms** defined by the NC-tensor K (or by the horizontal 1-form μ in the case of μ-twisted prolongation (comp.(33.7) in Sec.33).

$$\sum_{|I|\geqq 0}(-1)^{|I|}d_I\frac{\partial L}{\partial y_I^\mu} = -\sum_{I||I|>0}K_{\mu I}^\nu L_{,y_I^\nu} = -K_{\mu I}^\nu\pi_\nu^I,\ \mu=1,\dots,m, \tag{25.8}$$

where $\pi_\nu^I = L_{,y_I^\mu}$.

For a Lagrangian problem of **second degree** with $L = L(x^i, y^\mu, y_i^\mu, y_{ij}^\mu)$, with momenta of the first (π_μ^i) and second (π_ν^{ij}) degree , K-twisted Euler-Lagrange system has the form

$$\frac{\delta L}{\delta y^\mu} = \frac{\partial L}{\partial y^\mu} - d_i\frac{\partial L}{\partial y_i^\mu} + d_{ij}\frac{\partial L}{\partial y_{ij}^\mu} = -K_{\nu i}^\mu\pi_\nu^i - K_{\mu ij}^\nu\pi_\nu^{ij},\ \mu=1,\dots,m. \tag{25.9}$$

Example 16. Consider an example of classical Einstein equation:

$$Ric_{\mu\nu} = \lambda T_{\mu\nu},\ i,j = 0,1,2,3. \tag{25.10}$$

Here G is the metric in R^4 of the signature $(1,3)$ and $Ric_{\mu\nu}$ is the Ricci tensor of metric G (see Appendix I, (67.8)). $T_{\mu\nu}$ is the energy momentum tensor, see[79],Chapter XI.

Dynamical variables here are components $G_{\mu\nu}$ of the metric G and Euler-Lagrange equations are equations (see [80],Sec.XI)

$$\frac{\delta L}{\delta G_{\mu\nu}} = 0,\ \mu,\nu = 0,1,2,3.$$

Being written with the usage of the Energy-Momentum Tensor T_{ij} of the considered physical system (See [80], Ch.XI, Sec.94-95), Euler-Lagrange system takes the form 25.10.

In the situation of non-commuting variations defined by a NC-tensor with components $(K_{i\alpha\beta}^{\mu\nu}, K_{\alpha\beta ij}^{\mu\nu})$, Einstein equations takes the form

$$Ric_{\alpha\beta} = \lambda T_{\alpha\beta} - K_{\alpha\beta i}^{\mu\nu}\pi_{\mu\nu}^i - K_{\alpha\beta ij}^{\mu\nu}\pi_{\mu\nu}^{ij},\ \alpha,\beta = 0,\dots,3. \tag{25.11}$$

Notice the second and third terms in the right side of the equation. These terms correspond to the additional inputs of the "world" to the dynamics of the physical system.

25.2. **Degenerate Lagrangian and higher order dissipation.** Any Lagrangian of order k can be considered as one of order $q > k$ by lifting Lagrangian density Ldv to the bundle $J^q(\pi)$. Now, let K be a Ehresmann connection in the bundle $\pi_{s0} : J^q(\pi) \to Y$. This connection allows us to lift vector fields from Y to $J^q(\pi)$ (with the connection coefficients $K_{\mu I}^\nu$ *depending arbitrarily on the jet variables up to order q.*

Example 17. In the next example we take $n = 1$, $m = 3$. We define the connection K in such a way that the coefficients $K_{\nu 0}^\mu$ depend on third derivatives of the dynamical variables y^μ. Then, although the left side of the Euler-Lagrange equation is the standard second order ODE, the force in the right side will depend on $y_{,ttt}^\mu$.

Consider a point $\mathbf{r}(t)$ electrically charged with the charge e, moving under the action of the Lorentz force $F_0 = e[\mathbf{E} + \frac{v}{c} \times \mathbf{B}]$, where $\mathbf{E}$ is the electrical field and

B - magnetic field, **v** - velocity of the charge and *with the reaction of radiation has the form* ([48])

$$m\ddot{\mathbf{r}} = \mathbf{F_0}(t,x) + \frac{2e^2}{3c^2}\frac{d^3}{dt^3}\mathbf{r}(t). \qquad (25.12)$$

Here $\mathbf{r}(t)$ is the position of electrical charge e at moment t, $\mathbf{F_0} = e\mathbf{E_0} + \frac{e}{c}[\dot{\mathbf{r}} \times \mathbf{H_0}]$ is the Lorentz force and the second term on the right side is the radiation reaction force, [48].

Let $\mathbf{A} = \phi dt - A_I dx^I = \phi dt - \mathbf{A}_{(3)}$ be the 4-potential 1-form generating electrical and magnetic fields (see [79]). Without the reaction force, the system of equations (14.6) is the Euler-Lagrange system with the Lagrangian

$$L = -mc^2\sqrt{1 - \frac{\|v\|^2}{c^2}} + \frac{e}{c}\mathbf{v} \cdot \mathbf{A} - e\phi. \qquad (25.13)$$

Momenta of this Lagrangian are $\pi_i = mg_{ij}v^j + \frac{e}{c}A_{(3)i}$. Here (X,g) is the 4-dim Poincare space-time with the Lorentz metric g. Dynamical variables here are coordinates $x^i, i = 1,2,3$ of the charged point. As a result, $n = 1, m = 3$. Lifting the Lagrangian density to the 3-jet bundle $J^3(\pi)$, we now choose *canonical NC-tensor*

$$K^i_j = \chi^i_0 \frac{2e^2}{3c^2}\frac{d^3}{dt^3}(g_{jk}x^k(t)).$$

All other components of tensor K (see (14.5)) are zero. It is clear now that the equation (14.6) is the Euler-Lagrange system of equations with Lagrangian (14.7) and the chosen NC-tensor K.

Chapter 4. Twisted prolongations and the NC-variations.

26. Introduction.

In the year 2001, Spanish mathematicians C.Murial and J.Romero, introduced the new procedure of prolongations of vector fields from the space Y, of a configuration bundle $\pi : Y \to R$ to the jet spaces $J^k(\pi)$ (λ- **prolongations**), new class of "symmetries" (λ-**symmetries**) for the ordinary differential equations ([102]) and the new method of the **symmetry reduction** of the ordinary differential equations with the λ-symmetries corresponding to the λ- prolongations.

In the next several years, approach of Muriel and Romero was extended by the group of Italian mathematicians (E.Pucci and G.Saccommandi and then, by G.Gaeta, G.Cicogna and P.Morando, to name a few). The method of λ-prolongations, notion of λ-symmetries and the corresponding method of symmetry reduction was extended to partial differential equations and systems of partial differential equations under the name of μ-twisted prolongations or, simply μ- **prolongations and** μ-**symmetries**.

The main goal of their development was to introduce and study a new class of "symmetries" for the differential equations and systems of DE and to use these, "twisted symmetries" for the "symmetry reduction" of the corresponding DE and their systems.

In this Chapter we present a short exposition of basic results of the method of "twisted" prolongations and twisted symmetries close to our approach to the non-commutative variations. We will follow the presentation in the papers . For more detailed results of this very elegant mathematical development, including results of the last years (relation between λ and μ-prolongation with the gauge transformations in the tower of jet bundles, σ-symmetries, etc.) we refer the reader to the papers of authors cited above, esp. papers of C.Muriel, J.Romero, G.Gaeta, P.Morando and their collaborators, especially, the latest works [42, 43, 103, 37].

27. Twisted prolongation of vector fields in Y to the jet bundles.

In this section we introduce a class of twisted prolongations of vector fields in the space Y to the jet bundles $J^k(\pi)$. This class is specified, in comparison to the general $K - twisted$-prolongations (Chapter 3, (17.1) (Definition 8) by two conditions. The first condition is the expression of the horizontal component K^α_{ik} of the Christoffel coefficients of connection K (see Sec.17 or Appendix II, Sec.77) in terms of the coefficients of the vertical component:

$$K^\alpha_{ik} = y^\beta_k K^\alpha_{i\beta}. \tag{27.1}$$

Second condition is, essentially, the condition that the connection K in the bundle $J^1\pi \to Y$ has **zero vertical part of curvature**: (see (see Theorem 15, Sec.31) below.

Remark 36. We will mostly deal with the case where $k = 1$ as it is the most important for applications.

. In order to define twisted prolongations of vector fields it is convenient to use total derivatives "twisted" by a tensor $K^\alpha_{\beta i}$ of NC-type:

$$(\nabla_i)^\alpha_\beta = d_i \delta^\alpha_\beta + K^\alpha_{i\beta}. \tag{27.2}$$

© Springer International Publishing Switzerland 2016
S. Preston, *Non-commuting Variations in Mathematics and Physics*,
Interaction of Mechanics and Mathematics,
DOI 10.1007/978-3-319-28323-4_4

These "covariant total derivatives" appear naturally in the works of G.Garcia and his collaborators on the twisted symmetries as well as in the constructions of Chapter 2 above.

27.1. **K-twisted prolongations of order one.**

Definition 10. *Let K be an Ehresmann connection on the 1-jet bundle π_{10} : $J^1(\pi) \to Y$ such that*

$$K_{ki}^\alpha = y_k^\beta K_{\beta i}^\alpha.$$

. Introduce the horizontal $End(V(\pi))$-valued 1-form $\mu = K_{\beta i}^\alpha dx^i$ in $J^1(\pi)$ defined by a NC-tensor $K_{\beta i}^\alpha$ (of vertical connection).
    *A **first order K-prolongation of a vector field***

$$\xi = \xi^i \partial_i + \xi^\alpha \partial_\alpha$$

in the space Y is the vector field in $J^1(\pi)$

$$Pr_\mu^1 \xi = \xi + [(d_i\delta_\beta^\alpha + K_{\beta i}^\alpha)\xi^\beta - (y_k^\alpha d_i + K_{ki}^\alpha)\xi^k]\partial_{y_i^\alpha} = \xi + [(\nabla_i)_\beta^\alpha \xi^\beta - y_m^\beta(\nabla_i)_\beta^\alpha \xi^m]\partial_{y_i^\alpha}.$$
$$(27.3)$$

Remark 37. The μ-prolongation procedure (see (31.10) below) is a special case of (27.3), but it has some important properties that the general procedure (2.12) is lacking.

28. λ-TWISTED PROLONGATIONS - CASE $m = n = 1$

Our presentation of the μ -prolongation scheme starts with the case of $\lambda-$ (or, in other term,λ- twisted)-prolongation of C.Murial and J.Romero, where $n = m = 1$ and the scalar dynamical field $y(x)$ is a function of one independent variable x. To compare their prolongation with the conventional prolongation of vector fields we refer to Appendix II, Section 75).

A configurational bundle $\pi : Y \to X$ has $X = R$ as the base and the 2-dim bundle space Y (one can take $Y = R_x \times R_y$ unless the space Y has more complex topology or/and singularities, for instance $\pi : S^2 \to (-a, a)$).

In this case, pioneered by Muriel and Romero, [101], a twisted prolongation is defined by one scalar function $\lambda \in C^\infty(J^1(\pi))$. Then, this λ-**prolongation** of a vector field $\eta = \xi(x, y)\partial_x + \phi(x, y)\partial_y$ to $J^1(\pi)$ has the form

$$Pr_\lambda^1 \eta = \eta + [(d_x + \lambda)\phi - \dot{y}(d_x + \lambda)\xi)]\partial_{\dot{y}}. \qquad (28.1)$$

Remark 38. To compare λ-prolongation with the general K-twisted prolongation defined by a connection K (see Ch.2) notice that in this case (m=n=1) K-prolongation (21.2) of vector fields is defined by two scalar functions K_y, K_x, so that

$$Pr_K^1 \eta = \eta + [(d_x + K_Y)\phi - \dot{y}(d_x + K_X\eta)]\partial_{\dot{y}}.$$

28.1. λ-prolongations of higher order. Let $\lambda \in C^\infty(J^1(\pi))$ as above. λ-prolongation of vector fields $\eta = \xi(x,y)\partial_x + \phi(x,y)\partial_y$ to the k-jet space $J^k(\pi)$ is defined by the formula

$$\begin{cases} Pr^k_\lambda \eta = \eta + \sum_{p=1}^k \xi_p \partial_{y^{(p)}}, \text{ where } \xi_0 = \phi, \text{ and} \\ \text{for all } 1 \leqq p \leqq k, \ \xi_p = (d_x + \lambda)\xi_{p-1} - y^{(p+1)}(d_x + \lambda)\xi(x,y), \end{cases} \tag{28.2}$$

similar to the one for the formula (28.10 for the case $k = 1$.

Remark 39. The choice $\lambda \in C^\infty(J^1(\pi))$ guarantees that the λ-prolongation $Pr^k_\lambda \eta$ of a vector field η is the conventional (*not generalized*, see [106]) vector field on $J^k(\pi)$ for all $k \geq 1$.

28.2. Characteristic of λ-prolongation in terms of contact structure. Contact structure on the jet bundles plays an important role in the study of differential equations and systems of such equations. We refer our readers to the sources[106, 71] and other sources on the Global analysis.

In Appendix II, Sec.76 we present the basic result characterizing vector fields on the Jet bundles $J^k(\pi)$ of a bundle $\pi : Y \to X$ that are prolongations of order k of vector fields in the bundle space Y as the vector fields preserving the ideal of contact forms in $J^k(\pi)$.

In the article ([115]) E.Pucci and G.Saccomandi introduced the modified condition of "**invariance of contact structure**" satisfied by the λ-twisted prolongations of vector fields. These conditions were later extended to the case of PDE and to the systems of PDE.

In the next Proposition, the property 2) of λ-prolonged vector fields represent the modified form of the conservation of the contact structure in $J^1\pi$, see(In Appendix II, Sec.75). This condition was introduced by E.Pucci and G.Saccomandi in [115].

Proposition 4. *Let (Y, π, X) be a bundle over the real line $X = R$ with fiber Y_x. Let ξ^k be a vector field on the k-jet space $J^k(\pi)$ which projects to the vector field ξ in Y. Then the following properties of vector field ξ^k are equivalent:*

(1) *Vector field ξ^k is the λ-prolongation of the vector field ξ,*

(2) ***For any contact 1-form*** $\theta \in C\Omega(J^k(\pi))$, ***the 1-form*** $\mathcal{L}_{\xi^k}\theta + (i_{\xi^k}\theta)\lambda dx$ ***is contact:***

$$\mathcal{L}_{\xi^k}\theta + (i_{\xi^k}\theta)\lambda dx \in C\Omega(J^k(\pi)). \tag{28.3}$$

(3) *For any contact form $\theta \in C\Omega(J^k(\pi))$,*

$$i_{[d_x,\xi^k]}\theta = \lambda i_{\xi^k}\theta. \tag{28.4}$$

(4) *For any contact form $\theta \in C\Omega(J^k(\pi))$,*

$$[d_x, xi^k] = \lambda \xi^k + h d_x + \eta, \tag{28.5}$$

where $h \in C^\infty(J^1(\pi))$ and η is the $\pi_{k(k-1)}$-vertical vector field in $J^k(\pi)$.

Proof. **Equivalence of (1) and (2).** Write a general vector field on $J^k(\pi)$ as $\xi^k = \xi\partial_x + \sum_{p=0}^k \psi_p \partial_{y_p}$. With the notation $y_m = d_x^m y$, contact forms in the case where $n = 1$ (one independent variable) are $\theta_p = dy_p - y_{p+1}dx, p = 0, i, \ldots, k-1$. By explicit computations we have

$$\mathcal{L}_{\xi^k}\theta + (i_{\xi^k}\theta)\lambda dx = [-\psi_{k+1} + d_x\psi_k - y_{k+1}d_x\xi + \lambda(\psi_k - y_{k+1}\xi)]dx + \sigma,$$

where σ is contact form. Then, the relation (28.3) is valid if and only if the coefficient ψ_k is given by the prolongation formula (28.1,2).

Equivalence of (3) and (1). In the previous proof we see that $i_{[d_x,\xi^k]}\theta$ is given by formula (4) specialized to the case of one independent variable. With the notation $y_m = d_x^m y$ we have

$$[d_x,\xi^k] = -\psi_{p+1} + (d_x\psi_p - y_{p+1}d_x\xi);$$

on the other hand, $i_{\xi^k}\theta_p = \psi_p - y_{p+1}\xi$. combining these two calculations we see that condition (3) is equivalent to the λ-prolongation formula (27.1,2).

Proof of the equivalence of statements (1) and (4) is similar to the proof of similar statements in Proposition 2, Sec.18 for the **conventional prolongations**.

$\square$

29. λ-SYMMETRIES.

Definition 11. *Let $\Delta = 0$ be a k-th order ODE for a function $u = u(x)$ in a domain $x \in X \subset \mathbf{R}$. Let $\pi : Y = U \times X \to U$ be the corresponding (trivial) configurational bundle.*

*Let a vector field $\zeta = Pr^k\xi \in X(J^k(\pi))$ be the λ-prolongation of the vector field ξ in Y. Vector field ζ is called a λ-**symmetry of the equation** $\Delta = 0$ if it is tangent to the submanifold $\Delta = 0$ of the k-jet manifold $J^k(\pi)$. Locally this condition is equivalent to the existence of a smooth function $\Phi \in C^\infty(J^k(\pi))$ such that $\mathcal{L}_\zeta\Delta = \Phi\Delta$ (compare to [106]).*

Here we present only **two** examples of ordinary differential equations the λ-symmetry and the corresponding reduction by $\lambda - symmetry$ of the differential equations (see Section 77 for definition of the reduction of a differential equation with symmetry).

For more examples of λ- symmetries we refer the reader to the cited articles on the λ--prolongations, especially to the pioneering work ([101]).

Example 18. In the monograph [106],p.182 P.Olver considered the ordinary differential equation

$$u_{xx} = [(x + x^2)e^u]_x. \qquad (29.1)$$

This equation can be integrated in quadratures. At the same time, this equation has no nontrivial Lie group symmetries. This equation belongs to the class of equations of the form

$$u_{xx} = D_xF(x, u) \qquad (29.2)$$

which has the simple order reduction by integration:

$$u_x = F(x, u) + C, \ C \in R. \qquad (29.3)$$

No equations 29.2 can have non-trivial Lie symmetries.

At the same time, C.Muriel and J.Romero proved the following

Theorem 13. *A second order differential equation of the form (29.2) admits the λ-symmetry $\xi = \frac{\partial}{\partial u}$ with $\lambda = F_u(x, u)$. Reduction presented above is the λ-reduction defined by the function λ, see [101]. Theorem 4.1.*

Proof. Using λ-prolongation to the 2-jet bundle with the function λ introduced in Theorem, we obtain the vector field

$$V = \frac{\partial}{\partial u} + F_u\frac{\partial}{\partial u_x} + F_u^2 + u_xF_{uu} + F_{xu}\frac{\partial}{\partial u_{xx}}. \qquad (29.4)$$

By straightforward calculation, one checks that the vector field Pr_λ^2 satisfies the equation

$$Pr_\lambda^2(u_{xx} - D_x F(x, u)) = 0.$$

In particular, this is valid at the points where $u_{xx} - D_x F(x, u) = 0$. As a result, vector field $v = \frac{\partial}{\partial u}$ is the λ-symmetry of equation (29.2) with $\lambda = F_u(x, u)$.

Next, we notice that two functionally independent invariants for the vector field Pr_λ^2 constructed above are $z = x$ and $w = u_x - F(x, u)$.

Another invariant of this vector field can be obtained by differentiation

$$w_z = u_{xx} - D_x F(x, u).$$

In terms of obtained 2 invariants, equation (29.2) takes the form of trivial 1-st order equation

$$w_z = 0.$$

Its general solution $w = C - const$ allows us to obtain a general solution of (29.2) by solving for $u(x)$ the first order equation

$$u_x = F(c, u) + C. \tag{29.5}$$

$\square$

Example 19. Second example of an ordinary differential equations without Lie symmetries but having λ-symmetries and, due to this circumstance allows reduction to the explicitly integrable equation is the second order equation

$$u_{xx} = -\left(\frac{x^2}{4u^3} + u + \frac{1}{2u} \right). \tag{29.6}$$

It is proved in the paper [101], that this equation has no Lie (usual) symmetries. Then the authors shows that for the function $\lambda = \frac{x}{u^2}$, the vector field $v = u\frac{\partial}{\partial u}$ is a λ-symmetry of the equation (29.6). Introducing new variables

$$y = x, w = -\left(\frac{u_x}{u} + \frac{x}{2u^2} \right),$$

that serves as the functionally independent invariants for the first prolongation of the vector field v. In terms of variables y, w, w_y the starting equation (29.6) takes the form

$$1 + w^2 - w_y = 0 \tag{29.7}$$

This equation is easily integrated and provides the general solution of equation (29.6):

$$u = \frac{+}{-}\cos(x + c_1)\sqrt{-\log(\cos(x + c_1))} - x \cdot tan(x + c_1) + c_2,$$

where c_1, c_2 are two arbitrary constants.

30. μ-prolongations and μ-symmetries.Case of one PDE (m=1).

As the next step in our presentation of the formalism of$\lambda-$ and μ-prolongations we consider, following the works [13, 14, 38], the case of one partial differential equation of arbitrary order k for one function $u(x^1, \ldots, x^m)$. In our presentation we will mark the simplifications that take place for the the first order prolongations (i.e. when $k = 1$).

In the case where $m = 1$, the configurational bundle $\pi : Y \to X$ is the bundle with 1-dim fibers. In most cases, fibers of these bundles are either the real line R

or the circle S^1. As before we fix the configurational space $\pi : Y \to X$ and denote by $J^1(\pi)$ corresponding 1-jet space (see Section 73).

Introduce a **semibasic** 1-form

$$\mu = \lambda_i(x,y)dx^i \tag{30.1}$$

on the k-jet space $J^k(\pi)$.

We assume that the 1-form μ satisfies to the conditions of compatibility with the contact structure in $J^k(\pi)$ introduced in the next proposition (see also Definition 12 below).

Proposition 5. *The following **compatibility conditions** for a semibasic 1-form $\mu = \lambda_i dx^i$ in the k-jet space $(J^k(\pi))$ are equivalent:*

(1) *Differential $d\mu$ is contact form:*

$$d\mu \in C\Omega(J^k(\pi)),$$

(2)

$$d_i\lambda_j = d_j\lambda_i, i \neq j$$

or, what is equivalent, $D\mu = 0$ where D is the total exterior derivative operator: $D = \sum_i x^i d_i$. Hence, locally, a 1-form $\mu = \lambda_i dx^i$ satisfying to this condition is the differential

$$\mu = D\Phi$$

for some smooth real function $\Phi \in C^\infty(X)$.

(3) *For the twisted total derivatives $\nabla_i = d_i + \lambda_i I$ and for all $i,j = 1,\ldots,n$,*

$$[\nabla_i, \nabla_j] = 0.$$

Proof. Since λ_i are functions on $J^1(\pi)$, we calculate

$$d\mu = \frac{\partial \lambda_j}{\partial x^i \frac{\partial \lambda_j}{\partial u}} du \wedge dx^j + \frac{\partial \lambda_j}{\partial u_i} du_i \wedge dx^j,$$

i.e.,

$$d\mu = \left[\frac{\partial \lambda_j}{\partial x^i} + u_i \frac{\partial \lambda_j}{\partial u} + u_{ik} \frac{\partial \lambda_j}{\partial u_k} \right] dx^i \wedge dx^j + \frac{\partial \lambda_j}{\partial u} \omega_0 \wedge dx^j + \frac{\partial \lambda_j}{\partial u_i} \omega_i \wedge dx^j. \tag{30.2}$$

The last two terms in the final expression are, obviously, contact. The first term is contact only if it is zero. Thus, the first condition in the Proposition is satisfied if and only if expression in brackets vanish for all i,j. Since this expression is the total derivative $d_i\lambda_j$, it is easy to check that it vanishes if and only if second condition of Proposition is fulfilled. Equivalence of the second condition with the third one follows from the definition of ∇_i. $\square$

Remark 40. In the case of the first order prolongation, i.e. when $k = 1$, conditions of this Proposition are fulfilled automatically. Namely, in this case we get, using the lift to the second jet bundle $J^2(\pi)$,

$$d\mu = \frac{\partial \lambda}{\partial u} du \wedge dx + \frac{\partial \lambda}{\partial u_x} du_x \wedge dx = \frac{\partial \lambda}{\partial u} \omega \wedge dx + \frac{\partial \lambda}{\partial u_x} \omega^1 \wedge dx \in C\Omega(J^2(\pi)),$$

where $\omega = du - u_x dx$, $\omega^1 = du_x - u_{xx}dx$ are basic contact forms.

Now we define (in the case $m = 1$) the μ-**twisted prolongations** of vector fields $\xi \in \mathcal{X}(Y)$ of order k.

To characterize μ-prolongations of vector fields in terms of contact structure we define, following [38, 98], the adequate notion of invariance of contact structure under a one-parameter group of transformations defined by a vector field ζ.

Definition 12. *Let $\mu = \lambda_i dx^i$ be a 1-form on $J^k(\pi)$ satisfying the compatibility conditions of Proposition 5. We say that a vector field $\zeta \in X(J^k(\pi))$ μ-**preserves the contact structure** if for each $\theta \in C\Omega(J^k(\pi))$,*

$$\mathcal{L}_\zeta(\theta) + (i_\zeta\theta) \wedge \mu \in C\Omega(J^k(\pi)). \tag{30.3}$$

Definition 13. *A vector fields Y in $J^k(\pi)$ which projects to the vector field X in the space Y and which μ-preserves the contact structure in $J^k(\pi)$ is called the μ-**prolongation (of order k) of vector field** X.*

In the next Theorem, explicit prolongation formula for the vector fields ξ

Theorem 14. *let ξ be a vector field in the configurational space Y and $Pr^k_\mu\xi$ - its μ-**twisted prolongation of order** k. Then the vector field $Pr^k_\mu\xi$ has the coefficients satisfying to the next μ-prolongation formula*

$$Pr^k_\mu\xi = \xi + \sum_{0 \leq |J| \leq k} \psi_J \partial_{u_J}, \tag{30.4}$$

where coefficients ψ_J are defined by the recurrent formula

$$\psi_{J,i} = (d_i + \lambda_i)\psi_J - u_{J,m}(d_i + \lambda_i)\xi^m, \Psi_0 = \phi. \tag{30.5}$$

Proof. The basic contact forms in the k-jet space $J^k(\pi)$ are $\theta_J = du_J - u_{J,i}dx^i$, with $0 \leq |J| \leq k - 1$. In the proof of the prolongation formula for the conventional prolongation of vector fields (see),

$$\mathcal{L}_\xi(\theta_J) = (-\Psi_{J,i} + d_i\Psi_J - u_{J,m}d_i\xi^m)dx^i + \Theta,$$

where Θ is the contact form. Therefore,

$$\mathcal{L}_\xi(\theta_J) + (i_\xi\Theta_J)\mu = [(-\Psi_{J,i} + d_i\Psi_J - u_{J,m}d_i\xi^m) + \lambda_i 9\Psi_J - u_{J,m}\xi^m)]dx^i + \Theta.$$

This is a contact form if and only if all the coefficients of the differentials dx^i vanish, i.e. if the formula (30.5) is satisfied. $\qquad\square$

In the next Proposition we collect several equivalent conditions of a vector fields in the k-jet space $J^k(\pi)$.

Proposition 6. *Let $\mu = \lambda_i dx^i$ be a semibasic 1-form in the k-jet bundle space $J^k\pi$ satisfying the conditions of Proposition 5. Then, for a vector field $\hat{\zeta} \in X(J^k(\pi))$, the following conditions are equivalent:*

(1) *Vector field $\hat{\zeta}$ is the μ-twisted prolongation of order k of a vector field $\xi = \xi^i\partial_i + \phi\partial_\alpha \in \mathcal{X}(Y)$,*

(2) *Vector field $\hat{\zeta}$ is compatible with the contact structure in the sense of Definition 12,*

(3) *For any contact form θ and for all i,*

$$i_{[d_i,\xi]}\theta = \lambda_i i_\xi\theta.$$

(4) $[d_i, \xi] = \lambda_i \xi + h_i^m d_m + V$, *where* λ_i, h_i^m *are scalar functions on* $J^1(\pi)$ *and* V *is a* $\pi_{k(k-1)}$*-vertical vector field in the k-jet space* $J^k(\pi)$.

Proof. To prove the equivalence of statements 1) and 3), we use the calculation from the proof of theorem 6 giving

$$i_{[d_i,\xi]}\omega_J = -\xi_{J,i} + d_i\xi_J - y_{J,m}d_i\xi^m.$$

Since $i_\xi\omega_J = \xi_J - y_{Jm}\xi^m$, statement 2) of this Proposition is equivalent to the prolongation formula (27.3).

Equivalence of the last statement and the prolongation formula (30.4-5) follows from direct calculation. $\square$

To finish this section we present the relation between the coefficients of μ-prolonged vector field and the corresponding standard prolongation of this vector field. Let $Pr^k\xi = \xi^i\partial_i + \sum \Phi_J\partial_{u_J}$ be the standard prolongation of a field $\xi = \xi^i\partial_i + \phi\partial_u$, so $\Phi_0 = \phi$. Let

$$Pr_\mu^k\xi = \xi^i\partial_i + \sum \Psi_J\partial_{u_J}$$

be the μ-prolongation of the same field ξ. We have $\Psi_J = \Phi_J + F_J$. It was proved in [38] that

$$F_{J,i} = (d_i + \lambda_i)F_J + \lambda_i d_J Q, \tag{30.6}$$

where $Q = \phi - u_i\xi^i$ is the characteristic of vector field ξ (see Appendix III, Section 77 for the definition of vector fields characteristics).

Remark 41. Let X be a vector field in Y and let $I_X = \{d_J Q = 0,\ 0 \leqq |J| < k\} \subset J^k(\pi)$ be the X-invariant subspace. Then, **on this subspace**, the μ-prolongation of vector fields **coincide with the standard prolongation** (see [39].

31. μ-prolongation and μ-symmetries for the systems of PDE.

In the general case of (systems) of partial differential equations it is natural to consider 1-form μ as the $End(V(\pi)) = \mathfrak{gl}(m, R)$-**valued horizontal 1-form**

$$\mu = (\Lambda_i)^\alpha_\beta dx^i = K^\alpha_{i\beta} dx^i, \tag{31.1}$$

where the μ-form is defined both in terms of $1, 2$-tensor Λ as well as in terms of tensor field $K^\alpha_{i\beta}$ that was used to define K-twisted prolongations in Chapter 2. Such 1-form μ allows to introduce the matrix valued (or, in our context, K-covariant) derivatives:

$$\nabla_i = d_i \delta^\alpha_\beta + (\Lambda_i)^\alpha_\beta. \tag{31.2}$$

We *assume* that operators ∇_i commute between themselves:

$$[\nabla_i, \nabla_j] = 0, \tag{31.3}$$

This requirement is clarified in the next Theorem.

31.1. Compatibility condition for the 1-form $\mu = \Lambda_i dx^i$. Formula (30.4-5) is the recurrent relation that allows us to construct prolongations of vector fields of higher order from those of lower order. Yet, **in order for the coefficients ψ^α_J of the prolongations of order two and higher be defined uniquely**, 1-form μ should satisfy the additional conditions expressed in the next result:

Theorem 15. *The following properties of the 1-form $\mu = (\Lambda_i)^a_b dx^i$ are equivalent:*
   (1) $d_i\Lambda_j - d_j\Lambda_i + [\Lambda_i, \Lambda_j] = 0,$
   (2) $d\mu \in C\Omega,$
   (3) $[\nabla_i, \nabla_j] = d_i\Lambda_j - d_j\Lambda_i + [\Lambda_i, \Lambda_j] = 0, i \neq j,$
   (4) $D\mu + \frac{1}{2}[\mu, \mu] = 0,$ *i.e. 1-form μ satisfies to the horizontal Maurer-Cartan equation.*

Equivalence of the last condition with the first two was noticed in ([13]). For the proof of this Theorem we refer reader to ([38]).

To compare the compatibility condition(s) presented in Theorem 15 with the curvature of vertical connections (20.6) and the twisted bracket (20.8) we will write in the case $k = 1$ (first order prolongation) compatibility conditions presented in the last Theorem in explicit form. We write $(\Lambda_i)^\alpha_\beta = K^\alpha_{i\beta}$. Then,

$$(\nabla_i)^\alpha_\beta = \delta^\alpha_\beta d_i + K^\alpha_{i\beta},$$

and, acting by the twisted total derivative ∇_i on a vertical vector field $\xi = \xi_\mu \partial_\mu$ we get

$$((\nabla_i)^\alpha_\beta \xi)^\mu = d_i\xi^\mu + K^\mu_{i\beta}\xi^\beta. \tag{31.4}$$

As a result,

$$\nabla_i\nabla_j\xi = (d_i(\nabla_j)^\alpha + K^\alpha_{\beta i}(\nabla_j\xi)^\beta = d_i[d_j\xi^\alpha + K^\alpha_{j\beta}\xi^\beta] + K^\alpha_{\beta i}(d_j\xi^\beta + K^\beta_{j\gamma}\xi^\gamma), \tag{31.5}$$

Writing a similar equality with i and j exchanged we see the coincidence of the curvature defined by the covariant derivatives ∇_i and the curvature of the vertical connection in the bundle $\pi_{10} : J^1 \to Y$ defined by the NC-tensor $K^\alpha_{\beta i}$ corresponding to the 1-form μ.

$$[(\nabla_i\nabla_j - \nabla_j\nabla_i)\xi]^\alpha = (d_i K^\alpha_{j\beta} - d_j K^\alpha_{i\beta})\xi^\beta + (K^\alpha_{\beta i}d_j - K^\alpha_{\beta j}d_i)\xi^\beta + (K^\alpha_{\gamma i}K^\gamma_{\beta j} - K^\alpha_{\gamma j}K^\gamma_{\beta j})\xi^\beta. \tag{31.6}$$

Right side of obtained relation is the curvature of the vertical connection in the bundle $J^1(\pi) \to Y$ defined by the NC-tensor $K^\alpha_{i\beta}$. As a result, we prove the following statement:

Proposition 7. *Let* $\mu = (\Lambda_i)^\alpha_\beta$ *be the 1-form used for* μ*-prolongation of vector fields and defining the* $\mu - $ *symmetries of the Italian authors whose results are presented in this Chapter and let* $K^\alpha_{i\beta} = (\Lambda_i)^\alpha_\beta$ *be corresponding NC-tensor defining the* K*-twisted prolongation of vector fields (see Chapters 2,3,etc). Then, the 1-form* μ *satisfies to the compatibility conditions of Theorem 15 if and only if the curvature of the vertical connection defined by the NC-tensor* K *vanishes.*

Remark 42. Notice that for a given system of PDE $\Delta : \Delta_p = 0$, $p = 1,\ldots,r$, condition (31.3) has to be fulfilled only on the sub-manifold $S_\Delta \subset J^k(\pi)$ (generically of codimension r) defined by the equations of system Δ, see [41].

Definition 14. *Let* $\mu = (\Lambda_i)^\alpha_\beta$ *be a horizontal 1-form satisfying to the conditions of Theorem 12. Then, the* μ*-**prolongation of order** k of a vector field* $\xi = \xi^i \partial_i + \phi^\alpha \partial_\alpha \in \mathcal{X}(Y)$ *is defined by the following formula:*

$$Pr^k_\mu(\xi) = \xi^i \partial_i + \phi^\alpha \partial_\alpha + \sum_{\alpha, J||J|\leqq k} \Psi^\alpha_J \partial_{,y^\alpha_J},$$

where the coefficients Ψ^α_J *of the consecutive orders are determined by the the recurrent formula:*

$$\Psi^\alpha_{J,j} = (\nabla_i)^\alpha_\beta \Psi^\beta_J - y^\beta_{J,m}[(\nabla_i)^\alpha_\beta \xi^m = [d_j + \Lambda_j]\Psi^a_J - y^a_{J,m}[d_j + \Lambda_j]\xi^m.$$

Here we have used the twisted (matrix) operators $\nabla_i = d_i \delta^\alpha_\beta + (\Lambda_i)^\alpha_\beta$, *see (30.3).*

Remark 43. For a vertical vector field $\xi = \phi^\alpha \partial_\alpha$ in Y, the μ-prolongation formula **of the first order** (i.e. prolongation to the vector field in $J^1(\pi)$) takes the following simple and useful form:

$$Pr^1_\mu(\xi) = \xi + \psi^\alpha_i \partial_{y^\alpha_i}, \tag{31.7}$$

where

$$\psi^\alpha_i = (\nabla_i)^\alpha_\beta \phi^\beta = d_i \phi^\alpha + (\Lambda_i)^\alpha_\beta \phi^\beta. \tag{31.8}$$

Formula (31.8) coincides with formula (27.3) provided $K^\alpha_{i\beta} = (\Lambda_i)^\alpha_\beta$.

For a horizontal vector field $\xi = \xi^i \partial_i$, first prolongation has the form

$$Pr^1\xi = \xi + \Psi^\alpha_j \partial_{,y^\alpha_j} = \xi^i \partial_i - u^\alpha_m[d_j + \Lambda_j]\xi^m \partial_{u^\alpha_j}. \tag{31.9}$$

Adding these formulas we get the formula for the **first order** μ**-twisted prolongation** of the general vector field in the space Y: for $\xi = \xi^i \partial_i + \xi^\alpha \partial_\alpha$,

$$Pr^1\xi = \xi + [d_i \phi^\alpha + (\Lambda_i)^\alpha_\beta \phi^\beta - u^\alpha_m[d_j + \Lambda_j]\xi^m]\partial_{y^\alpha_j}. \tag{31.10}$$

For the **second order** μ**-prolongation** we have, for $\xi = \xi^i \partial_i + \phi^\alpha \partial_\alpha$,

$$Pr^2_\mu(\xi) = (\xi^i \partial_i + \phi^\alpha \partial_\alpha) + \Psi^\alpha_i \partial_{u^\alpha_i} + \Psi^\alpha_{ij} \partial_{u^\alpha_{ij}} = Pr^1_\mu \xi + \Psi^\alpha_{ij} \partial_{u^\alpha_{ij}}. \tag{31.11}$$

Here $\Psi^\alpha_{ij} = [d_j + \Lambda_j]\Psi^a_J - u^\alpha_{J,m}[d_j + \Lambda_j]\xi^m.$

31.2. Twisted invariance condition of the contact structure preservation at the μ-prolongation: general case. Here we introduce, in the general case of systems of PDE, the modified condition of invariance of contact structure under the twisted prolongation of vector fields. We will follow the works [38, 39, 40] with the appropriate modifications. Below we will present another form of these conditions suggested by P.Morando. This form of the condition will be a deformed version of the condition (see Sec.17, Proposition 2) of conservation of contact structure (or, equivalently, conservation of the Cartan distribution).

In order to formulate recurrent relations in this, general, case, it is convenient to collect the contact forms into the **vector-valued (R^m-valued) 1-forms**. Denote by Θ^m the space of vector-values contact 1-forms (module over the $m \times m$ square functions) generated by the basic 1-forms

$$\theta_J^\mu = dy_J^\mu - y_{J,i}^\mu dx^i - \tag{31.12}$$

We get the space of 1-forms $\eta \in R^m \otimes \Lambda^1(X)$ with components

$$\eta^\alpha = (R^J)_\beta^\alpha \theta_J^\beta, \ \alpha = 1,\ldots,m, \tag{31.13}$$

i.e. with values in the vertical bundle $V(\pi)$. Matrix function $(y) \to R^J(y)$ take values in the bundle of linear endomorphisms $End(V(\pi))$ of vertical bundle: $R^J(x,y) \in End(V_y(\pi))$.

Fundamental form $\mu = \Lambda^i dx^i$ defining the twisted prolongations (see above) is also considered to take values in the fibers of the bundle $End(V(\pi)) \to Y$.

Using local (fibred) charts one may consider Λ_i and tensors R^J taking values in the Lie algebra $gl(m, R)$ or in some Lie subalgebra of this Lie algebra.

Theorem 16. *Let $\hat{\xi}$ be a vector field in $J^k(\pi)$ projecting to the vector field $\xi \in \mathcal{X}(Y)$. The following properties of vector field $\hat{\xi}$ are equivalent:*

(1) *Vector field $\hat{\xi}$ is the μ-prolongation of the vector field ξ, see definition 14.*
(2) *Vector field $\hat{\xi}$ μ-**preserves the vector contact structure** Θ^m **in the following sense**: for any collection of matrix functions R^J, 1-form $\eta = \{\eta^\alpha\}$ defined by (30.9) satisfies the condition*

$$\mathcal{L}_{\hat{\xi}}\eta^\alpha + i_{\hat{\xi}}[(\Lambda_i)_\beta^\alpha \eta^\beta]dx^i = \widehat{\eta^\alpha} \in \Theta^m. \tag{31.14}$$

Proof. For the proof it is sufficient to show that, in terms of the coefficients of vector field ξ - projection of $\hat{\xi}$ in Y, vector field $\hat{\xi}$ satisfying condition 2) has the form 1). This is done by straightforward calculation similar to that in Proposition 11. $\qquad\square$

Remark 44. (see [38]) Let $\xi = \xi^i \partial_i + \xi^\mu \partial_\mu \in \mathcal{X}(Y)$ be a vector field and let

$$Pr^k \xi = \xi + \sum \Phi_J^\mu \partial_\mu^J$$

be its standard (flow) prolongation and

$$\hat{\xi} = \xi + \sum \Psi_J^\mu \partial_\mu^J$$

- its μ-prolongation of order k: $Pr_\mu^k \xi$ defined by a 1-form μ (see Definition 14). Then

$$\Psi_J^\mu = \Phi_J^\mu + F_J^\mu$$

for some tensor field F_J^μ. Then, the coefficients F_J^μ satisfy to the following recurrent formula

$$F_{J,i}^\mu = [\delta_\nu^\mu d_i + (\Lambda_i)_\nu^\mu] F_J^\nu + (\Lambda_i)_\nu^\mu d_J Q^\nu, \tag{31.15}$$

where $Q^\nu = \xi^\nu - u_k^\nu \xi^k$ is the characteristic of vector field ξ.

Proof of this statement is similar to the proof of the similar statement in the case where $m = 1$, see [38], Theorem 3.

32. μ-SYMMETRIES AND REDUCTION OF PDE SYSTEMS.

Having defined the notion of μ-prolongation of vector fields to the k-jet bundle we can now introduce the class of symmetries of a differential equation or of a system of such equations and study the class of "conservaion laws" corresponding to μ-symmetries. Thus, we define

Definition 15. *Let $\pi : Y \to X$ be a vector bundle over an n-dim manifold X. Let*

$$\Delta : \quad \Delta_\mu = 0, \quad \mu = 1, \ldots, m \tag{32.1}$$

be a system of r PDE of order k for the fields $u^\mu(x), \mu = 1, \ldots, m$. Let $S_\Delta \subset J^k(\pi)$ be the sumanifold in the k-jet space $J^k(pi)$ corresponding to the system Δ.

Let $\xi \in \mathcal{X}(Y)$ be a vector field in Y and let μ be - a $gl(m, R)$-valued semibasic 1-form satisfying the compatibility condition (Theorem 15). Let $Pr_\mu^k \xi$ be the μ-prolongation of order k of the vector field ξ.

(1) *We say that vector field ξ defines (or simply is) the μ **infinitesimal symmetry of system** Δ if, being restricted to the submanifold S_Δ, vector field $Pr_\mu^k \xi$ is tangent to S_Δ. Equivalently, see [106], under usual conditions of non-degeneracy, ζ is a μ-symmetry if there is a smooth matrix function $R : J^k(\pi) \to Mat(m, R)$ such that*

$$\zeta(\Delta_\mu) = R_\mu^\nu \Delta_\nu, \quad \mu = 1, \ldots, m. \tag{32.2}$$

(2) *Let $F \in C^\infty(J^k(\pi)$ be a smooth function on the k-jet space $J^k(\pi)$. We say that F is μ-invariant under the vector field ξ if*

$$\mathcal{L}_{Pr_\mu^k} F = 0.$$

It is interesting that the k-jet space contains the submanifold where conventional and μ-prolongations of vector field ξ coincide - *invariant submanifold of vector field ξ.*

Definition 16. *Let $\xi \in \mathcal{X}(Y)$ be a vector field on the manifold Y with the characteristic $Q^\mu \partial_\mu$. The ξ-**invariant submanifold of vector field** ξ **in** $J^k(\pi)$ is the set*

$$I_\xi = \{z \in J^k(\pi) | d_J Q^\mu(z) = 0, \ \forall \mu = 1, \ldots, m, J, \ 0 \leqq |J| \leqq k - 1.\} \tag{32.3}$$

Theorem 17. *In the conditions above (fulfillment of conditions of Theorems 10,11), let Pr_μ^k be the μ-prolongation of vector field $\xi \in \mathcal{X}(Y)$. Then this prolongation coincides with the standard (flow) prolongation Pr^k, on the ξ-invariant submanifold I_ξ.*

We refer to [38], Thm.8, for the proof of this statement.

Remark 45. Conventional symmetries and μ-symmetries of a systems of DE are different. In particular, μ-symmetries do not map solutions of DE to the solutions. More then this, μ-symmetries do not form a Lie algebra (comp. Sec.18 for the same property of the K-twisted prolongations). Considerations of Sections (18-21) show that the obstructions for this are "curvature of vertical connection" and the deformed "Poisson bracket, see Sec.20, Ch.3.

Now we turn to the "conservation (an balance) laws" corresponding to a NC-tensor $K - (seeChapter2)$ or μ-symmetries. We notice that for the Euler-Lagrange system corresponding to a Lagrangian L there are four possibilities:

(1) cc-case: To use *conventional* prolongation for variational vector fields and the *conventional symmetries*. This leads to the standard Noether conservation laws, see [106], or Appendix III, Sec.80-81 ,

(2) cK-case: To use conventional variations and the μ-symmetries,

(3) Kc-case: To use K-twisted or μ-prolongations and conventionally defined symmetries (in this case we get new types of "conservation" or "balance" laws for a given Euler-Lagrange system of differential equations.

(4) KK-case: To use $K - twisted$- or μ-prolongations and, correspondingly, $K - twisted$- and μ-symmetries.

Remark 46. For a general system of differential equations, the four situations introduced above lead to different results: If a system Δ of PDE $\Delta_j = 0$, $j = 1,\ldots,r$ of order k has a symmetry vector field $\zeta \in \mathcal{X}(J^k(\pi))$ (symmetry in the conventional sense (see Appendix III), it may or may not be a prolongation or μ-twisted, or K-twisted prolongation of a vector field $\xi \in \mathcal{X}(Y)$. This condition of such a prolongation is the fulfillment of twisted condition of invariance of contact structure (in a case of systems of PDE - condition of Theorem 16).

In general, for each 1-form μ, there may or may not be some μ-symmetries for the system Δ. On the other hand, a π_{k0}-projectable vector field ζ tangent to S_Δ (a symmetry of the system Δ)), should satisfy to the equation (31 .2). This equation can be considered as the condition for 1-forms μ (i.e. for matrices of functions Λ_i) such that a π_{k0}-projectable vector field ζ is the μ-prolongation of the projection of ζ to the space Y.

So, the notion of μ-symmetries refer to the vector fields in the configurational space Y rather then to the vector fields in the jet space $J^k(\pi)$. In general the symmetry Lie algebra $\mathfrak{sym}(\Delta) \subset \mathcal{X}(J^k(\pi))$ of a PDE system Δ contains the Lie subalgebra $\mathfrak{psym}$ of π_{k0}-projectable vector fields. This subalgebra contains, for each 1-form μ, the linear subspace $\mu - sym(\Delta)$ of infinitesimal μ-symmetries of the system Δ - the Lie subalgebra of conventional geometrical symmetries of system Δ.

33. μ-CONSERVATION LAWS.

Let μ be the 1-form taking values in a Lie algebra $\mathfrak{g}$ (usually, $\mathfrak{g}$ is the Lie subgroup of $gl(m, R)$).

A *standard conservation law* has the form of the relation

$$d_i P^i = 0,$$

valid for all solutions of the corresponding system of partial differentiable equations. Here $\mathcal{P}^i$ is the n-tuple of functions in $J^k(\pi)$. For the "conservation laws" corresponding to the μ-symmetries, we introduce the following type of $\mu-conservation laws$:

Definition 17. *Let $\mathcal{P} : J^k(\pi) \to g \subset gl(m, R)$ be a matrix valued vector. The μ-conservation law defined by $\mathcal{P}$ is the relation*

$$Tr(\nabla_i \mathcal{P}^i) = 0 \tag{33.1}$$

valid for all solutions of the system of PDE. In components it reads,

$$(\nabla_i)^a_b (P^i)^b_a \equiv (\nabla^a_b)_i (P^b_a)^i = 0. \tag{33.2}$$

This "μ-conservation law" does not emply conservation of P or $Tr(P)$ in the usual sense: Defining $\mathbf{P}^i = Tr(\mathcal{P}^i)$ and using definition of operators ∇_i we **get the balance law**

$$d_i \mathbf{P}^1 = d_i(\mathcal{P}^i)^a_a = -(\Lambda_i)^a_b (\mathcal{P}^i)^b_a = -Tr(\Lambda_i \mathcal{P}^i). \tag{33.3}$$

34. NOETHER THEOREM FOR μ-SYMMETRIES.

In the paper [102], C.Muriel, J.Romero and P.Olver proved the Noether type Theorem for λ-symmetries. In the article [14], the authors formulated and proved the Noether type Theorem for μ-symmetries.

We start with the Noether type Theorem for a Lagrangian of the first order.

Theorem 18. *([14]) Let $L(y^\alpha, y^\alpha_x)$ be a first order Lagrangian and let $\xi = \phi^\alpha \partial_{y^\alpha}$ be a vertical vector field in Y. Define*

$$(P^a_b)^i = \phi^\alpha \pi^i_\beta$$

where $\pi^i_\beta = L_{,u^\alpha_i}$ are momenta.

Then, the vector field ξ is a μ-symmetry if and only if the matrix valued vector function $\mathcal{P}^i$ is a μ-conserved vector, i.e, if

$$(\nabla_i)^a_b (P^i)^b_a \equiv (\nabla^a_b)_i (P^b_a)^i = 0.$$

Proof. For the first order Lagrangians one can check by direct calculation that the condition of μ-conservation in this Theorem for a vertical vector field ξ has the form

$$(\nabla_i)^\alpha_\beta (P^i)^\beta_\alpha = d_i Tr(\mathcal{P}^i) + (\Lambda_i)^\alpha_\beta \phi^\beta \frac{\partial L}{\partial y^\alpha_i} = d_i \mathcal{P}^i + Tr(\Lambda_i \mathcal{P}^i) = 0 \tag{34.1}$$

Now, let $\xi = \phi^\alpha \partial_{y^\alpha}$. Then, the μ-prolongation of this vector field of the first order is equal to

$$Pr^1_\mu \xi = \phi^\alpha \partial_{y^\alpha} + [d_i \phi^\alpha + (\Lambda_i)^\alpha_\beta \phi^\beta] \partial_{,y^\alpha_i}.$$

Acting on the Lagrangian L by this vector field and integrating by parts we get

$$Pr^1_\mu \xi L = \phi^\alpha \left(L_{,y^\alpha} - d_i L_{,y^\alpha_i} \right) + d_i(\phi^\alpha_i \pi^i_\alpha) + (\Lambda_i)^\alpha_\beta \phi^\beta \pi^i_\alpha \tag{34.2}$$

At the solutions of the Euler-Lagrange system $E_L(s(x)) = 0$, the first term vanishes and the result simplifies to

$$Pr^1_\mu \xi L = d_i(\phi^\alpha_i \pi^i_\alpha) + (\Lambda_i)^\alpha_\beta \phi^\beta \pi^i_\alpha = (\nabla_i)^\alpha_\beta \pi^i_\alpha \phi^\beta. \tag{34.3}$$

This shows that the symmetry condition $Pr_\mu^1 \xi L = 0$, fulfilled on the solutions of Euler-Lagrange system implies condition (33.1) of μ-conservation law for $\mathcal{P}^i$. Following these calculations in the opposite order we prove that, from the μ-conservation of $\mathcal{P}^i$ the symmetry condition follows. $\square$

Noether Theorem for μ-symmetries also holds for higher order Lagrangians. We present the result here without proof refering reader to paper [102] where the recursive procedure was used for proving this result in the case $m = n = 1$. Similar arguments prove the result of Theorem 14 in the general form.

Theorem 19. *Let $L \in C^\infty(J^k(\pi))$ be a Lagrangian of order k. Let ξ be a vector field. Then,*

(1) *ξ is a μ-symmetry for L (i.e. $Pr_\mu^k \xi L = 0$) if and only if there exist a matrix valued vector function $(\mathcal{P}^i)_\beta^\alpha$ satisfying to the μ-conservation law*

$$(\nabla_i)_\beta^\alpha (\mathcal{P}^i)_\alpha^\beta = 0. \tag{34.4}$$

(2) *Matrix valued vector $(\mathcal{P}^i)_\beta^\alpha$ is obtained in the following way: write the usual current density vector $(P_0)^i$ as determined by the Lagrangian L and by the vector field ζ considered as the **standard symmetry** for L (i.e. with $\mu = 0$). Then replace each term $d_J \phi^\alpha$ appearing in $(P_0)^i$ with $(\nabla_J \phi)^\alpha$ for all $J, |J| \geqq 0$.*

(3) *The conservation law (34.4) is obtained by replacing the total divergence operator d_i with $(\nabla_i)_\alpha^\beta$.*

Example 20. For a second order Lagrangian, the μ-conserved vector $P_\beta^{\alpha i})$ has the form

$$P_\beta^\alpha)^i = \phi^\beta \frac{\partial L}{\partial y_i^\alpha} + ((\nabla_j)_\gamma^\beta \phi^\gamma) \frac{\partial L}{\partial y_{ij}^\alpha} - \phi^\beta d_j \frac{\partial L}{\partial y_{ij}^\alpha}. \tag{34.5}$$

34.1. Conservation laws for μ-symmetries. Let

$$A_D(s) = \int_D L(y^\mu, y_{x^i}^\mu) dv \tag{34.6}$$

be an action with Lagrangian of the first order and let $\mu = (\Lambda_i)_\beta^\alpha dx^i = K_{i\beta}^\alpha dx^i$ be a 1-form μ satisfying the compatibility condition (Theorem 15). Using this NC-tensor K to define the 1-prolongation of vertical vector fields we get the $K-$ (or $\mu-$)-**modified Euler-Lagrange system** of equations (6.4), in the notation of this Chapter -

$$\frac{\partial L}{\partial y^\alpha} - \frac{\partial L}{\partial y_j^\alpha} = -\frac{\partial L}{\partial y_i^\beta}(\Lambda_i)_\alpha^\beta = -(\Lambda_i^T)_\alpha^\beta \pi_\beta^i. \tag{34.7}$$

Now, let Lagrangian L admit a μ-symmetry being a π-vertical vector field $\xi = \phi^\alpha \partial_\alpha$.

μ-prolongation of ξ is $Pr_\mu^1 \xi = \xi + [(\nabla_i)_\beta^\alpha \phi^\beta] \partial_{y^\alpha_i}$. The invariance condition of the Lagrangian has the form $\mathcal{L}_{Pr_\mu^1 \xi} L = 0$.

Using $(\nabla_i)_\beta^\alpha = \delta_\beta^\alpha d_i + (\Lambda_i)_\beta^\alpha$, we write the invariance condition as follows

$$\phi^\alpha \frac{\partial L}{\partial y^\alpha} + (d_i \phi^\alpha + (\Lambda_i)_\beta^\alpha \phi^\beta) \frac{\partial L}{\partial y_j^\alpha} = 0.$$

Theorem 20. *Let L be a first order Lagrangian admitting the vector field $\xi = \phi^\alpha \partial_\alpha$ as a μ-symmetry for some 1-form μ satisfying the compatibility conditions. Then, the vector-function $P^i = \phi^\alpha \pi_\alpha^i$ defines the standard conservation law*

$$d_i P^i = 0 \tag{34.8}$$

for the solutions of associated μ-twisted Euler-Lagrange system (34.7).

Proof. Integrating the condition of invariance above by parts, we get

$$\phi^\alpha \frac{\partial L}{\partial y^\alpha} + d_i \left(\phi^\alpha \frac{\partial L}{\partial y_i^\alpha} \right) - \phi^\alpha d_i \frac{\partial L}{\partial y_i^\alpha} + ((\Lambda_i)_\alpha^\beta \phi^\alpha) \frac{\partial L}{\partial y_i^\beta} = 0.$$

Collecting terms with ϕ^α we get

$$\phi^\alpha \left[\frac{\partial L}{\partial y^\alpha} - d_i \frac{\partial L}{\partial y_i^\alpha} + \frac{\partial L}{\partial y_i^\beta} (\Lambda_i)_\alpha^\beta \right] + d_i \left(\phi^\alpha \frac{\partial L}{\partial y_i^\alpha} \right) = 0.$$

The terms in the brackets vanish on the solutions of μ-modified EL system and we see that the symmetry condition above implies the conservation law in its standard form

$$d_i \left(\phi^\alpha \right) \frac{\partial L}{\partial y_i^\alpha} = 0 \tag{34.9}$$

as claimed in Theorem. $\qquad\qquad\qquad\qquad\qquad\qquad\qquad\qquad\qquad\qquad\qquad\qquad\square$

Remark 47. Last result shows that **using a μ-prolongation both for defining variations of derivatives y_i^μ and for prolongation of a vertical (and geometrical) invariance vector field ξ** in the Noether formalism leads **to the conventional symmetry law.**

In contrast to this, combining μ-(or $K-$) prolongations of variations with the conventional symmetry defined by a geometrical ($\xi \in \mathcal{X}(Y)$) vector field we modify the conservation law as well.

This leaves open the last possibility - to combine conventional flow prolongation of variational vector fields with the μ-symmetry corresponding to the vector field $\xi \in \mathcal{X}(Y)$- infinitesimal generator of the symmetry.

Remark 48. Notice that in presenting μ-symmetries and using $\mu-twisted$-prolongations for modifying Euler-Lagrange Equations, the authors have dealt almost exclusively with the π-vertical vector fields. At the same time, there are interesting conservation laws (energy-momentum and angular momenta balance laws are examples) also containing the "horizontal" (by ∂_{x^i}) - components.

35. Gauge transformations and comparison of μ- and conventional prolongations.

35.1. Exponential vector fields and the μ-symmetries. Let $\mu = \lambda_i dx^i$ be a horizontal 1-form satisfying the compatibility condition of Proposition 5. Let X_0 be a vector field in Y. The exponential vector field corresponding to μ and X_0 has, in general, the formal expression

$$X = e^{\int \mu} X_0. \tag{35.1}$$

This notion was introduced by P.Olver (see [106], p.181) for presenting nonlocal symmetries.

If $\mu = (D_i P)dx^i$, the compatibility conditions of Proposition 5 are automaticallly satisfied and $X = e^P X_0$.

Let μ be a general 1-form (satisfying to the conpatibility condition) and let Δ be a PDE such that the compatibility condition $D_i \lambda_j = D_j \lambda_i$, equivalent to the condition 1) of the Proposition 15, is satisfied on the subset S_Δ defined by the equation Δ of the corresponding jet space.

Then the following result is valid: Vector field X is a (in general, nonlocal) symmetry for the PDE Δ if and only if the the vector field X_0 is μ-symmetry for Δ. We refer to the paper [13] for the proof and for the generalization of this result.

35.2. Gauge transformations and the action of Linear Group . We start with a case where the configurational bundle $Y = X \times F \to X$ **is trivial**. Manifold F is the standard fiber of the bundle π.

Let $G \times F \to F$ be a smooth action of the Lie group G at the standard fiber manifold F of the bundle π.

Then, a connected finite dimensional Lie group G acts fiberwisely (say, from the right) smoothly on the space $Y = X \times F$.

Introduce the gauge group $\Gamma = C^\infty(Y \to G)$, a - group of smooth mappings $\phi : Y \to G$. Γ has the structure of an infinite-dimensional Lie group with the Lie algebra being the Loop algebra. Now we define the (pointwise) action of the group Γ at the space Y:

$$\Gamma \times Y \to Y : \quad (\phi, y) \to \phi(y) \cdot y. \tag{35.2}$$

This action induces the (pointwise) action of the gauge group Γ on the smooth sections $s : X \to Y$: $(s(x), \phi) \to \phi(s)(x) = \phi(s(x)) \cdot s(x)$. In other words, the element $\phi \in \Gamma$ acts on a section $s : X \to Y$ by producing section s^ϕ - sending a base point $x \in X$ to the point $\phi(s(x)) \cdot s(x)$.

More then this, introduced action of the gauge group Γ on the points of Y and the sections of the bundle π defines the following actions:

(1) *Induced push forward* of the π-vertical vector fields in Y: $\mathcal{X}(Y) \ni \xi(x, y) \to \phi_{*(x,y)}\xi(x, y) \in \mathcal{X}_{\phi(x,y)}(Y)$.

   Let γ be an element of Γ - smooth function $Y \to G$.

(2) *Induced action on the k-jets $j^k s(x)$ of sections $s : X \to Y$* - by jet prolongation of sections and,

(3) By the flow prolongation the action of the group Γ extends to the action at the k-jet bundle $\ni \Gamma \ni \phi \to j^k \phi : J^k \pi \to J^k \pi$. Denote the group of prolongations of elements of the group Γ to the k-jet space by Γ_k. This is the group of smooth mappings $\Gamma : J^k \pi \to G$. Its elements act on the k-jet bundle by vertical transformations (evolutional in terminology of [106]).

   A vertical vector field in the bundle $J^k \pi \to X$ has the form

$$Y = \sum_{a, J/|J| \leqq k} \Psi^a_J \frac{\partial}{\partial u^a_J}.$$

An element γ of the gauge group Γ_k acts on the vertical vector field Y linearly as follows:

$$\gamma \cdot Y = [\gamma_b^a \Psi_J^b] \frac{\partial}{\partial u_J^a}. \tag{35.3}$$

(4) Last defined action defines, for each element $\phi \in \Gamma$, the action of the group G of the vector fields on the jet bundles: $\Gamma^k \phi : \mathcal{X}(J^k \pi) \to \mathcal{X}(J^k \pi)$.

Example 21. Let the configurational bundle $\pi : Y \to VX$ be a vector bundle and let the linear group $G = GL(m, R)$ act on the right by differentiable action $(g \to T_g)$ on the fibers Y_x, $x \in X$ of the configurational bundle $\pi : Y \to X$. This action (called gauge transformations or gauge group action) can be prolonged in the standard (flow) way to the right action of G on the fibers of the jet spaces $J^k \pi$ commuting with the projections $\pi_{pq} : J^p \pi \to J^q \pi$.

Definition 18. *Let Y, W be two π_k-vertical vector fields in $J^k \pi$. We say that the vector fields Y, W are G-equivalent if there exists a function $\gamma \in \Gamma_k$ such that $Y = \gamma E$ globally in $J^k \pi$.*

We say that the vector fields Y, W are G-equivalent locally if for any point $z \in J^k \pi$ there exists an neighborhood A_z and a function γ_z defined in the neighborhood A_z such that $W = \gamma_z Y$.

Remark 49. Below we will be using, together with a function $\gamma : J^k \to G$, the function $\gamma^{-1} \in \Gamma_k$. The mapping γ^{-1} is the G-valued function in $J^k \pi$ such that at each point $(x, u^k) \in J^k \pi$, take as its value the element inverse to the element of $\gamma(x, u^k) \in J^k \pi$.

35.3. Darboux derivative and the local presentation of the form μ.

Remark 50. Based on this condition and using a result of Marvan on the zero curvature representation of partial differential equations ([?]) one can prove that locally, in any contractible neighborhood $W \subset J^k(\pi)$ there exists a matrix-valued function $F : W \to GL(R, m)$ such that (in W) the 1-form μ is the **Darboux derivative of** F. As a result, due to the conditions on the form μ, presented in Section 31, (see Theorem 15), any μ-prolonged vector field is **locally gauge-equivalent** to a vector field prolonged in the standard way (by the *flow-prolongation*).

We start with the definition of Darboux derivative.

Definition 19. *Let G be a Lie group with the Lie algebra $\mathfrak{g}$. Let ω be the Maurer-Cartan form on the group G. Let M be a smooth manifold and $f : M \to G$ be a smooth mapping. The (left) Darboux derivative of mapping f is the $\mathfrak{g}$-values 1-form $\omega_f = f^* \omega_G$.*

The mapping f is called an integral of primitive of the form ω. Due to the naturality of differential d, the 1-form ω_f satisfies the analog of the structural equation

$$d\omega_f(X, Y) + [\omega_f(X), \omega_f(Y)] = 0. \tag{35.4}$$

The next property of Darboux derivative is the uniqueness of the primitive.

Theorem 21. *([121], Theorem 5.2)). Let M be a connected smooth manifold and $f_1, f_2 : M \to G$ are two smooth mappings such that*

$$\omega_{f_1} = \omega_{f_2}.$$

Then, there is an element $c \in G$ ("constant of integration") such that for all $x \in M$, $f_2(x) = C \cdot f_1(x)$.

Next we present the fundamental result of E.Cartan used below. For the proof of this Theorem we refer the reader to the monograph [121] (Theorem 6.1).

Theorem 22. *Let G be a Lie group with Lie algebra $\mathfrak{g}$. Let ω be a $\mathfrak{g}$-valued 1-form on the smooth manifold M satisfying the structural equation*

$$d\omega + \frac{1}{2}[\omega, \omega] = 0.$$

Then, any point $x \in M$ has a neighborhood U and a smooth mapping $f : U \to G$ such that the restriction of the form ω to the neighborhood U coincide with the Darboux derivative of the mapping f:

$$\omega_U = \omega_f. \tag{35.5}$$

Lemma 3. *The compatibility condition (see Proposition 5 for one PDE equation and Proposition Theorem 15 for systems of PDE) for the 1-form $\mu = \Lambda_i dx^i$:*

$$D_i\Lambda_j - D_j\Lambda_i + [\Lambda_i, \Lambda_j] = 0 \Leftrightarrow [\nabla_i, \nabla_j] = 0 \tag{35.6}$$

is equivalent to the horizontal Maurer-Cartan equation

$$D\mu = \frac{1}{2}[\mu, \mu] = 0. \tag{35.7}$$

Equivalently, it states that the Maurer-Cartan equation is satisfied up to a form

$$\rho = -\frac{\partial \Lambda_j}{\partial u^a_K} dx^j \wedge \omega^a_K$$

,i.e.,

$$d\mu + \frac{1}{2}[\mu, \mu] = \rho \in C\Omega.$$

Proof. For the proof we will use some properties of $\mathfrak{g}$-valued exterior forms (see [121, ?]). Let $\{e_i\}$ be a basis of Lie algebra $\mathfrak{g}$. Then, any $\mathfrak{g}$-valued k-form ω can be presented in the form

$$\omega = e_i \otimes \omega^i$$

for some k-forms ω^i.

Given two $\mathfrak{g}$-values forms $\alpha = e_i\alpha^i$ and $\beta = e_j \otimes \beta^j$, we define the bracket

$$[\alpha, \beta] = [e_i, e_j] \otimes (\alpha^i \wedge \beta^j). \tag{35.8}$$

Denote by D the total exterior derivative. For a horizontal form $\alpha = A_\sigma dx^\sigma$ (this is the shorthand of the presentation $\alpha = A_{\sigma_1\sigma_2...\sigma_k} dx^{\sigma_1 \wedge ... \wedge \sigma_k}$) action of D on the form α is $D\alpha = (D_i A_\sigma)dx^i \wedge dx^\sigma$. Using these definitions, we have

$$D\mu + \frac{1}{2}[\mu, \mu] = (D_i\Lambda_j + \frac{1}{2}[\Lambda_i, \Lambda_j] \otimes (dx^i \wedge dx^j) = \frac{1}{2}(D_i\Lambda_j - D_j\Lambda_i) + [\Lambda_i, \Lambda_j] \otimes (dx^i \wedge dx^j),$$

as claimed. Using the form of basic contact forms () we have

$$D\mu = d\mu - \frac{\partial \Lambda_j}{\partial u^a_K}\theta^a_K \wedge dx^j,$$

which proves the equivalence of statements. $\qquad\square$

The next Proposition is the form of the local Fundamental Theorem of E.Cartan (see Theorem 18 above) necessary for our goals (see appendix for the full form of the Fundamental Theorem on the Darboux derivatives.

Proposition 8. *Let M be a smooth manifold. Let G be a Lie group with Lie algebra $\mathfrak{g}$. Let μ be a $\mathfrak{g}$-valued one form in M satisfying the structural (Maurer-Cartan) equation:*

$$d\mu + \frac{1}{2}[\mu, \mu] = 0. \tag{35.9}$$

Then, for any point $z \in M$, there is a neighborhood $A_z \subset M$ of the point z and a smooth function $\gamma_z : A_z \to G$ such that in the domain A_z,

$$\mu = \gamma_z^{-1} d\gamma_z.$$

Remark 51. If ω in the Proposition above is the Maurer-Cartan form on M, this result guarantees that any $\mathfrak{g}$-valued 1-form satisfying the structural equation is (locally) the Darboux derivative of som (local) G-valued function γ.

The next statement covers the bundle form of the local presentation of a $\mathfrak{g}$-valued 1-form satisfying the structural equation. Its proof follows from Cartan Theorem 18 in the same way as the previous proposition.

Proposition 9. *Let $\pi : E \to M$ be a fiber bundle, G a Lie group with the Lie algebra $\mathfrak{g}$ and μ a horizontal $\mathfrak{g}$-valued 1-form satisfying the horizontal Maurer-Cartan equation $D\mu + \frac{1}{2}[\mu, \mu] = 0$. Then, for each point $z \in E$ there is a neighborhood A_z and a locally defined function $\gamma_z : A_z \to G$ such that in A_z, $\mu = \gamma_z^{-1} D\gamma_z$.*

35.4. Comparison of flow- and μ-prolongations.

Lemma 4. *Let $\mathfrak{g}$ be the Lie group of the Lie group G and let μ be a $\mathfrak{g}$-valued horizontal 1-form on the space $J^k(Y)$ of the k-jet bundle $\pi_k : J^k \to X$ having, in local fiber coordinates, the form $\mu = \Lambda_i dx^i$. Then, the 1-form μ satisfies the compatibility condition (AAA) if and only if for any $z \in J^k(Y)$, there is a neighborhood $A_z \subset J^k(Y)$ and a smooth mapping $\gamma_z : A_z \to G$ such that*

$$\Lambda_i = \gamma_z^{-1} D_i \gamma_z, \ i = 1, \ldots, n. \tag{35.10}$$

Proof. If 1-form $\mu = \Lambda_i dx^i$ satisfies the condition (35.10), we have $D_i \Lambda_j = (D_i \gamma_z^{-1})(D_j \gamma_z) + \gamma_z^{-1}(D_i D_j \gamma_z)$; moreover, $\Lambda_i \Lambda_j = \gamma_z^{-1}(D_i \gamma_z) = (D_i \gamma_z^{-1} \gamma_z) = -(D_i \gamma_z^{-1})(D_j \gamma_z)$. This shows that the compatibility condition () is satisfied provided (35.10) holds. T proof of the converse statement follows directly from Lemma 4 and Proposition 8. $\qquad \square$

Remark 52. Notice that if a function γ_1 satisfies the condition (34.10), any function of the form $\gamma_2 = h\gamma_1$ with a constant $h \in G$ also deliveres the presentation (34.10).

Theorem 23. *Let Y be a π_k- vertical vector field in the space $J^k(Y)$, $Y = \Psi_J^a \frac{\partial}{\partial u_J^a}$ that is the μ-prolongation of the (possibly generalized) vector field $X = \Psi_0^a \cdot \frac{\partial}{\partial u^a} \in V(\pi)$. Suppose that the 1-form μ has the form*

$$\mu = \gamma^{-1}(D\gamma) = \gamma^{-1}(D_i \gamma) dx^i,$$

for a smooth G-values function $\gamma : Y \to G$, i.e. $\gamma \in \Gamma_n$. Then the vertical vector field $W = \gamma \cdot Y \in V_k$ is the standard prolongation of (possibly generalized) π-vertical vector field $\tilde{X} = \gamma \cdot X \in \mathrm{X}(\pi)$.

Conversely, let $W = \tilde{X}^{(k)} = \tilde{\Phi}^a_J \frac{\partial}{\partial^a_J}$ *be the standard prolongation of the evolutional vector field* $\tilde{X} = \tilde{\Phi}^a_0 \frac{\partial}{\partial u^a} \in V(\pi)$, *let* $\gamma \in \Gamma_k$ *be a smooth function. Then, the vector field* $Y \in V(\pi_k)$ *defined by* $Y = \gamma^{-1} \cdot W$ *is the* μ-*prolongation of the (possibly generalized) vector field* $X = \gamma^{-1} \cdot \tilde{X}$, *with* $\mu = \gamma^{-1}(D\gamma)$.

Proof. Vector field $W = \tilde{X}^{(k)}$ is a flow prolongation if (and only if) $\tilde{\Phi}^a_{J,i} = D_i(\tilde{\Phi}^a_J)$. On the other hand, $W = \gamma \cdot Y$, and, therefore, $\tilde{\Phi}^a_K = \gamma^a_b \Psi^b_K$ for any multiindex K, see (). In vector notation

$$\tilde{\Phi}^a_{J,i} = -\gamma^a_b \Psi^b_{J,i} = \gamma^a_b(D_i \Phi^b_J) + \gamma^b_c[(\gamma^{-1})^b_c(D_i \gamma^c_m)]\Psi^m_J.$$

Matrices $\gamma(x, u, u^{(k)})$ are invertible. This gives us

$$\Psi^a_{J,i} = [\delta^a_b D_i + (\gamma^{-1})^a_k]\Psi^b_J,$$

which is the formula of μ-prolongation with the identification $\Lambda_i = \gamma^{-1}(D_i\gamma)$, as claimed in the statement. In order to prove the converse, it suffices to perform the computation in the opposite direction, i.e., starting from $\Psi^a_{J,i} = \nabla_i(\Psi^a_J)0$. $\square$

For the general case, where μ is not known to have the form $\mu = \gamma^{-1}(D\gamma)$, we have

Theorem 24. *Let* Y *be a* π_k- *vertical vector field in the space* $J^k(Y)$, *that is the* μ-*prolongation of the (possibly generalized) vector field* $X = \Psi^a_0 \cdot \frac{\partial}{\partial u^a} \in V(\pi)$ *with the 1-form* μ *satisfying to the structural equation (2.2). Then, for any* $z = (x, u, u^{(k)}) \in J^k(Y)$ *there is a neighborhood* $A_z \subset J^k\pi$ *and a local mapping* $\gamma : A_z \to \Gamma_k$ *such that locally at* z,

(1) *I. The* **g**-*valued 1-form* $\mu \in \Lambda^1(J^k(Y, \mathbf{g})$ *is given by* $\mu = \gamma_z^{-1}(D\gamma_z)$,

(2) *II. Vector field* Y *is G-equivalent to the vector field* $W = \gamma_z \cdot Y$, *which is the standard (flow-) prolongation of the* π-*vertical vector field* $\tilde{X} = \gamma_z \cdot X$.

Proof. This Theorem merely states that Theorem (23) that deals with the case $\mu = \gamma^{-1}(D\gamma)$ for some mapping $\gamma \in \Gamma_k$, actually holds locally for all 1-forms μ satisfying the conditions of (Theorem 15), or, equivalently, condition (35.6). Lemma (4) guarantees that any such 1-form μ can be locally written in the form $\mu = \gamma^{-1}(D\gamma)$. $\square$

Theorem 25. *Let* $\mu = \Lambda_i dx^i$ *be a* **g**-*valued 1-form on the k-jet space* $J^k\pi$ *satisfying () or, equivalently,() . Let* $Y = \Psi^a_J \frac{\partial}{\partial u^a_J}) \in V(\pi_k)$ *be the* μ-*prolongation of the (possibly generalized) vector field* $X = \Psi^a_0 \frac{\partial}{\partial u^a} \in V(\pi)$. *Then the following statements are equivalent:*

(1) *There exists a vector field* $W \in V_k$ *which is G-equivalent to* Y *and which is the standard prolongation of a vector field* $\tilde{X} \in V_{(\pi)}$. *The vector fields* W *and* $\tilde{X}$ *are given by* $W = \gamma Y m$ $\tilde{X} = \gamma \cdot X$ *with* $\Lambda_i = \gamma^{-1}(\gamma^{-1}D(\gamma))$..

(2) *1-form* μ *is the horizontal Darboux derivative of a* $\gamma \in \Gamma_k, \mu = \gamma^{-1}(D\gamma)$.

Proof. This result follows at once from Lemma 4 and Theorem 20. $\square$

Denote by $\mathcal{M}$ the set of all $\mathfrak{g}$-values horizontal 1-forms satisfying the horizontal Maurer-Cartan equation (35.9) and by $\mathcal{D}$ the set of 1-forms that are Darboux derivatives of a mapping $\gamma \in \Gamma_k$. By Proposition 7, $\mathcal{D} \subset \mathcal{M}$.

We define $\mathcal{H} = \mathcal{M}/\mathcal{D}$. This amounts to factorization with respect to gauge equivalence. Space (in general, sheaf bundle) $\mathcal{H}$ depends on the topology of $J^k(Y)$, in particular, on the 1-st homotopy group. If the fibers F of the bundle $\pi : Y \to X$ (and hence, those of the bundle $J^k(Y) \to X$) are contractible, properties of $\mathcal{H}$ are defined by the topology of the base X, see [122].

Corollary 2. *If $\mathcal{H} = 0$, then any μ-prolonged vector field, for any $\mu \in \mathcal{M}$, is globally G-equivalent to a standard prolongation. In particular, it holds if the jet space $J^k(Y)$ is contractible.*

The following **commutative diagram** illustrates the relation of standard prolongation and the twisted prolongation delivered by the prolonged action of the gauge group.

$$
\begin{array}{ccc}
Y \in \mathcal{V}(J^k(Y)) & \longrightarrow & W = \gamma(Y) \in \mathcal{V}(J^k(Y)) \\[6pt]
\Big\uparrow {\scriptstyle Pr^k_\mu} & & \Big\uparrow {\scriptstyle Pr^k} \\[6pt]
X \in \mathcal{V}(Y) & \longrightarrow & \tilde{X} = \gamma \cdot X \in \mathcal{V}(Y)
\end{array}
\qquad .
$$

36. Applications and Examples.

In this Section we discuss some aspects of μ-symmetries and μ-conservation laws together with some applications. We also give examples of applications.

we start with the remark that the μ-symmetries often appears as the perturbation of conventional Noether symmetry and, as a result, the μ-conservation law is the modification of usual conservation law.

As already pointed out, one of the main properties of μ-symmetries is that the conventional Noether conservation law is present in the modified form. In some cases μ-symmetries reduces to the standard ones. An important aspect of the presence of μ-symmetry is the possibility of performing the **procedure of reduction** of order of differential equation (or a system of such equations. We refer to the book by P.Olver [106], Ch.3 for the detailed presentation of the method of reduction defined by an action of Lie group at the configuration space of a physical system

. In particular,in some cases it is possible to integrate explicitly the μ-conservation law. In some cases this will help to find other solutions from a given one.

Example 22. Let $X = u\partial_u + \partial_v$ be a vector field on the plane (u,v). Here $u = u(x,y), v = v(x,y)$ are two dependent variables.

Take the μ-form to be $\mu = \Lambda_1 dx + \Lambda_2 dy$, where $\Lambda_1 = \begin{pmatrix} = 0 & 0 \\ u_x & 0 \end{pmatrix} dx$ and, where $\Lambda_2 = \begin{pmatrix} 0 & 0 \\ u_y & 0 \end{pmatrix} dy$.

It is easy to check that the Lagrangian $L = \frac{1}{2}(u_x^2 + u_y^2) - \frac{1}{u}(u_x v_x + u_y v_y) + u^2 \cdot exp(-2v)$ is μ-invariant (but not not invariant in the usual sense) under the vector field X.

The M-vector $(P)_a^b$ defining the form of corresponding conservation law has the components

$$(P^1)^b_a = \begin{pmatrix} uu_x - v_x & -u_x \\ u_x - \frac{v_x}{v} & \frac{-u_x}{u} \end{pmatrix},$$

$$(P^2)^b_a = \begin{pmatrix} uu_y - v_y & -u_y \\ u_y - \frac{v_y}{v} & -\frac{u_y}{u} \end{pmatrix}.$$

M-vector $(P)^b_a$ satisfies to the μ-conservation law $Tr(\nabla_i P^i) = 0$, which here takes the form

$$d_i P^i = d_x \left(uu_x - v_x - \frac{u_x}{u} \right) = d_y(uu_y - v_y - \frac{u_y}{u}) = u_x^2 + u_u^2. \tag{36.1}$$

For more detail about this and other similar examples see [14].

Example 23. In this example, we deal with one independent variable (t) and two dependent - $q_1(t)$, $q_2(t)$.

Vector field of symmetries is

$$X = q_1 \frac{\partial}{\partial q_1} + \frac{\partial}{\partial q_2}.$$

We choose 1-form μ to be: $\mu = \Lambda dt$, where $\Lambda = \lambda I$ with $\lambda = q_1$ and I to be 2-dim unit matrix.

Consider now the Lagrangian

$$L = \frac{1}{2} \left(\frac{\dot{q}_1}{q_1} - q_1 \right)^2 + \frac{1}{2}(\dot{q}_2 - q_1)^2) \tag{36.2}$$

, **μ-invariant with respect to the vector field** X.

It is easy to check the existence of the μ-conservation law $\nabla_t \mathbf{P} = 0$ for the quantity $\mathbf{P} = \left(\frac{\dot{q}_1}{q_1 + \dot{q}_2 - 2q_1} \right)$.

If we introduce the μ-invariant quantities $\alpha = \frac{\dot{q}_1}{q_1} - q_1$, $\beta = \dot{q}_2 - q_1$ as the new dependent variables,the Euler-Lagrange equations (system of two ODE) takes tyhe form

$$\begin{cases} \alpha_t + q_1(\alpha + \beta) = 0, \\ \beta_t = 0. \end{cases} \tag{36.3}$$

In this case, the μ-symmetry with respect of vector field X is lost together with the μ-invariance. On the other side, this becomes an example of "**partial reduction**" in the sense of the work [102]. In this case, this system admit the transformation to the system of first order equations $\alpha = \beta = 0$. This immediately allows to obtain the solutions

$$\begin{cases} q_1 = \frac{c_1}{1 - c_1 t}, \\ q_2 = c_2 - log(1 - c_1 t) \end{cases} \tag{36.4}$$

with arbitrary constants c_1, c_2.

Example 24. As the last example of a partial differential equation with the μ-symmetry we consider the euler equation

$$u_t + uu_x = 0. \tag{36.5}$$

Let $X = \xi\partial_x + \tau\partial_t + \phi\partial_u \in \mathcal{X}(Y)$ be a potential μ- symmetry vector field for a 1-form $\mu = \alpha dx + \beta dt$. In order for 1-form μ to sat5isfy to the conditions of Proposition 5, it is necessary and sufficient that $d_x\beta = d_t\alpha$ when $u_t + uu_x = 0$.

Condition for vector field X to be a μ-symmetry for euler equation is that

$$\phi u_x + \alpha u\phi + \beta\phi + u^2 u_x \alpha\tau + uu_x\beta\tau - uu_x\alpha\xi - u_x\beta\xi + \phi_t + uu_x\tau_t -$$
$$- u_x\xi_t + u\phi_x + u^2 u_x\tau_x - uu_x\xi_x = 0. \quad (36.6)$$

With the ansatz $\alpha = \alpha(x, t, u)$, $\beta = \beta(x, t, u)$, obtained equation splits into two equations:

$$\begin{cases} (\alpha u + \beta)\phi + \phi_t + u\phi_x = 0, \\ \phi + (\alpha u^2 + \beta u)\tau - (\alpha u + \beta)\xi + u\tau_t - \xi_t + u^2\tau_x - u\xi_x = 0. \end{cases} \quad (36.7)$$

These are nonlinear equations for $\alpha, \beta, \xi, \tau, \phi)$. A special solution is provided as follows:

$$\begin{cases} \alpha = u, \\ \beta = -\frac{u^2}{2}, \\ \xi = 0, \\ \tau = [B(u) - \frac{A(u)t}{u}]exp[-(u^2/2)t], \\ \phi = A(u)exp[-\frac{u^2}{2}] \end{cases} \quad (36.8)$$

Notice that for this choice of 1-form μ, compatibility condition $d_t\alpha = d_x\beta$ is satisfied only on the manifold of solutions S_Δ. Obtained μ-symmetry corresponds to **a nonlocal ordinary symmetry Z of exponential type. In fact,**

$$Z = e^{\int udx - (u^2/2)dt}X.$$

37. Deformation of exterior differential, Lie derivatives and the μ-prolongation (by P.Morando,[98]).

In the article [98], Paola Morando introduced the deformation (gauging) of Lie derivative in terms of closed 1-forms μ. Using the obtained picture, Dr.Morando described μ-prolongations and their basic properties in an elegant geometrical fashion. In my opinion, this approach may be useful in other problems of geometrical theory of differential equations. That is why we present basic structures and results of the cited paper here.

37.1. **Witten's gauging and the deformed operators d and $\mathcal{L}_\xi$.** Let M be a smooth manifold and $f \in C^\infty(M)$ be a smooth function.

Definition 20. *Define the deformed differential d^{df} by*

$$d^{df}\beta = e^{-f}d(e^f\beta) = d\beta + df \wedge \beta), \quad (37.1)$$

for all exterior forms $\beta \in \Lambda^*(M)$.

Lemma 5. *The deformed exterior differential is the first order differential operator satisfying*

$$d^{df} \circ d^{df} = 0. \quad (37.2)$$

The proof of Lemma is a straightforward computation using the facts $d(df) = 0$ and $df \wedge df = 0$. Next we define the deformed Lie derivative along a vector field $X \in \mathcal{X}(M)$:

Definition 21. *Let M be a smooth manifold, $f \in C^\infty(M)$, $\beta \in \Lambda^*(M)$ and $x \in \mathcal{X}(M)$. Then,*

$$\mathcal{L}_X^{df}\beta = e^{-f}\mathcal{L}_{(e^f X)}\beta = \mathcal{L}_X\beta + df \wedge (i_X\beta). \quad (37.3)$$

Lemma 6. *The deformed Lie derivative L_X^{df} satisfies*

(1) $L_X^{df}(\beta_1 + \beta_2) = L_X^{df}\beta_1 + L_X^{df}\beta_2)$,

(2) $L_X^{df}(\beta_1 \wedge \beta_2) = L_X^{df}\beta_1 \wedge \beta_2 + \beta_1 \wedge (L_X^{df}\beta_2)$.

for any vector field X and arbitrary exterior forms β_1, β_2.

The proof of this lemma is an easy computation.

As the next step we extend this deformation to the lie derivative of vector fields.

Definition 22. *Let M be a smooth manifold and let $f \in C^\infty(M)$ and $X, Y \in \mathcal{X}(M)$ be two vector fields. We define*

$$\mathcal{L}_X^{df}(Y) = e^{-f}\mathcal{L}_{e^f X}Y = \mathcal{L}_X Y - i_Y dFX. \tag{37.4}$$

Lemma 7. *The deformed Lie derivative $\mathcal{L}_X^{df}$ satisfies*

$$L_X^{df}(i_Y \beta) = L_X^{df}(Y) + i_Y(L_X^{df}\beta) \tag{37.5}$$

for all vector fields X, Y in B and any exterior form β on M.

The proof is straightforward. The next statement is the (deformed) analog of a Cartan form for the Lie derivative of an exterior form.

Theorem 26. *Let M be a smooth manifold, $f \in C^\infty(M)$ and $X \in \mathcal{X}(M)$ and $\beta \in \Lambda^*(M)$. Then*

$$\begin{cases} L_X^{df}(\beta) = i_X d\beta + d^{df} i_X(\beta), \\ L_X^{df}(d\beta) = d^{df} L_X^{df}(\beta). \end{cases} \tag{37.6}$$

37.2. **Derformation of exterior differential and Lie derivative.**

Definition 23. *Let M be a manifold and μ be - a closed (i.e. such that $d\mu = 0$) 1-form in M. We define the μ-twisted differential d^μ and the μ-twisted Lie derivative of vector field X as follows:*

Differential d^μ:

On the differential forms $\beta \in \Lambda^*(M)$, μ-twisted differential is

$$d^\mu \beta = d\beta + \mu \wedge \beta. \tag{37.7}$$

Definition 24. μ-**twisted Lie derivative** $\mathcal{L}^\mu$:
On exterior forms, μ-twisted Lie $\mathcal{L}^\mu$ derivative along a vector field ξ is

$$\mathcal{L}_\xi^\mu \beta = \mathcal{L}_\xi \beta + \mu \wedge i_\xi \beta, \tag{37.8}$$

At the same time, μ-twisted Lie $\mathcal{L}^\mu$ derivative of the vector fields ζ is

$$\mathcal{L}_\xi \zeta = L_\xi \zeta - (i_\zeta \mu)\xi. \tag{37.9}$$

Example 25. In the special case where the 1-form μ is exact (i.e. $\mu = df$, $f \in C^\infty(M)$),introduced deformation of the exterior differential d has the form of Witten's gauging, ().

Introduced deformations of exterior differential and the Lie derivative have the following simple properties

Proposition 10. *Let M be a smooth manifold and let $\mu \in \Lambda^1(M)$ be a 1-form. Then, for any $\xi \in \mathcal{X}(M)$, and for arbitrary $\beta \in \Omega^*(M)$, the following properties are equivalent:*

(1) $d\mu = 0$,
(2) $d^\mu \circ d^\mu = 0$,
(3) $\mathcal{L}_\xi^\mu d\beta = d^\mu(\mathcal{L}_\xi^\mu \beta)$.

Proof. Equivalence of these statements follows from the arbitrariness of ξ and β and the following straightforward calculations:

$$d^\mu(d^\mu \beta)) = d^\mu(d\beta + \mu \wedge \beta) = d\mu \wedge \beta,$$

$$\mathcal{L}_\xi^\mu d\beta = d^\mu(i_\xi d\beta) = d^\mu(\mathcal{L}_\xi^\mu \beta) - d^\mu(d^\mu(i_\xi \beta)). \quad (37.10)$$

$\square$

37.3. Modifications for the case of the 1-jet bundle. Let μ is a horizontal 1-form $\mu = \Lambda_i(x^i, y^\mu, y_i^\mu)dx^i$ on the 1-jet space $J^1(\pi)$ of a bundle $\pi : Y \to X$. We assume that $d\mu \in C\Omega(J^1(\pi))$ is the contact 1-form. Modify the previous definition as follows.

Definition 25. *Let M be a manifold and let μ be a horizontal 1-form $\mu = \Lambda_i(x^i, y^\mu, y_i^\mu)dx^i$ on the 1-jet space $J^1(\pi)$ of a bundle $\pi : Y \to X$. We assume that $d\mu \in C\Omega(J^1(\pi))$ is the contact 1-form. We define the μ-twisted differential d^μ and the μ-twisted Lie derivative of vector field ξ on the k-jet bundles $J^k(\pi)$*

(1) On the differential forms $\beta \in \Lambda^*(J^k(\pi))$ to be

$$d^\mu \beta = d\beta + \mu \wedge \beta, \quad (37.11)$$

and

(2)

$$\mathcal{L}_\xi^\mu \beta = \mathcal{L}_\xi \beta + \mu \wedge i_\xi \beta, \quad (37.12)$$

on the vector fields ζ,

$$\mathcal{L}_\xi \zeta = L_\xi \zeta - (i_\zeta \mu)\xi. \quad (37.13)$$

In this case it is not guaranteed that $d^\mu \circ d^\mu = 0$. Somewhat weaker properties are guaranteed by the following

Proposition 11. *Let $\pi : Y \to X$ be a fiber bundle and let μ be a horizontal 1-form $\mu = \Lambda_i(x^i, y^\mu, y_i^\mu)dx^i$ on the 1-jet space $J^1(\pi)$. Then, for any $\xi \in \mathcal{X}(Y)$, and for arbitrary $k \geq 1$ and $\beta \in \Omega^*(J^k(\pi))$, following properties are equivalent:*

(1) $d\mu \in C\Omega^*(J^1(\pi))$,
(2) $d^\mu \circ d^\mu \in C\Omega^*(J^1(\pi))$,
(3) $\mathcal{L}_\xi^\mu d\beta - d^\mu(\mathcal{L}_\xi^\mu \beta) \in C\Omega^*(J^1(\pi))$.

Proof. Direct calculation using the Definition (25) gives us

$$d^\mu \circ d^\mu \beta = d^\mu(d\beta + \mu \wedge \beta) = (d\mu + \mu \wedge \mu) \wedge \beta.$$

As $\mu \wedge \mu = 0$ and using the fact that β is an arbitrary exterior form, we see equivalnece of statements 1) and 2). To prove the equivalence of 2) and 3) we write $\mathcal{L}_\xi^\mu \beta = i_\xi d\beta + d^\mu(i_\xi \beta)$, using Definition (). Then we have

$$\mathcal{L}_\xi^\mu(d\beta) = i_\xi dd\beta + d^\mu(i_\xi d\beta) = d^\mu(i_\xi d\beta).$$

Considering the d^μ differential of $L_\xi^\mu(\beta)$ we find

$$d^\mu(\mathcal{L}_\xi^\mu(\beta)) = d^\mu(i_\xi d\beta) + d^\mu(d^\mu(i_\xi\beta)).$$

Then, $\mathcal{L}_\xi^\mu(s\beta) = d^\mu(\mathcal{L}_\xi^\mu(\beta)) - d^\mu d^\mu(i_\xi d\beta)$. Statement of the Proposition now follows from the arbitrariness of ξ and β. $\qquad\qquad\qquad\qquad\qquad\Box$

Remark 53. Recall the variational bicomplex obtained by the decomposition of DeRham complex into the component with fixed horizontal and vertical degrees. In variational bicomplex, whose terms are $\Lambda^{p,q}(J^\infty(M))$, exterior differential splits into the horizontal and vertical components:

$$d = d_H + d_v.$$

For these components we have

$$d_H f = D_i f dx^i, \ d_H dx^i = 0, \ d_H \omega_J^\alpha = -\omega_{J.i}^\alpha \wedge dx^i).$$

We can associate with the ν-twisted differential d^μ, a cohomological operator

$$d_H^\mu(\beta) = d_H\beta + \mu \wedge \beta, \qquad\qquad\qquad (37.14)$$

for all exterior forms β. Thus, in some sense d^μ "defines the horizontal cohomology.

Now we apply the deformed differential and Lie derivatives to the contact characterization of μ-prolongations $\hat\xi \in \mathcal{X}(J^k(\pi))$ of vector fields $\xi \in \mathcal{X}(Y)$.

Theorem 27. *(P.Morando,[98], Theorems 4,5) Let $\pi : Y \to X$ be a fiber bundle and let $\mu = \Lambda_i dx^i \in \Lambda_{hor}^1(J^1\pi)$ be a π-horizontal 1-form such that $d\mu \in C\Omega^*(J^1\pi)$ is contact. Then, for a vector field $\hat\xi \in \mathcal{X}(J^k(\pi))$ the following statements are equivalent*

    (1) *Vector field $\hat\xi$ is the μ-prolongation of the vector field $\xi \in \mathcal{X}(Y)$, i.e. that a)vector field $\hat\xi$ is (π_{k0}-)projectable to Y with the projection ξ and b)*

$$\hat\xi = \xi + \sum_{a,J|0<|J|\leqq k} \Psi_J^a \partial_{y_J^a}, \ \Psi_{J,j}^a = [d_j + \Lambda_j]\Psi_J^a - y_{J,m}^a[d_j + \Lambda_j]\xi^m, \ \Psi_0^a = \phi^a.$$

    (2) *Vector field $\hat\xi$ is (π_{k0}-)projectable to Y with the projection ξ and $\mathcal{L}_\xi^\mu(\theta) \in C\Omega^*(J^k(\pi))$ for all contact forms $\theta \in C\Omega(J^k(\pi))$.*

    (3) *Let Ca^k be the Cartan distribution in the k-jet space $J^k(\pi)$ (linear span of reduction to $J^k(\pi)$ of total derivatives d_i). Then a)Vector field $\hat\xi$ is (π_{k0}-)projectable to Y with the projection ξ and b) for all Lie fields $\eta \in Ca^k$ in $J^k(\pi)$, $\mathcal{L}_\xi^\mu\eta \in Ca^k$.*

Proof. To prove the equivalence of first two statements we use basic contact forms $\theta_J^\alpha] = dy_J^\alpha - y_{J,i}^\alpha dx^i$ and represent vector field $\hat\xi$ in the basis:

$$\hat\xi = \xi^i\partial_i + \xi^\mu\partial_\mu + \psi_J^\mu\partial_{y_J^\mu}.$$

Then, the direct calculation shows that the fulfillment of the second condition on the basic forms:

$$\mathcal{L}_{\hat\xi}^\mu(\theta_J^\alpha) \in C\Omega^*(J^k(\pi)) = L_{\hat\xi}$$

leads to the λ (or respectively μ)-prolongation formula introduce above in this Section(Section 30) for the coefficients of the vector field $\hat\xi$.

For the proof of equivalence of statements 3) and 2) we recall that

$$\mathcal{L}_{\hat\xi}^\mu(d_i) = \mathcal{L}_{\hat\xi}(d_i) - (i_{d_i}\mu)\hat\xi = \mathcal{L}_{\hat\xi}(d_i) - -\Lambda_i\hat\xi$$

and use arguments in the proofs of previous statements. $\square$

38. Variational λ- and μ-symmetries and the Noether Theorem for λ and μ-symmetries.

In this section we discuss the relation between the variational λ- and μ- symmetries and the conservation laws for a first order regular Lagrangians. A geometrical interpretation of the variational λ- and μ symmetries allows us to realize the corresponding conservation law in the form of the Noether equation (compare to [102]).

All results for ODE can be extended to the case of higher order Lagrangians using corresponding Poincare-Cartan form, see [75]. In the case of PDE some difficulties could arise if we consider Lagrangians of order higher then 2 for in this case one can not define uniquelly the corresponding Poincare-Cartan form.

38.1. **Case of ODE.** . Let $\pi_0 : M \to B$ be a fiber bundle with 1-dim base B. Introduce fibred coordinates $(x, u^\alpha, u_x^\alpha)$ and let $L : J^1\pi \to R$ be a first order regular Lagrangian function. Introduce the Poincare-Cartan 1-form

$$\Theta = Ldx + \frac{\partial L}{\partial u_x^\alpha}\omega^\alpha \qquad (38.1)$$

corresponding to L.

Definition 26.   (1) *Let $\mu = \lambda(x, u^\alpha, u_x^\alpha)dx$ be a horizontal 1-form on $J^1(M)$. A vector field $X_0 \in \mathcal{X}(M)$ is a **variational λ-symmetry** for Lagrangian L if and only if the first order λ- prolongation $X \in \mathcal{X}(J^1(M))$ is such that $L_X^\mu(\Theta) \in \mathcal{C}\Omega$, i.e., if μ-twisted Lie derivative of the Poincare-Cartan form Θ is **is contact**.*

   (2) *A vector field $X_0 \in \mathcal{X}(M)$ is called a divergence variational λ-symmetry for the Lagrangian L if its first order λ-prolongation $X \in \mathcal{X}(J^1(M))$ satisfies*

$$\mathcal{L}_X^\mu(\Theta) - d^\mu R \in \mathcal{C}\Omega, \qquad (38.2)$$

   for a suitable function $R \in C^\infty(M)$.

Lemma 8. *Condition () is equivalent to*

$$X(L) + L(D_x + \lambda)\xi = (D_x + \lambda)(R), \qquad (38.3)$$

where X is the λ-prolongation of a vector field X_0 on M.

For the proof we refer to the paper [98].

As the last step we will show that the geometrical characteristic of the introduced λ-symmetries leads to the corresponding Noether Theorem. We recall that in our geometrical framework, a conservation law for an ODE $\Delta = 0$, is a function P such that on the equation $\Delta = 0$, $dP \in \mathcal{C}\Omega$ is contact 1-form.

For a system of Euler-lagrange Equations, this condition is equivalent to the requirement that $D_x P = 0$ on the solutions of Euler-Lagrange system. Next is the extension of the notion of conservation law to the case of "deformed operators."

Definition 27. *A λ-conservation law for a given ODE $\Delta := F(x, u^a, u_n^a) = 0$ is a function P such that $d^\mu P = 0$ on the subset $F = 0$ in the jet space. In particular, $i_{D_x} d^\mu P = D_x(P) + \lambda P = 0$ on $F = 0$.*

The next result (Noether Theorem for λ-symmetries) completely agrees with the conservation law in [102].

Theorem 28. *Let X_0 be a divergence variational λ-symmetry for a first order regular Lagrangian L. If Θ is the Poincare-Cartan 1-form associated with L, we have the conservation law for the solutions of the system of Euler-Lagrange equations:*

$$D_x(i_{X_0}\Theta - R + \lambda(i_{X_0}\Theta - R) = 0. \tag{38.4}$$

Proof. Recall that X_0 is a divergence variational λ-symmetry if and only if its first-order λ-prolongation satisfies to the relation

$$\mathcal{L}_X^\mu(\Theta) - d^\mu R = i_X d\Theta + d^\mu i_X \Theta - R) \in \mathcal{C}\Omega(J^1(M)) \tag{38.5}$$

for a suitable function $R \in C^\infty(M)$. On the solutions of Euler-Lagrange Equations, we have $i_X d\Theta = 0$ and we get the Noether relation (relation 38.4) by using that for the form Θ, $i_X\Theta = i_{X_0}\Theta$. $\qquad\square$

38.2. The case of PDE.. Consider a fibred bundle($\pi_0 : M \to B$) with an n-dimensional connected orientable base B (endowed with a riemannian metric g and corresponding volume n-form ω_0). Let $J^1\pi_0$ be the 1-jet bundle of the (configurational) bundle π_0 and $L(x^i, u^\alpha, u_i^\alpha)$ - be regular first order Lagrangian.

Poincare-Cartan n-form, corresponding to the Lagrangian L, is ([10])

$$\Theta = L\omega_0 + \frac{\partial L}{\partial u_i^\alpha}\theta^\alpha \wedge \omega_i, \tag{38.6}$$

where $\omega^\mu = du^\mu - u_i^\mu dx^i$ are basic contact 1-forms and $\omega_i = i_{\partial_{x^i}}\omega_0, i = 1, \ldots, n$.

Definition 28. *Let $\Lambda_i(x^k, u^\alpha, u_k^\alpha)dx^i$ be a horizontal 1-form on $J^1\pi_0$ satisfying $d\mu \in \mathcal{C}\Omega$ and Θ be the Poincare-Cartan n-form associated with L. A vector field X_0 on M is a variational μ-symmetry for the Lagrangian L if its first order μ-prolongation $X \in \mathcal{X}(J^1(M))$ satisfies $\mathcal{L}_X^\mu\Theta \in \mathcal{C}\Omega^n$.*

A vector field $X_0 \in \mathcal{X}(M)$ is called a divergence variational μ-symmetry for the Lagrangian L if its first order prolongation $X \in \mathcal{X}(J^1(M))$ satisfies

$$\mathcal{L}_X^\mu\Theta \in \mathcal{C}\Omega^n \tag{38.7}$$

for a suitable $(n-1)$-form ρ. If we present ρ in the form $\rho = R^i\omega_i + \sigma$ with contact form σ, it is an easy computation (see Lemma 8) to prove that the last condition for vector field X can be written, in local (fibred) coordinates, as

$$X(L) + L(D_i + \Lambda_i)\xi^i = (D_i + \Lambda_i)R^i. \tag{38.8}$$

Finally, we will show that in the case of Euler-Lagrange equations (variational PDE), we associate with any variational μ-symmetry, a μ-conservation law (see above).

Recall that a conservation law for a PDE $\Delta = 0$, can be defined as an exterior $(n-1)$-form π such that $d\pi$ is a contact n-form on the surface defined by Δ. Generalizing this definition to the case of deformed differential equations we come up with the following

Definition 29. *A μ-conservation law for a given PDE $\Delta := F(x^i, u^\alpha, u_i^\alpha) = 0$ is a (n-1)-form π such that $d^\mu\pi$ is contact on the solutions of PDE Δ. In particular, this means that $i_{D_i}d^\mu\pi = 0$ for $i = 1, \ldots, n$ on the solutions of PDE $\Delta = 0$.*

Theorem 29. *Let X_0 be a divergence variational μ-symmetry for a first order regular Lagrangian L, and let Θ be the Poincare-Cartan n-form associated with L. Then, the following μ-conservation law holds:*

$$i_{D_i} d^\mu (i_{X_0} \Theta - \rho) = 0 \tag{38.9}$$

Proof. The proof of this result is similar to the proof of Theorem 18
. In particular, on the solution of Euler-Lagrange equations the μ-conservation law (36.9) explicitly read

$$D_i \left(\frac{\partial L}{\partial u_i^\alpha} (\Phi^\alpha - u_i^\alpha \xi^k) + \xi^i L - R^i \right) + \Lambda_i \left(\frac{\partial L}{\partial u_i^\alpha} (\Phi^\alpha - u_i^\alpha \xi^k) + \xi^i L - R^i \right) = 0. \tag{38.10}$$

$\square$

39. Appendix:Darboux derivatives.

In this Section we present, following [121], definitions and the basic result (Cartan Theorem, Theorem 26) about Maurer-Cartan forms and the Darbough Derivatives used in this Chapter.

Definition 30. *Let G be a Lie group. Then, the left-invariant Maurer-Cartan form $\omega_G : T(G) \to \mathfrak{g}$ is defined by*

$$\omega_G(v(= L_{g^{-1}*}(v) \tag{39.1}$$

for tangent vectors $v \in T_g(G)$.

This form is invariant under the left translations. More specifically, let $v \in T_g(G)$. Then, $L_{h*} \in T_{hg}(G)$ and

$$(L_{h*}\omega_G)v = \omega_G(L_{h*}(v))) = L_{(hg)^{-1}*}(L_{h*}(v)) = L_{g^{-1}*}(v) = \omega_G(v).$$

Let G be a Lie group and $\mathfrak{g}$ be the Lie algebra of the group G. Calculating the differential of the Maurer-Cartan form leads to the formula called the **structural equation for the Maurer-Cartan form, see** [121], **Ch.3, Sec.3**: For arbitrary left-invariant vector fields X, Y in the Lie group G,

$$d\omega_G(X, Y) + [\omega_G(X), \omega_G(Y)] = 0. \tag{39.2}$$

Since any tangent vectors u, v at a point $g \in G$ can be presented as the values, at the point g, of left-invariant vector fields on G, the structural equation is valid **for any pair of vector fields** in G. The structural equation has also another form:

$$d\omega_G + \frac{1}{2}[\omega_G, \omega_G] = 0. \tag{39.3}$$

Definition 31. *Let G be a Lie group with the Lie algebra $\mathfrak{g}$. Let ω_G be the Maurer-Cartan form on G. Let now M be a smooth manifold and $f : M \to G$ be a smooth mapping.*

The (left) Darboux derivative of the mapping f is the $\mathfrak{g}$-valued 1-form $\omega_f = f^*(\omega_G)$ on the manifold M.

The map f is called the primitive of ω. From the naturality of exterior differential d it follows that the form ω_f satisfies the analog of structural equation

$$d\omega_f(X, Y) + [\omega_f(X), \omega_f(Y)] = 0. \tag{39.4}$$

The next basic result is the fundamental Theorem of E.Cartan that characterized $\mathfrak{g}$-valued 1-forms ω on a manifold M that are Darboux derivatives of smooth mappings $f : M \to G$.

Theorem 30. *Let G be a Lie group with Lie algebra $\mathfrak{g}$. Let ω be a $\mathfrak{g}$-values 1-form on the smooth manifold M satisfying to the structural equation $d\omega + \frac{1}{2}[\omega, \omega] = 0$. Then, for each point $p \in M$, there is a neighborhood U of p and a smooth mapping $f : U \to G$ such that*

$$\omega_{|U} = \omega_f. \tag{39.5}$$

Chapter 5. Applications: Holonomic and non-Holonomic Mechanics, H.Kleinert Action principle,

Uniform Materials, Non commutative variations and the Dissipative potentials .

In this part we present five different situations in Mechanics, holonomic and non-holonomic, in Continuum Thermodynamics and in material Science where non-commuting variations appears as a necessary tool for the formulation of Variational Equations. In all five situations considered in this chapter tensor K is related to (or even defined by) some geometrical structure present in the configurational bundle -affine connection, non-holonomic frames, absolute parallelism and its torsion. In fourth of the presented examples NC-tensor K is defined by the torsion of the present connection and directly determine the non-conservative, dissipative sources.

40. Nonconservative forces in Holonomic Mechanics.

Successes of Differential Geometry at the end of XIX and the beginning of XX century, and, in particular, description of the inertia motions on the ellipsoids by K.K.Jacoby led specialists in Mechanics and Geometry to the following question: is it possible to write down Newton equations of motion of a mechanical system as the equations for motion **along geodesics** of some metric or a linear (affine) connection. This problem, called the **geometrization** problem, was studied in the second half of XIX century and in the first half of XX century by different scientists who had offered its solutions in different form.

40.1. Geometrization via affine connection. One of the ways to solve this problem was suggested by B.Vujanovich and Dj.Djukis in the papers [133, 134]. They considered a holonomic nonconservative mechanical system with the configurational space Q, generalized coordinates $q^i, i = 1\ldots, m$ and the kinetic energy

$$T = \frac{1}{2} g_{ij} \dot{q}^i \dot{q}^j. \tag{40.1}$$

Here g_{ij} is the symmetrical nondegenerate positive definite (0,2)-tensor - **fixed metric in the configurational space** Q.

We assume that this system is subject to the action of *nonconservative* forces $f_i(t, q^i, \dot{q}^j)$. Newton equations of motion in contravariant form (metric g_{ij} is used for lifting of indices) are

$$^g\nabla_t \frac{dq^i}{dt} = \ddot{q}^i +^g \Gamma^i_{jk} \dot{q}^j \dot{q}^k = f^i, i = 1\ldots, m. \tag{40.2}$$

Here $^g\Gamma^i_{jk}$ are Christoffel coefficients of the Levi-Civita connection of the metric g.

B.Vujanovich and Dj.Djukis have used the affine (linear) connection (see Appendix I, Sec.69-70 or [69], Vol. I) defined by the metric g and the 1-form force $\omega = f_i dq^i$, where $f_j = f^i g_{ij}$.

This connection in the space Q is defined as follows

$$\Gamma^i_{jk} =^g \Gamma^i_{jk} + (\frac{1}{2T}(-f_j \delta^i_k + f_k \delta^i_j - f^i g_{jk}). \tag{40.3}$$

© Springer International Publishing Switzerland 2016
S. Preston, *Non-commuting Variations in Mathematics and Physics*,
Interaction of Mechanics and Mathematics,
DOI 10.1007/978-3-319-28323-4_5

Remark 54. Notice that unless forces f_j have a special form (homogeneous by velocities $\dot{q}^j$), connection coefficients Γ^i_{jk} are singular at the equilibria points where $T = 0$.

Remark 55. Geometrization equations suggested by B.Vujanovich and Dj.Djukis are equations of geodesic curves of the connection Γ. These geodesics differ from the geodesics of the Levi-Civita connection of the metric g. This follows from the expression of an arbitrary affine connection having the same geodesic curves as the Levi-Civita connection ${}^g\Gamma^i_{jk}$

$$L^i_{jk} = {}^g\Gamma^i_{jk} + 2\delta^i_j\psi_k \tag{40.4}$$

where $\psi_k dy^k$ is an arbitrary 1-form in the space Q (see [21],Sec.12, p.30 or Appendix I, Sec.69).

Proof of the following statement is straightforward.

Proposition 12. *Equations of motion (40.2) are identical with the equations of* motion along geodesics *of the affine connection Γ:*

$$\nabla_t \frac{dq^i}{dt} = \ddot{q}^i + \Gamma^i_{jk}\dot{q}^j\dot{q}^k = 0, i = 1\ldots, m. \tag{40.5}$$

Decompose the connection Γ into symmetrical and antisymmetrical parts (recall that Levi-Civita connection ${}^g\Gamma$ is symmetrical: ${}^g\Gamma^i_{jk} = {}^g\Gamma^i_{kj}$)

$$\Gamma^i_{jk} = \Gamma^i_{(jk)} + \Gamma^i_{[jk]} = ({}^g\Gamma^i_{jk} - \frac{1}{2T}f^i g_{jk}) + S_{[j}\delta^i_{k]}, \tag{40.6}$$

where $S_j = -\frac{f_j}{T}$ is the (1,2)-tensor field. Tensor

$$T^i_{jk} = \Gamma^i_{[jk]} = -\frac{f_{[j}}{2T}\delta^i_{k]} \tag{40.7}$$

in the space Q is the **torsion** of the *affine connection Γ.*

Notice that the connection Γ is not a metric one, in particular metric g is not invariant under the parallel translation defined by connection Γ. The measure of "non-metricity" - **non-metricity tensor** of metric g under the parallel translation by the connection Γ is

$$\nabla^\Gamma_i g_{jk} = \frac{1}{T}f_i g_{jk} \tag{40.8}$$

is not zero (See Appendix I, Sec.69). notice that for a fixed metric g, this tensor vanishes if and only if $f_i = 0$ for all i.

Remark 56. In the terminology of J.Schouten, Γ is the semi-symmetric, semi-metric affine connection ([120], pp.126,133).

The symmetrical part Δ of the connection Γ,

$$\Delta^i_{jk} = {}^g\Gamma^i_{jk} - \frac{1}{2T}f^i g_{jk} \tag{40.9}$$

is the torsionless affine connection ($\Delta^i_{jk} = \Delta^i_{kj}$) that has *the same geodesics* as the connection Γ since geodesics of an affine connection do not depend on the antisymmetrical part of connection (i.e. on the torsion of the connection). This follows immediately from the equation for geodesics $\gamma(t) = \{x^i(t)\}$ of an affine connection L^i_{JK} written in the form (see [21],Sec.7)

$$\frac{dx^j}{dt}\left(\frac{d^2x^i}{dt^2} + L^i_{kl}\frac{dx^k}{dt}\frac{dx^l}{dt}\right) - \frac{dx^i}{dt}\left(\frac{d^2x^j}{dt^2} + L^j_{kl}\frac{dx^k}{dt}\frac{dx^l}{dt}\right) = 0. \qquad (40.10)$$

This proves the following

Proposition 13. *Let (Q, g) be a manifold with the Riemannian metric g and Levi-Civita connection $^g\Gamma$. Equations of motion (40.2):*

$$^g\nabla_t\frac{dq^i}{dt} = \ddot{q}^i +^g \Gamma^i_{jk}\dot{q}^j\dot{q}^k = f^i, i = 1\ldots,n.$$

in the manifold Q with the Lagrangian given by the kinetic energy $L = T = \frac{1}{2}g_{ij}\dot{q}^i\dot{q}^j$, under the action of forces f_j can be presented as the motion along geodesics of the symmetrical affine connection Δ in Q. This symmetric connection is unique:

$$\Delta^i_{jk} =^g \Gamma^i_{jk} - \frac{1}{2T}f^i g_{jk}. \qquad (40.11)$$

Tensor of non-metricity of connection Δ has the form

$$D^\Delta(g)_{ijk} = \frac{1}{2T}(g_{ij}f_k + g_{ik}f_j). \qquad (40.12)$$

Connection Δ is metric if and only if $f_j = 0$ for all j.

Proof. The first statement was proved above and the uniqueness of Δ follows from Appendix, Sec.69. Expression for the **non-metricity tensor** of the connection Δ is determined by the formula (69.13) where the tensor presenting connection Δ in terms of the Levi-Civita connection $^g\Gamma$ has the form $R^i_{jk} = -\frac{1}{2T}f^i g_{jk}$. The last statement follows immediately from the expression for non-metricity tensor. $\qquad\square$

40.2. Description of a mechanical system with non-conservative forces via NC-variations.

Here we show that a mechanical system with non-conservative forces can be presented as the Euler-Lagrange equations of the variational problem with non-commutative variations.

If we write down the Euler-Lagrange equations for the kinetic energy (arc length) $T = \frac{1}{2}\|\dot{q}\|^2$ and use, for obtained action functional

$$S(\gamma) = \int_{t_0}^{t_1}\frac{1}{2}T(q^i(t), \dot{q}^i(t))dt, \qquad (40.13)$$

the non-commutativity rule defined by (Definition 8,(17.1)), we get the Newton equations (40.2) with the non-conservative forcers f_j.

Conditions of non-commutativity (k-twisted prolongations) of variations ξ^i of q^i and the time derivative have, in this situation, the form

$$\delta\dot{q}^i = \dot{\xi}^i + \frac{1}{2T}\dot{q}^i f_j\xi^j. \qquad (40.14)$$

Comparing the system (40.5) with obtained Euler-Lagrange system with the Lagrangian (40.4) and the NC-variations (40.14) we see that system (40.5) has the form (6.4) with the NC-tensor

$$K^i_{\cdot j} = \frac{\dot{q}^i}{2T}f_j = \frac{\dot{q}^i}{\|\dot{q}\|^2}f_j. \qquad (40.15)$$

Force	Tensor K
$f_j = f_j(q^k)$	$K^i_{\cdot j} = \frac{\dot{q}^i}{\|\dot{q}\|^2} f_j(y)$
$f_j = a_{ji}\dot{q}^i$	$K^i_{\cdot j} = \frac{a_{jk}\dot{q}^k \dot{y}^i}{\|\dot{q}\|^2}$
$f_j = a_{jkl}\dot{q}^k \dot{q}^l$	$K^i_{\cdot j} = \frac{\dot{q}^i}{\|\dot{q}\|^2} a_{jkl}\dot{q}^k \dot{q}^l.$
$m = 1, \ f = a\dot{y}^p$	$K_{\cdot} = a\dot{q}^{p-1}.$

TABLE 1. Tensor K for different forces $\{f_j\}, j = 1, \ldots, m$.

Using the expression $\pi_i = g_{ij}\dot{q}^j$ we see that the non-commutativity tensor $K^i_{\cdot j}$ here *has the canonical form* (9.8).

Finally, comparing expression (40.7) for the torsion T of connection Γ with the expression for tensor K we find that

$$T^i_{jk} = \frac{1}{2}\left[\frac{K^s_j}{\dot{q}^s}\delta^i_k - \frac{K^s_k}{\dot{q}^k}\delta^i_j\right]. \tag{40.16}$$

Notice that in this formula there are no summation by s.

This representation establishes the direct relation of non-commutativity tensor K with the connection Γ.

At the same time we have the expression of tensor R^i_{jk} participating in the definition of "canonical" connection Δ

$$R^i_{jk} = -\frac{f^i}{2T}g_{jk} = \frac{K^s_p}{\dot{q}^s}g^{ip}_{jk}. \tag{40.17}$$

In particular, we notice that the three conditions: $f_j = 0$, $K^i_j = 0$ and $\Delta = \Gamma =^g \Gamma$ **are equivalent**.

In the next table we present the form of tensors K of non-commuting variations for different types of forces f_j in Newton equations 40.2.

Example 26. ([134]) The following example demonstrates the application of the NC-variations to a simple problem of the type considered in this section where the extremal of variational problem with the non-commuting variations delivers the minimum of action.

Consider the following nonlinear equation

$$\ddot{x} = -k\dot{x}^3, \tag{40.18}$$

with the initial conditions $x(0) = 0, \ x(T) = A$.

Solution of the equation with these initial conditions is

$$x(t) = \frac{1}{1}v_0 k\left[\sqrt{1 + 2v_0^2 kt} - 1\right], \tag{40.19}$$

and $A = x(T)$.

Consider a variation of solution x(t): $y(t) = x(t) + h(t)$. Thus, $\delta x = h, \frac{d}{dt}\delta x = h(t)$. For the variation of velocity we have

$$\dot{x} = \frac{v_0}{\sqrt{1 + 2v_0^2 kt}} + \dot{\beta}(t),$$

and

$$\delta\dot{x} = \dot{\beta}(t).$$

According to the scheme of NC-variations, equation (40.18) can be obtained as the Euler-Lagrange equation with the Lagrangian $L = \frac{\dot{x}^2}{2}$ and the action $I(x) = \int_0^T \frac{\dot{x}^2}{2} dt$. The non-commutativity rule here has the form

$$\dot{x}\delta\dot{x} = \dot{x}d_t\delta x - k\dot{x}^3\delta x \Leftrightarrow \delta\dot{x} = d_t\delta x - k\dot{x}^2\delta x. \tag{40.20}$$

Using the expression for variations we have

$$\delta\dot{x} = \dot{\beta} = \dot{h} - \frac{kv_0^2}{\sqrt{1 + 2v_0^2 kt}}h. \tag{40.21}$$

for the varied motion we have

$$I + \delta I = \int_0^T \frac{\dot{y}^2}{2} dt = \int_0^T \left(\frac{kv_0^2}{\sqrt{1 + 2v_0^2 kt}} + \dot{\beta} \right) dt.$$

This gives

$$\delta I = \int_0^T \left(\frac{\dot{\beta}^2}{2} + \frac{2v_0\dot{\beta}}{\sqrt{1 + 2v_0^2 kt}} \right) dt.$$

Inserting expression (40.21) for $\dot{\beta}$ into the last expression and integrating by parts we get

$$\delta I = \frac{kv_0^2}{\sqrt{1 + 2v_0^2 kt}}\Big|_0^T + \int_0^T \frac{\dot{\beta}^2}{2} dt = \frac{1}{2}\int_0^T \frac{\dot{\beta}^2}{2} dt. \tag{40.22}$$

As a result, $\delta I > 0$ if $\dot{\beta} \neq 0$. Thus, comparing all the motions in the neighborhood of $x(t)$ with the solution $x(t)$ (see (40.18)) we see that $x(t)$ delivers the minimum of the action $I(y)$ in the presence of expression (40.21) for the velocity variations.

40.3. **Fundamental quadrilateral and non-commuting variations, see more.** Here we present the "fundamental quadrilateral" illustrating notion of non-commuting of variations in general situation and, in particular, in the studied here geometrization in nonconservative Mechanics. The motion of a holonomic nonconservative mechanical system can be considered as the motion of a point in the configurational space L. Let C_i be the path of a point M_i under the true motion while C_i^* - variated path. Mark positions M_i and M' of the point M_i at the moments t and t' respectively (see Fig 0). If r is the vector of the point M_i, then then the radius vector of the point M_i' is $r_i + dr_i$. Let M_i^* and $M_i^{*'}$ - be positions of the point on the variated system configurations at the same time instants.

Variation of the radius vector on the real path is $M_iM_i' = dr$ while on the variated path $M_i^*M_i^{*'} = dr' + d\delta r$. Here vector dr' is obtained by the parallel displacement of the vector dr from the point M_i to the point M_i^*. The variation of vector dr_i i.e. the vector δdr is defined in such a way that $M'M*' = \delta r + \delta dr$. From the figure 0 it is clear that the following vector equation is valid

$$\delta r_i + \delta r_i^* + d\delta r = dr_i + \delta r_i + \delta dr_i. \tag{40.23}$$

Introduce, in the configurational space L, a Riemannian metric $ds^2 = 2T(dt)^2 = g_{ab}dq^a dq^b$. Here ds is the length of the vector dr. Combining covariant derivatives defined by the connection (38.3) with this metric we obtain

$$d\mathbf{r}* = (1 + \frac{1}{2T}f_i\delta q^i)d\mathbf{r}.$$

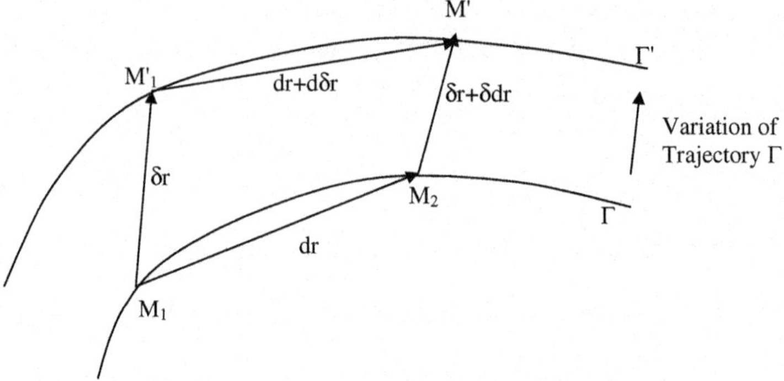

FIGURE 1. A sketch of fundamental quadrilateral.

Substituting this into the equation 38.23and using the relation

$$d\delta\mathbf{r} - \delta d\mathbf{r} = \frac{\partial\mathbf{r}}{\partial q^a}(d\delta q^a - \delta dq^a),\tag{40.24}$$

see [88], p.29, we get

$$\frac{\partial\mathbf{r}}{\partial q^a}(d\delta q^a - \delta dq^a + \frac{1}{2T}q^a f_b q^b) = 0$$

and, finally,

$$\delta dq^a = d\delta q^a + \frac{1}{2T}q^a f_b q^b.\tag{40.25}$$

This result shows that in the configurational space L with the connection 40.3 operator of variation δ and the operator of differentiation d **to not commute**. If the force here is absent $(= 0)$, commutativity will be restored.

41. Variational methods in the Nonholonomic Mechanics and Boltzmann connection.

In this section we mostly follow the "Analytical Dynamics" by E.T.Whittaker ([140]) published in 1904 (see Sec.30.), and the monographs [105, 88] to introduce the quasi-coordinates and formulation of Euler-Lagrange equations of a nonholonomic systems in quasi-coordinates. We will be using the notation $\{q^i\}$ for the dynamical variables of a mechanical system.

The birth of the nonholonomic mechanics and, more generally, nonholonomic dynamics occurred at the time when the variational formalism of Euler and Lagrange was found to be unapplicable to a variety of mechanical problems starting with the dynamics of rigid bodies rolling without slipping on a plane. Rolling systems attracted the attention of a variety of eminent scientists at the second part of XIX and the first half of XX century - Hamel, Herz, Zhukovsky, Appel, Chaplygin to name a few.

In mechanics the configurational bundle has the form of the direct product $Y = T \times Q \to T$, where T is the time axis and Q is the m-dim configurational space of the mechanical system. Denote by $q^\mu, \mu = 1, \ldots, m$, the generalized coordinates of a mechanical system. A kinematical constraint is the relation of the form

$$f_\sigma(\dot{q}^1, \ldots, \dot{q}^m;\ q^1, \ldots, q^m; t) = 0,\ \sigma = 1, \ldots, k, \tag{41.1}$$

that is imposed on the mechanical system.

Sometimes a system (41.1) of these constraints can be *integrated* to the equivalent system of relations of the form $\phi_\alpha(t, q^k) = 0$ where functions ϕ_α, defining obtained constraints, depend on the coordinates q^i, t but not on the velocities $\dot{q}^i$ (such constraints are called **geometrical**).

The system (41.1) of constraints is called **integrable** (or *holonomic*) if it can be reduced this way to the system of geometric constraints. Otherwise, it is called **non-integrable**.

In 1894, H.Herz introduced the distinction between two classes of constraints and the mechanical systems - he called the mechanical systems with non-integrable kinematical constraints (that can not be reduced to the geometrical constraints) - a **nonholonomic systems**.

It often happens that the non-integrable constraints encountered in Mechanics depend linearly on the generalized velocities $\dot{q}^i$.

Remark 57. It might happen that a non-integrable constraint can be transformed into equivalent geometrical (integrable one) using the other constraint equations of the same system. This means that the nonholonomicy is a property of the system of constraints of mechanical system as a whole, rather then the sum of properties of separate constraints.

Example 27. Consider the system of two constraints in R^3 with cartesian coordinates x, y, z)

$$\begin{cases} (x^2 + y^2)dx + xzdz = 0, \\ (x^2 + y^2)dy + yzdz = 0. \end{cases}$$

Each of these constraints is not integrable. Yet, taken together and using simple transformations one can transform this system into the integrable one

$$\begin{cases} (x^2 + y^2)d(x^2 + y^2 + z^2) = 0, \\ (d\ln\frac{x}{y} = 0. \end{cases}$$

A key property of nonholonomic mechanical system, making it impossible to use conventional variational formalism, is the presence of nonholonomic constraints imposing the restrictions that we take, for simplicity, to be linear by velocities

$$a_i^\alpha(q)\dot{q}^i = 0, \alpha = 1, \ldots, r. \tag{41.2}$$

Here $a_i^\alpha(q)$ are functions of (holonomic) generalized coordinates q^i.

We assume that r conditions (41.2) are linearly independent at all points $q \in Q$ and that $r \leqq m$.

It is convenient to consider these constraints as the *conditions of nullity* $\dot{\pi}^\alpha = 0$, $\alpha = 1, \ldots, r$ for new variables - **"quasi-velocities"**. In a case where $r = m$, quasi-velocities are defined by the equivalent relations

$$\dot{\pi}^\alpha = a_i^\alpha(q)\dot{q}^i, \; j = 1, \ldots, m \Leftrightarrow d\pi^\alpha = a_i^\alpha dq^i \leftrightarrow \dot{q}^i = b_\beta^i(q)\dot{\pi}^\beta. \tag{41.3}$$

Equalities (41.3) assume that the "matrix" a_i^α is non-degenerate and that

$$a_k^\alpha b_\beta^k = \delta_\beta^\alpha.$$

If $r < m$, in order to achieve the introduction of m quasi-velocities and perform the related non-holonomic transformation, we have to complete the set of covectors $d\pi^\alpha = a_i^\alpha(q)dq^i = a_i^\alpha(q)\dot{q}^i dt$ with $m - r$ linearly independent combinations of differentials dq^i (one can simply rearrange q^i and take $dq^j, j = (r + 1), \ldots, m$) so that the covectors $\dot{\pi}^\alpha, \alpha = 1, \ldots, r; dq^{r+1}, \ldots, dq^m$ form the **(non-holonomic) coframe** in Q.

Second of the conditions (41.3) can be interpreted as the realization, along the solutions $q(t) = \{q^i(t)\}$, of the non-holonomic transformation of the cotangent bundle $T^*(Q) \to Q$:

$$d\pi^\alpha = a_i^\alpha dq^i, \tag{41.4}$$

i.e., performance of the non-holonomic transformation of the cotangent bundle $T^*(Q) \to T$.

Unless the integrability conditions

$$a^\beta_{i,q^j} = a^\beta_{j,q^i}, i \neq j, \tag{41.5}$$

are fulfilled, functions $\dot{\pi}^\alpha$ ARE NOT the time derivatives of any functions $\pi^\alpha(q)$. In such a case, there is no change from coordinates q^i to new coordinates π^α whose time derivatives are defined by (41.3). Yet, we will be using notation π^α having in mind that only their "differentials" (and their derivatives - variations, dual to the "differentials" (see below)) are defined by (41.3) as the **coframe** in the space Q.

Remark 58. In some situations one can define the quasi-coordinates as follows:

$$\pi_i = \int_0^t \dot{\pi}_i dt.$$

In general this definition is formal, it is practical if one can solve for dynamical variables.

From the relation (41.3) it follows that

$$\frac{\partial \dot q^i}{\partial \dot\pi^\alpha} = b^i_\alpha \equiv \frac{\partial q^i}{\partial \pi^\alpha}, \tag{41.6}$$

where the right equality is the definition of $\frac{\partial q^i}{\partial \pi^\alpha}$. More generally **we define**, for a function f,

$$\frac{\partial f}{\partial \pi^\alpha} \equiv^{def} \frac{\partial f}{\partial q^i}\frac{\partial q^i}{\partial \pi^\alpha} = b^i_\alpha \frac{\partial f}{\partial q^i}. \tag{41.7}$$

Variations of coordinates q^i and of the quasi-coordinates π^α are related by the condition of the form (41.3):

$$\delta q^s = b^s_\alpha \delta \pi^\alpha.$$

In this relation we recognize the **nonholonomic transformation of the tangent bundle** $T(Q) \to Q$.

We assume that the mechanical system is *under no action of any forces* and, therefore, **its Lagrangian coincide with the kinetic energy**: $L = T(q, \dot q^i)$. Corresponding Euler-Lagrange equations (in the D'Alambert-Lagrange form) (see [140],Sec.30), are

$$(d_t T_{,\dot q^i} - T_{,q^i})\delta q^i = 0. \tag{41.8}$$

Next, we transform Euler-Lagrange equations to the *non-holonomic coordinates* i.e. we write down EL equations in terns of $q^i, \dot\pi^\alpha$ and the "formal" derivatives (39.3-4). Notice that, in contrast to the variables $(q^i, \dot q^i)$, the variables $q^i, \dot\pi^\alpha$ are **functionally independent**.

Replacing the variations we write the last equation in the form

$$(d_t T_{,\dot q^i} - T_{,q^i})b^i_\alpha \delta\pi^\alpha = 0 \Leftrightarrow (d_t T_{,\dot q^i} - T_{,q^i})b^i_\alpha = 0, \ \forall\alpha, \tag{41.9}$$

due to the independence of variations $\delta\pi^\alpha$.

Write the kinetic energy as the function of coordinates q^i and quasi-velocities $\dot\pi^\alpha$:

$$T^*(q^i, \dot\pi^\alpha) = T(q^i, \dot q^i)|_{\dot q^i = b^i_\alpha(q)\dot\pi^\alpha}.$$

We have $T_{,\dot q^i} = T^*_{,\dot\pi^\alpha} a^\alpha_i$. Using this in (41.9), calculating the total time derivative of the functions a^α_i and substituting $\dot q^m = b^m_\beta \dot\pi^\beta$, we get

$$d_t T_{,\dot q^i} = a^\alpha_i d_t \frac{\partial T^*}{\partial \dot\pi^\alpha} + T^*_{,\dot\pi^\alpha}\frac{\partial a^\alpha_i}{\partial q^j}b^j_\sigma \dot\pi^\sigma.$$

Using this, we transform (41.9) to

$$d_t(T^*_{,\dot\pi^\sigma}) + \frac{\partial a^\alpha_i}{\partial q^m}b^m_\beta b^i_\sigma \dot\pi^\beta T^*_{,\dot\pi^\alpha} - b^i_\sigma T_{,q^i} = 0, \sigma = 1\ldots,m. \tag{41.10}$$

Then, we have

$$\frac{\partial T}{\partial q^i} = \frac{\partial T^*}{\partial q^i} + \frac{\partial T^*}{\partial \dot\pi^\beta}\frac{\partial \dot\pi^\beta}{\partial q^i} = \frac{\partial T^*}{\partial q^i} + \frac{\partial T^*}{\partial \dot\pi^\beta}\frac{\partial a^\beta_k}{\partial q^k}\dot q^k = \frac{\partial T^*}{\partial q^i} + \frac{\partial T^*}{\partial \dot\pi^\beta}\frac{\partial a^\beta_k}{\partial q^k}b^k_\delta \dot\pi^\delta.$$

Substituting this into the equations (41.8) we obtain the **Euler-Lagrange equations for non-holonomic mechanical system in quasi-coordinates**(see [88, 105])

$$d_t T^*_{,\dot\pi^\beta} - T^*_{,\pi^\beta} + \gamma^\sigma_{\alpha\beta}T^*_{,\dot\pi^\sigma}\dot\pi^\alpha = 0 \Leftrightarrow d_t T^*_{,\dot\pi^\beta} - T^*_{,\pi^\beta} = -\gamma^\sigma_{\alpha\beta}T^*_{,\dot\pi^\sigma}\dot\pi^\alpha = -\gamma^\sigma_{\alpha\beta}\tilde p_\sigma \dot\pi^\alpha, \ a = 1,\ldots,m, \tag{41.11}$$

where

$$\gamma^\sigma_{\alpha\beta} = b^s_\alpha b^l_\beta \left(\frac{\partial a^\sigma_s}{\partial q^l} - \frac{\partial a^\sigma_l}{\partial q^s} \right) \tag{41.12}$$

is the "**Boltzmann tensor**", $[8, 88, 105, 140]$ and $\tilde{p}_\sigma$ are components of momentum in the quasi-coordinates $\tilde{p}_\sigma = T^*_{,\dot\pi^\sigma}$.

Obtained system of equations 41.11 **has the form of an Euler-Lagrange system (in the non-holonomic coordinates) with the non-commuting variations.** For the NC-tensor we have

$$\frac{\delta T^*}{\delta \pi^\beta} = -\gamma^\sigma_{\alpha\beta} T^*_{,\dot\pi^\sigma} \dot\pi^\alpha = -K^\sigma_\beta p_\sigma.$$

Here $p_\sigma = T^*_{,\dot\pi^\sigma}$ are the momenta in (non-holonomic) quasi-coordinates. As a result, NC-tensor K has the form

$$K^\sigma_\beta = \gamma^\sigma_{\alpha\beta} \dot\pi^\alpha. \tag{41.13}$$

Now we **define** the zero curvature connection (that could be naturally called the "**Boltzmann connection**") in the space Q as follows:

$$\Gamma^k_{ij} = b^k_\alpha \frac{\partial a^\alpha_i}{\partial x^j}. \tag{41.14}$$

From this it follows that $\frac{\partial a^\alpha_i}{\partial x^j} = a^\alpha_k \Gamma^k_{ij}$.

Substituting this into the formula for $\gamma^\sigma_{\alpha\beta}$ we get

$$\gamma^\sigma_{\alpha\beta} = b^i_\alpha b^j_\beta a^\sigma_k (\Gamma^k_{ij} - \Gamma^k_{ji}) = 2 b^i_\alpha b^j_\beta a^\sigma_k S^k_{ij}, \tag{41.15}$$

where

$$S^k_{ij} = \frac{1}{2} \left(b^k_\alpha \frac{\partial a^\alpha_i}{\partial x^j} - b^k_\alpha \frac{\partial a^\alpha_j}{\partial x^i} \right). \tag{41.16}$$

is the **torsion of the AP-connection Γ defined by the non-holonomic frame** $\delta\pi^\alpha$ **(41.3)** (see Sec.38).

Remark 59. Connection Γ and its torsion are defined naturally in the holonomic frame, see (Sec.67 in Appendix I). At the same time, Boltzmann tensor γ is defined in the non-holonomic frame and (up to the constant 2) coincide with the torsion tensor in the non-holonomic frame.

From (41.12) and (41.15) we obtain the following relation of the torsion S of the AP-connection Γ and the NC-tensor K:

$$K^\sigma_\beta = 2 b^i_\alpha b^j_\beta a^\sigma_k S^k_{ij} \dot\pi^\alpha. \tag{41.17}$$

There are situations where non-holonomic relations (41.1) are such that *for some values of β* commutativity relations (41.4) are valid for all i, j. If this is true for a fixed value β_0 of index β. Then, the corresponding quasi-coordinate π^{β_0} is a function of generalized coordinates q^i and the corresponding quasi-velocity - time derivative $\dot\pi^{\beta_0} = \frac{d\pi^{\beta_0}}{dt}$.

In this case, $\gamma^{\beta_0}_{\alpha\sigma} = 0$ for all values of α, σ, see (41.11-12). Acting on (41.12) (in the form $\gamma^{\beta_0}_{\alpha\sigma} = 2 b^i_\alpha b^j_\sigma a^{\beta_0}_k S^k_{ij}$) with $a^\alpha_s a^\sigma_l$ we get the equivalent relation in terms of torsion S:

$$\delta^i_s \delta^j_l a^{\beta_0}_k S^k_{ij} = 0 \Leftrightarrow a^{\beta_0}_k S^k_{sl} = 0 \tag{41.18}$$

for all s, l. Thus, the covector $a^{\beta_0}_k$ belongs to the kernel of the torsion (1,2)-tensor S^k_{ij}.

Remark - Exercise.

Equalities (41.3-4) allow us to define derivatives of functions by non-holonomic coordinates as the coefficients of decomposition of differential $d\phi$ by the non-holonomic coframe:

$$d\phi = \frac{\partial \phi}{\partial \pi^\alpha} d\pi^\alpha.$$

Continuing this, one can define the higher order derivatives by non-holonomic co-ordinates. We suggest as a nice exercise, to prove the following interesting formula for the commutation of second order derivatives

$$\frac{\partial^2 \phi}{\partial \pi^\alpha \partial \pi^\beta} - \frac{\partial^2 \phi}{\partial \pi^\beta \partial \pi^\alpha} = \gamma^\sigma_{\alpha\beta} \frac{\partial \phi}{\partial \pi^\sigma}. \tag{41.19}$$

Example 28. To illustrate described Lagrangian approach to the non-holonomic mechanical systems, we present the following example.

Consider a circular disc of radius R with a sharp age that rolls on the xy-plane without sliding. Dynamical variables describing the rolling of this disk (see Picture ()) are: two linear variables - coordinates x, y of the point M of the contact of disc with the plane and three angles: angle ψ measured from a chosen point of the rim to the point M of contact, angle ϕ between the tangent to the disc at the point M and the Ox axis; and the angle of inclination θ between the plane of the disc and the plane of rolling.

The disc rolls without sliding and, as a result, variations of five variables can not be arbitrary. More specifically, coordinates x, y of the point of contact can be arbitrary. We refer to [105], Sec.1.1. for an explanation of how the disc can be brought from any fixed position $(x_0, y_0, \psi_0, \phi_0, \theta_0)$ to ANY prescribed position.

Geometry of the rolling shows that disc has to satisfy two conditions

$$\begin{cases} dx = Rcos(\phi)d\psi, \\ dy = Rsin(\phi)d\psi. \end{cases} \tag{41.20}$$

Dividing by dt we get, from the non-slicing condition, two **kinematical constraints**

$$\begin{cases} \dot{x} = R\dot{\psi}cos(\phi), \\ \dot{x} = R\dot{\psi}sin(\phi) \end{cases} \tag{41.21}$$

These constraints do not impose any conditions on the values of dynamical variables, but to the values of velocities. Such conditions (first recognized by Hertz in 1894 as different from the "holonomic constraints") were called **non-holonomic** .

Introducing quasi-velocities $\dot{\pi}_1 = \dot{x} - Rcos(\phi)\dot{\psi}$, $\dot{\pi}_2 = \dot{y} - Rsin(\phi)\dot{\psi}$ and introducing three quasi-coordinates $\pi_3 = \phi, \pi_4 = \phi, \pi_5 = \theta$ we can use arguments presented in this section.

In particular, we get for equality $\dot{\pi}^\alpha = a_i^\alpha \dot{q}^i$,

$$a_i^\alpha \cdot \langle \dot{q}^i \rangle = \begin{pmatrix} 1 & 0 & -Rcos(\phi) & 0 & 0 \\ 0 & 1 & -Rsin(\phi) & 0 & 0 \\ 0 & 0 & 1 & 0 & 0 \\ 0 & 0 & 0 & 1 & 0 \\ 0 & 0 & 0 & 0 & 1 \end{pmatrix} \cdot \begin{pmatrix} \dot{x} \\ \dot{y} \\ \dot{\psi} \\ \dot{\phi} \\ \dot{\theta} \end{pmatrix}. \tag{41.22}$$

Inverse tensor has the form

$$
b^i_\beta = \begin{pmatrix}
1 & 0 & R\cos(\phi) & 0 & 0 \\
0 & 1 & R\sin(\phi) & 0 & 0 \\
0 & 0 & 1 & 0 & 0 \\
0 & 0 & 0 & 1 & 0 \\
0 & 0 & 0 & 0 & 1
\end{pmatrix}
\tag{41.23}
$$

It follows that only nonzero components $\Gamma^i_{j\phi}$ of the zero curvature connection (41.14) are two components given here:

$$
\Gamma^i_{j\phi} = \begin{pmatrix}
0 & 0 & R\sin(\phi) & 0 & 0 \\
0 & 0 & -R\cos(\phi) & 0 & 0 \\
0 & 0 & 0 & 0 & 0 \\
0 & 0 & 0 & 0 & 0 \\
0 & 0 & 0 & 0 & 0
\end{pmatrix} .
\tag{41.24}
$$

i.e. only two Christoffel coefficients are nonzero: $\Gamma^1_{3\phi} = R\sin(\phi)$, $\Gamma^2_{3\phi} = -R\cos(\phi)$.

For the torsion tensor $S^i_{jk} = \frac{1}{2}(\Gamma^i_{jk} - \Gamma^i_{kj})$ the only non-zero coefficients are (using the $\phi = \pi_4$),

$$
S^1_{34} = \frac{R}{2}\sin(\phi), \; S^1_{43} = -\frac{R}{2}\sin(\phi); \; S^2_{34} = \frac{-R}{2}\cos(\phi), \; S^2_{43} = \frac{R}{2}\cos(\phi).
$$

We refer the reader to the monographs [88, 105] for more detailed analysis of the role of non-commuting variations in the Non-holonomic Mechanics and for more examples of calculations and physical meaning of Boltzmann tensor γ (and of the related NC-tensor K).

42. Gauge transformations, torsion and H.Kleinert's Action Principle.

In this section we present, with some modifications, the scheme developed by H.Kleinert and his collaborators (P.Fiziev, A.Pelster to name a few) by using a non-holonomic transformations in constitutional space in order to formulate an Action principle in a Cartan space-time. This new action principle for the motion of a particle in the Cartan space-time (**space-time** had to take into account the *curvature and torsion* in the space endowed with the metric and an affine connection. **Anholonomic** (gauge, nonholonomic) transformations in the space of fields (in the fibers Y_x of the configurational bundle $\pi : Y \to X$, in our terminology) allows us to construct both metric and zero curvature connection ω in the way similar to the construction of material metric and material connection in the theory of uniform materials, see [22]. Lifting these transformations to the 1-jet space allows to get the form of variation of the fields and their derivatives that differs from the standard by the terms defined by the torsion of connection ω.

We formulate this scheme for the case of Field Theory using the language of jet bundles adopted in this work (see Appendix II).

42.1. Connection defined by an automorphism.

Let $\pi : y \to X$ be a configurational bundle over the base (typically a space-time) X.

We assume that a metric $x \to h_x$ is defined on the fibers Y_x of the bundle $\pi : Y \to X$. If π is a *natural bundle*,(ref), metric g on the base X defines the metric g_x on the fibers Y_x smoothly depending on x, [33]. Metric g_x defines the torsion-free Levi-Civita connections θ_x on the fibers Y_x, i.e., on the **vertical bundle** $T(Y_x) \simeq V_x(\pi) \to Y_x$.

Let now $y \to D_y$ be a **gauge automorphism of the vertical bundle** $\mathcal{V}(\pi)$ (thus, projecting to the identity diffeomorphism of Y). Tensor field $D(y)$ allows us to define another connection θ_D on the vertical bundle $\mathcal{V}(\pi) \to Y$. This connection is defined by the condition that if a frame $f_x = \{f_i(x, y) \in V_{x,y}\}$ on Y_x is h_x-orthonormal (and, therefore, θ-**horizontal**), then we declare the field of frames $\{\zeta_\alpha\} = \{D_\alpha^i f_i(x, y) \in V_{x,y}\}$ on Y_x to be θ_D-**horizontal**.

It is easy to check that this definition defines the connection θ_D correctly, i.e. independently on the choice of the frame **f**.

If the original metric h_y is flat (LC-connection θ has zero curvature), the same is true about connection θ_D.

Connection θ_D has zero curvature (being defined by the global frame, see (appendix, Sec.())) but has, in general, the non-zero torsion. Torsion tensor of the connection θ_D has, in the non-holonomic frame ζ_a, the form ([22])

$$S_{bc}^a = D_l^a \zeta_b \cdot D_c^l - D_l^a \zeta_c \cdot D_b^l. \tag{42.1}$$

42.2. Automorphism and the prolongation of variations.

Let D is a non-degenerate (1,1)-tensor field defining the non-holonomic transformation (authomorphism) of the vertical bundle $\mathcal{V}(\pi) : V(\pi) \to Y$. We interpret this transformation as the change of basis in the vertical spaces V_y, $y \in Y$ from the basis ∂_{y^k} induced by the local fibred chart (x^μ, y^k) to the non-holonomic frame ζ_a:

$$\zeta_a = D_a^k \partial_{y^k} \Leftrightarrow \partial_{y^k} = D_k^a \zeta_a.$$

Basis ζ_i in physical literature is identified with the derivatives ∂_{w^a} by the ("non existent") non-holonomic coordinates w^i. In what follows we will be using only frame δw^i and coframe dw^i dual to the frame $\delta w^i = \zeta_a$ but not the coordinates w^i themselves.

In terms of components of a vertical vector $v \in V_y(\pi)$,

$$v = \xi^k \partial_{y^k} = \eta^a \zeta_a,$$

where we using the notations similar to notations used in the works of H.Kleinert and his collaborators (): $\xi^k = \delta y^k$, $\eta_a = \delta w^i$ this transformation has the form

$$\delta y^k = D_a^k \delta w^i \Leftrightarrow \xi^k = D_a^k \eta^a. \tag{42.2}$$

In the dual space $V(\pi)_y^*$, we have the dual transformation

$$dy^k = D_a^k dw^a \Leftrightarrow^K y_{,\mu}^k = D_a^k(w) w_{,\mu}^a, \quad k, i = 1, \ldots, m. \tag{42.3}$$

Here dw^a are (basic) vertical 1-forms dual to the vector fields ζ^b: $\langle dw^a, \zeta^b \rangle = \delta_a^b$. Quantities $w_{,\mu}^a$ are coefficients in the decomposition of *basic covector fields* dw^a by the base coframe dx^μ: $dw^a = w_{,\mu}^a dx^\mu$. (This decomposition is valid and has sense if one take the pullback of this equality with any section of the bundle π). Symbols $w_{,\mu}^a$ play
the role of *"non-holonomic 1-jet variables"*.

We can collect introduced objects and correspondence of holonomic and non-holonomic quantities in the following form. Sign $\leftrightsquigarrow$ in this list means correspondence between holonomic and non-holonomic quantities.

$$\begin{cases} y^i \leftrightsquigarrow w^a, \\ \partial_{y^i} \leftrightsquigarrow \zeta_a = D_a^k \partial_{y^k}, \\ dy^i \leftrightsquigarrow dw^a = D_i^a dy^i, \\ For\ a\ vector\ field\ v = \xi^i \partial_{y^i} = \eta^a \zeta_a, \\ \zeta_a = D_a^k \partial_{y^k} \Leftrightarrow \partial_{y^j} = D_j^b \zeta_b, \\ \xi^k = \delta y^k \leftrightsquigarrow \eta^a = \delta w^a = D_i^a \xi^i = D_i^a \delta y^i. \end{cases}$$

In the calculations presented below we prefer to use physical rather then mathematical notations; their usage makes the meaning of calculations more transparent. In these calculations components D_i^k are considered as functions of variables w^a, or, more exactly, their variations and derivatives by x^μ are written through dw^a and δw^a.

Taking total derivative by μ in (42.2) we get

$$d_\mu \delta y^k = D_a^k d_\mu \delta w^a + (\partial_{w^b} D_a^k) w_{,\mu}^b \delta w^a. \tag{42.4}$$

Quantities $\partial_{w^b} D_a^k$ here are coefficients in the decomposition $dD_a^k = \partial_{w^b} D_a^k dw^b$ of the differential of functions D_a^k with respect to the non-holonomic coframe dw^b.

Taking variation in (42.3) we get, in components by dx^μ,

$$\delta y_{,\mu}^k = \delta(D_a^k(w) w_{,\mu}^a) = D_a^k \delta w_{,\mu}^a + (\partial_{w^b} D_a^k(w)) \delta w^b w_{,\mu}^a. \tag{42.5}$$

Subtract the last two formulas

$$d_\mu \delta y^k - \delta y_{,\mu}^k = D_a^k(d_\mu \delta w^a - \delta w_{,\mu}^a) + [(\partial_{w^b} D_a^k) w_{,\mu}^b \delta w^a - (\partial_{w^b} D_a^k) w_{,\mu}^a \delta w^b)].$$

Apply the inverse transformation D_k^c to this equality and moving the last two terms in the right to the left we get

$$d_\mu \delta w^c - \delta w^c_{,\mu} = D_k^c(d_\mu \delta y^k - \delta y^k_{,\mu}) - \left[D_k^c(\partial_{w^b} D_a^k) w^b_{,\mu} \delta w^a - D_k^c(\partial_{w^b} D_a^k) w^a_{,\mu} \delta w^b \right].$$

$$(42.6)$$

Let us rewrite this formula in more formal, mathematical, notation:

$$d_\mu \eta^c - \delta w^c_{,\mu} = D_k^c(d_\mu \xi^k - \delta y^k_{,\mu}) - \left(D_k^c(\partial_{w^b} D_a^k) w^b_{,\mu} \eta^a - D_k^c(\partial_{w^b} D_a^k) w^a_{,\mu} \eta^b \right). \quad (42.7)$$

Let us now use *the conventional rule for the prolongation of variation*: $\xi = \xi^k \partial_{y^k}$ for the field variations **in the holonomic frame** ∂_{y^k} . Then, the first term in the right side vanishes and the formula becomes

$$\delta w^c_{,\mu} = d_\mu \eta^c + \left[D_k^c(\partial_{w^b} D_a^k) w^b_{,\mu} \eta^a - D_k^c(\partial_{w^b} D_a^k) w^a_{,\mu} \eta^b \right] =$$

$$= d_\mu \eta^c + \left[D_k^c(\partial_{w^b} D_a^k) - D_k^c(\partial_{w^a} D_b^k) \right] w^b_{,\mu} \eta^a = d_\mu \eta^c + 2S^c_{ab} w^b_{,\mu} \eta^a. \quad (42.8)$$

Expression

$$S^c_{ab} = \left[D_k^c(\partial_{w^b} D_a^k) - D_k^c(\partial_{w^a} D_b^k) \right] \quad (42.9)$$

represent the **torsion of the zero curvature (AP) connection defined by the (1,1)-tensor field** D_a^k that corresponds to the automorphism D of vertical bundle $V(\pi) \to Y$, see (24.1).

In terms of vector fields on the bundle $J^1(\pi) \to X$ defining variation of variables w^a and their derivatives w^a_μ the last result has the form

$$\eta^{1D} = \eta^b \partial_{w^b} + \eta^c_\mu \partial_{w^c_\mu} = \eta^b \partial_{w^b} + \left(d_\mu \eta^c + S^c_{ba} w^b_\mu \eta^a \right) \partial_{w^c_\mu}. \quad (42.10)$$

This result is the geometrical multidimensional form of the *formula for variations after a non-holonomic transformation of the vertical bundle* $V(\pi)$ of H. Kleinert.

To determine the variations of derivatives y^i_μ we switch in the formula (40.5) to the holonomic frame ∂_{y^i}. Using transition formulas (42.8) we get (recall that $\delta y^i = \xi^i$)

$$\delta y^i_\mu = d_\mu \xi^i + 2S^i_{jk} y^j_\mu \xi^k, \quad (42.11)$$

where

$$S^i_{jk} = \frac{1}{2}(\Gamma^i_{jk} - \Gamma^i_{kj}) = \frac{1}{2}(D_c^i \partial_{y^j} D_k^c - D_c^i \partial_{y^k} D_j^c). \quad (42.12)$$

Notice that the variations of derivatives y^i_μ have the form (42.11) with the NC-variations tensor $K^i_{j\mu} = 2S^i_{kj} y^k_\mu$ being defined by the torsion of connection θ_D.

Return now to the notations of the main body of the paper with the fields denoted by $y^i(x)$ and their derivatives - by y^i_μ. Let L be a Lagrangian of the first order (function of x^μ, y^i, y^i_μ). Then the Euler-Lagrange equations (6.3) with the tensor (42.11) of non-commutativity have the form

$$\frac{\partial L}{\partial y^j} - d_\mu \left(\frac{\partial L}{\partial y^j_\mu} \right) = -2S^i_{kj} y^k_{,\mu} \frac{\partial L}{\partial y^i_\mu}, \quad j = 1, \ldots, m. \quad (42.13)$$

Thus, the action of the gauge automorphism θ_D in the vertical bundle $V(\pi) \to Y$ introduces the sources (forces in mechanics) defined by the torsion of the corresponding zero curvature connection θ_D and proportional to the momenta $\pi^\mu_i = \frac{\partial L}{\partial y^i_\mu}$.

Corresponding energy-momentum balance laws has, for the canonical EM-tensor $T^\mu_\sigma = L\delta^\mu_\sigma - (\Gamma^i_\sigma - y^i_\sigma)L_{,y^i_\mu}$, the form

$$d_\mu T^\mu_\sigma = -L_{,x^\sigma} - (\Gamma^i_\sigma - y^i_\sigma) 2y^k_\mu S^j_{ki} L_{,y^j_\mu}, \quad \mu = 0, 1, 2, 3. \quad (42.14)$$

Example 29. Model geometrical example ([67], Sec.10.2). Consider the 2-dim space $\mathbb{R}^2$ with a Riemannian metric g_{ij} and global coordinates x^1, x^2. A (classical) point is moving in this space under the action with the Lagrangian $L(x^i(t), \dot{x}^i(t)) = \frac{M}{2} \sum_i (g_{ij} \dot{x}^i \dot{x}^j)$ and the action $A(x(t)) = \int_{t_1}^{t_2} L dt$.

Now we state that the motion in the space with curvature (induced by the metric g) and the torsion can be described by non-holonomic change of coordinates $x^i \to q^a$ in the space $\mathbb{R}^2$.

This non-holonomic coframe is defined by the transformation

$$\begin{cases} dy^1 = dw^1, \\ dy^2 = dw^2 + \epsilon \frac{\partial \phi(w)}{\partial w^k} dw^k. \end{cases}$$

Second term in the second line is defined as the decomposition of 1-form $d\phi$ in terms of dw^i.

In matrix form

$$\begin{pmatrix} dy^1 \\ dy^2 \end{pmatrix} = \begin{pmatrix} 1 & 0 \\ \epsilon\phi_{,w^1} & 1 + \epsilon\phi_{,w^2} \end{pmatrix} \begin{pmatrix} dw^1 \\ dw^2 \end{pmatrix}, \qquad (42.15)$$

where $\phi(w) = tan^{-1}(\frac{w^2}{w^1})$. Substituting ϕ to () we get

$$\begin{pmatrix} dy^1 \\ dy^2 \end{pmatrix} = (D_a^k) \begin{pmatrix} dw^1 \\ dw^2 \end{pmatrix} = \begin{pmatrix} 1 & 0 \\ \epsilon\frac{-w^2}{(w^1)^2+(w^2)^2} & 1 + \epsilon\frac{w^1}{(w^1)^2+(w^2)^2} \end{pmatrix} \begin{pmatrix} dw^1 \\ dw^2 \end{pmatrix}. \qquad (42.16)$$

Introduce the notation $(D_a^k) = \begin{pmatrix} 1 & 0 \\ A & B \end{pmatrix}$. Then, for the inverse matrix we have

$$(D_k^c) = \begin{pmatrix} 1 & 0 \\ -\frac{A}{B} & B^{-1} \end{pmatrix}.$$

AAA Substituting this to the expression (40.9) for the torsion tensor we get

$$S_{ab}^c = D_1^c \left(\partial_{w^b} D_a^1 - \partial_{w^a} D_b^1 \right) + D_2^c \left(\partial_{w^b} D_a^2 - \partial_{w^a} D_b^2 \right). \qquad (42.17)$$

In 2-dim space, the torsion tensor looks as follows:

$$S = S_{ab}^c \partial_{w^c} \otimes dw^a \wedge dw^b = S_{12}^c \partial_{w^c} \otimes dw^1 \wedge dw^2. \qquad (42.18)$$

Thus, the torsion has two (apriori) non-zero components: $S_{12}^c, c = 1, 2$. Direct calculation (comp. [67], Sec.10.2) shows that the torsion is concentrated at the origin of w plane: $S_{12}^c = \epsilon\pi\delta(w)$, where $\delta(w)$ is the delta-function in the w-plane.

Remark 60. Non-holonomic transformation above defining the non-holonomic coframe is related to the construction of Burgers tensor in the dislocation theory, see [68],Sec.10.2.

42.3. Kustaanheimo-Stiefel transformation carries the Kepler-Coulomb problem to the Harmonic Oscillator. , [35], or [67],Sec.13.4.

We consider the Kepler problem in celestial mechanics. Here a body is moving in $X = \mathbb{R}^3$ with coordinates $x^i, i = 1, 2, 3$ and the radius-vector $\mathbf{r}(t)$. The Lagrangian is $L(x^i, \dot{x}^i) = \frac{1}{2}\|\dot{x}\|^2 + \frac{\alpha}{r}$, where $r = \sqrt{\sum_i x^{i^2}}$) and $\alpha = const$. Dynamical equations for the motion of the body are:

$$m\ddot{\mathbf{r}} + \frac{\alpha \mathbf{r}}{r^3} = 0.$$

Now we perform the Kustaanheimo-Stiefel transformation (taking "quaternionic square root") which transforms the Kepler Problem to the 4-dim Harmonic oscillator in $\mathbb{R}^4$ with the coordinates u^i. We introduce the 4-dim Euclidian vector space with the coordinates $x^i, i = 1, \ldots, 4$ and identify the original 3-dim space with the hyperplane in $\mathbb{R}^4$ defined by the condition (invariant constraint) $x^4 = 1$.

We define the non-holonomic transformation $dx^i = e^i_\mu(u)du^\mu$ setting for x^1, x^2, x^3:

$$\begin{cases} x^1 = 2(u^1 u^3 + u^2 u^4), \\ x^2 = 2(u^1 u^4 + u^2 u^3), \\ x^3 = (u^1)^2 + (u^2)^2 - (u^3)^2 - (u^4)^2, \end{cases} \tag{42.19}$$

for $\vec{u} = \begin{pmatrix} u^i \\ u^2 \\ u^3 \\ u^4 \end{pmatrix} \in \mathbb{R}^4.$

Next we introduce the dummy variable x^4 embedding the original 3-dim space $\mathbb{R}^3_x$ into the 4-dim space R^4 as the affine subspace $x^4 = const$. Then, introduce the *non-holonomic transformation* extending the mapping $R^4_u \to \mathbb{R}^3_x$ introduced above.

$$A(u) = \|e^i_\mu(u)\| = \begin{pmatrix} u^3 & u^4 & u^1 & u^2 \\ u^4 & -u^3 & -u^2 & u^1 \\ u^1 & u^2 & -u^3 & -u^4 \\ u^2 & -u^1 & u^4 & -u^3, \end{pmatrix} \tag{42.20}$$

Use the notation $\bar{d}x^k$ for the non-holonomic frame defined by (42.19) in the space $\mathbb{R}^4_x$. In particular, $\bar{d}x^4 = u^2 du^1 - u^1 du^2 + u^4 du^3 - u^3 du^4$. It is clear that $d(\bar{d}x^i) = 0$ for $i = 1, 2, 3$. but that $d(\bar{d}x^4) \neq 0$ and ,as a result, transformation $u \to x$ is not a coordinate but only non-holonomic transformation.

As a result, this non-holonomic transformation transforms the Euclidian geometry in $\mathbb{R}^4_x$ to the non-Euclidian geometry in $\mathbb{R}^4_u$ with curvature and torsion. Space $\mathbb{R}^4_u$ get the connection Γ with the torsion The only non-zero components of the torsion tensor are $S^1_{\mu 2} = S^3_{\mu 4} = 4 \left(u^2 \quad -u^1 \quad u^4 \quad -u^3 \right)$.

Introducing forth coordinate to the Lagrangian:

$$\hat{L} = L(x, \dot{x}) + \frac{m}{2}(\dot{x}^4)^2,$$

we get the new 4-dim problem in x^i coordinates that will be equivalent to the original Kepler problem if one enforces the constraint $x^4 = const$.

Introduced mapping sends the Lagrangian $\hat{L}$ to the Lagrangian in 4-dim space R^4_u:

$$\hat{L}(u, \dot{u}) = 2m\| \mathbf{u}\|^2 \cdot \|\dot{\mathbf{u}}\|^2 + \frac{\alpha}{\|\mathbf{u}\|^2}.$$

One can produce the correct dynamical equations in the space $\mathbb{R}^4_u$ if we are using the modification of the variational principle defined above with the help of the torsion of the connection defined by the non-holonomic coframe $\bar{d}x^i$. Variation of action $\mathcal{A}(u)$ leads to the equation

$$\frac{\delta A}{\delta u^\mu} + F_\mu = 0, \tag{42.21}$$

with the torsion force $F_\mu = m\dot{x}^4 \dot{S}^1_{\mu 2}$,[67]).

42.4. Motion of a point in Cartan space-time (by H.Kleinert, A.Pelster, P.Fiziev). In the works ([35, 65, 66, 67]), H.Kleinert and his collaborators introduced the Action Principle in spaces with the pseudo-Riemannian metric g (with the corresponding Levi-Civita connection ${}^g\Gamma$ and its curvature) and torsion. Torsion is introduced by the non-holonomic change of coordinates (frame) that leads to the new connection Γ having (in the absence of singularities) zero curvature and the connection S.

Consider the motion of a particle (body) in Euclidian space $\mathbb{R}^3$ by the law

$$\ddot{x}^i = 0, \ i = 1, 2, 3.$$

Changing variables $x \to q$ with the frame change defined by $e_i^\mu = \frac{\partial x^i}{\partial q^\mu}$ we rewrite equations of motions in the form

$$\ddot{q}^\mu + \Gamma^\mu_{\kappa\lambda}\dot{q}^\kappa\dot{q}^\lambda = 0, \ \mu = 1, 2, 3, \tag{42.22}$$

where

$$\Gamma^\mu_{\lambda\kappa} = -e^i_\kappa e^\mu_{i,\lambda} \tag{42.23}$$

are coefficients of affine connection Γ defined by the choice of variables.

Suppose now we make the non-holonomic change of variables by defining

$$dx^i = e^i_\mu(q)dq^\mu,$$

where components e^i_μ of the non-holonomic frame do not satisfy to the integrability condition $\partial_{q^\nu}e^i_\mu = \partial_{q^\mu}e^i_\nu$.

This non-holonomic frame generates the affine connection (see Appendix I) with the torsion

$$S^\mu_{\lambda\kappa} = \frac{1}{2}e^\mu_i \left(e^i_{\kappa,\lambda} - e^i_{\lambda,\kappa}\right).$$

Consider now the action for a point moving in R^3_q with the action $L(t, q, \dot{q})$:

$$A = \int_{t_1}^{t_2} L(t, q(t)\dot{q}(t))dt. \tag{42.24}$$

Calculating variation by q we use the non-commutativity rule (42.11) As a result we get the dynamical equation with the torsion force determined by the NC-tensor

$$K^\lambda_{\mu\nu} = 2S^\lambda_{\mu\nu}, \tag{42.25}$$

$$\frac{\partial L}{\partial q^\mu} - \frac{d}{dt}\frac{\partial L}{\partial \dot{q}^\mu} = -2S^\lambda_{\mu\nu}\dot{q}^\nu\frac{\partial L}{\partial \dot{q}^\lambda}. \tag{42.26}$$

In a special case where the action in Euclidian space R^3_x has the form $A = \int_{t_1}^{t_2}\frac{M}{2}\|\dot{q}\|_g^2dt$, with the metric $g_{\mu\nu}$ we obtain equation of motion

$$M[\ddot{q}^\mu + g^{\mu\kappa}(\frac{\partial g_{\lambda\kappa}}{\partial q^\nu} - \frac{1}{2}\frac{\partial g_{\nu\lambda}}{\partial q^\kappa}) - 2S^\mu_{\nu\lambda}]\dot{q}^\nu\dot{q}^\lambda = 0 \tag{42.27}$$

with the torsion force induced by the character of non-commuting variations of dynamical variables.

43. Elastic deformation of uniform materials.

In this section we describe the evolution equations for a **uniform material media** with the internal sources. Such sources are presented by the torsion of material connection corresponding to the inhomogeneity of the material (See Appendix I, Sec.68).

Usage of a material connection for presenting the internal structure of a material was pioneered by K.Kondo (1955), E.Kroner (1958) and B.A.Bilby, see [22] and the references therein. In the mid-sixties of XX century the theory of uniform materials got further development in the works of W.Noll, C.C.Wang and C.Truesdell as presented in the monograph ([137]) of C.Truesdell and C.C.Wang.

Later on, the study of uniform materials was continued in the works M.Epstein, M.de Leon, G.Maugin, M.Elzanowski and the author (see [23, 24, 25] and the references at [22]).

43.1. **Elasticity in material coordinates.** A material body is modeled by the 3-dim manifold B with the boundary ∂B and (material, Lagrange) coordinates $X^I, I = 1, 2, 3$ of the material points X in the body B.

Position of the body in physical Euclidian space $(E^3, h; x^i)$ (h being Euclidian metric, $x^i, i = 1, 2, 3$ are global coordinates in E^3) is determined by a **configuration** - diffeomorphic embedding $\phi_t : B \to E^3$. We fix the initial (non-stressed) configuration $\phi_0 : B \to E^3$. It is called the **reference configuration**.

Evolution of solid body is presented by the family of time-dependant configurations $\phi_t = \phi(t, X) = \{\phi^i(t, X^I), \ i = 1, 2, 3\}$ describing the motion and elastic deformation of the body B in the physical space. Motion of B in E^3 is characterized by the **velocity** - vector field $\bar{v}(t, X) = \{v^i = \frac{\partial \phi^i}{\partial t}\}, \ i = 1, 2, 3$ *over the embedding* ϕ_t. Elastic deformation of B is described by the **deformation gradient** - two point tensor $F_I^i = \frac{\partial \phi^i}{\partial X^I}$.

A reference configuration ϕ_0 allows us to introduce the (flat) *reference* metric $g_0 = \phi_0^* h$ in the body B obtained by pulling back physical (euclidian) metric h to the body B by the reference configuration mapping ϕ_0: $g_0 = \phi_0^* h$.

In the same way the current configuration ϕ_t defines in the body B the **Cauchy metric** $C(\phi) = \phi_t^* h$. In components, Cauchy metric has the form

$$C(\phi)_{IJ}(X) = h_{ij}(\phi_t(X))F_I^i F_J^j.$$

Deformation of the body is characterized by the **strain tensor** a measure of the decline of Cauchy metric $C(\phi_t)$ from the reference metric g_0:

$$E_J^I = \frac{1}{2}ln(g_0^{-1}C(\phi)) \approx \frac{1}{2}g_0^{-1}(C(\phi)-g_0), \ E_J^I = \frac{1}{2}ln(g_0^{IK}C(\phi)_{KJ)} \approx \frac{1}{2}g_0^{IK}(C(\phi)_{KJ}-g_{0KJ}),$$

$$(43.1)$$

where approximate value corresponds to the case of **Linear Elasticity** (see [80]).

Strain energy of a deformed body is the function $W = W(X, E_J^I)$ of the strain tensor E_J^I and, if the elastic properties of material varies from point to the point, of the material point. For the *linear elasticity*, $W = e_{AC}^{BD} E_B^A E_D^C$, where the (2,2)-elasticity tensor $e_{AC}^{BD}(X)$ characterizes the properties of material.

For a *homogeneous isotropic material*, $W = \frac{\lambda}{2}Tr(E^2) + \mu(Tr(E))^2$, where λ, μ are - **Lame constants**, [90], Chapter 4.

Lagrangian function L of the elasticity theory includes the kinetic energy density and the strain energy of the body

$$L(\phi) = \frac{\rho_0}{2}\|v\|^2 - W(E) \tag{43.2}$$

$\rho_0(X)$ here is the material density of the body. Denote by $\mathbf{b} = \{b_j\}$ the covector (1-form) of the bulk forces that may act on the body.

Elasticity equations in material (Lagrange) coordinates X^I are the Euler-Lagrange equations for the configuration ϕ with the action

$$A_B(\phi) = \int_{[t_1,t_2]\times B} Ldt \wedge d_{g_0}V. \tag{43.3}$$

Here $d_{g_0}V$ is the volume element in the body B corresponding to the reference metric g_0.

These equations for three components $\phi^j(t, X^I)$ of the configuration ϕ_t have the form

$$\rho_0 \frac{\partial v_j}{\partial t} + \frac{\partial}{\partial X^I} P_j^I = b_j, \; j = 1, 2, 3. \tag{43.4}$$

Here $P_j^I = \frac{\partial W}{\partial F_I^j}$ is the first Piola-Kirchoff stress tensor (see [90]).

43.2. **Uniform materials, material connections.** Simply speaking, a uniform material is the solid consisting of the same material at all points. Translating this into the geometrical language, we get the defining condition (of long distance parallelism or absolute parallelism, see Appendix I,Section 68)): for any couple of points $X, Y \in B$ there is the mapping $\xi_{XY} : T_X(B) \to T_Y(B)$ such that, for the strain energy function function $W \in C^\infty(T(B))$ of the material at the points X, Y, , we have the relation $W(Y, -) \circ \xi_{XY} = W(X, -)$.

Localization of this definition leads to the definition of a **uniform material** as the **media with the (non-holonomic) frame** $\zeta_\alpha, \alpha = 1.2.3$ **defined by a gauge transformation** D_α^I

$$\zeta_\alpha = D_\alpha^I \partial_{X^I}, \tag{43.5}$$

of the holonomic frame ∂_{X^I} in the material B generated by the Lagrange coordinates X^I.

Frame ζ_α defines the *zero curvature connection* (**absolute parallelism**, see Appendix I)

$$\Gamma_{ij}^k = D_s^k \frac{\partial D_i^{-1^s}}{\partial x^j} \tag{43.6}$$

with the **torsion** $S = \frac{1}{2}(\Gamma_{ij}^k - \Gamma_{ji}^k)$, see Appendix I, Sec. 68.

More then this, uniform structure defines the **(material) metric** g in the body B by requiring the frame ζ_α to be g-orthonormal. In the local frame ∂_{X^I},

$$g_{IJ} = \langle \partial_{X^I}, \partial_{X^J} \rangle = D_I^\beta D_J^\gamma g_{0\beta\gamma}.$$

Now we define the **strain tensor** in a uniform material (E_{un}) as the measure of decline of the Cauchy metric $C(\phi)$ of a configuration ϕ from the **material metric** g:

$$\mathbf{E}_{un} = \frac{1}{2}ln(g^{-1}C(\phi)) \Leftrightarrow E_J^I = \frac{1}{2}ln(g^{IK}C(\phi)_{KJ}).\frac{1}{2}g^{IK}(C(\phi)_{KJ} - g_{KJ}) \tag{43.7}$$

In the approximation of *linear elasticity* this definition takes the form

$$E_J^{lin\ I} = \frac{1}{2}g^{IK}(C(\phi)_{KJ} - g_{KJ}).$$

The **strain energy** of a uniform material with the frame ζ_α and material metric g is the function of the Strain tensor E_{un} and, the torsion of the connection Γ, see (43.6):

$$W = W(E^I_{un\ J}, S) \approx W(\frac{1}{2}g^{IK}(C(\phi_{KJ} - g_{KJ})), S). \tag{43.8}$$

Thus, *all the dependance of material properties on a point of material goes through the* point dependance of the non-holonomic frame $\zeta_\alpha = D^I_\alpha \partial_{X^I}$ (see ([22])), defining material connection Γ and material metric g.

Taking the Lagrangian of uniform material in the conventional form (41.2) (with the newly defined strain tensor $\mathbf{E}_{un}$) and following the scheme of the previous section in the defining variations of the derivatives of dynamical fields ϕ^i we get the Elasticity Equations **with the internal sources induced by the torsion S^i_{kj} of the connection Γ**

$$\frac{\partial L}{\partial \phi^\nu} - d_i \left(\frac{\partial L}{\partial \phi^\nu_{,i}} \right) = -2S^i_{kj}\phi^\sigma_{,s}\frac{\partial L}{\partial \phi^\mu_{,s}} = -2S^i_{kj}\phi^k_{,0}\frac{\partial L}{\partial \phi^i_{,0}} - 2S^i_{kj}J(\phi)\sigma^k_i, \tag{43.9}$$

where $\sigma^j_i = \phi^j_{,I}\frac{\partial L}{\partial \phi^i_{,I}} = J^{-1}\phi^j_{,I}P^I_i$ is the **Cauchy stress tensor** σ^i_j, J is the Jacobian of the deformation mapping ϕ measured in Euclidian metric h in the physical space and the reference metric $g_0 = \phi^*_0 h$ is the metric in the material body (see above)). Here $i, j = 1, 2, 3$.

Equation (43.9) is the Euler-Lagrange equation with the evolution induced by the NC-tensor

$$K^i_{j\nu} = 2S^i_{k\nu}\phi^k_{,j}. \tag{43.10}$$

Substituting here Lagrangian (43.2) and expressing (first) Piola-Kirchoff stress tensor P^I_i through the Cauchy stress tensor σ^i_j, we get the elasticity equations for a uniform material in the form

$$\rho_0\frac{\partial v_j}{\partial t} + \partial_{x^i}(\sigma^i_j) = -2S^i_{kj}\rho h_{is}\phi^k_{,0}\phi^s_0 - 2S^i_{kj}J(\phi)\sigma^k_i, \ j = 1, 2, 3. \tag{43.11}$$

Thus, as the works of G.Maugin and M.Epstein demonstrated (see [26]), it is the **torsion of the connection Γ** that produces the force of inelastic evolution of the uniform material. Notice that this inelastic evolution consists of two terms: the first one having a *kinematic nature* (depending on the velocity $\mathbf{v}$) and the second one having a dynamical nature (torsion S is paired with the stress in the material).

To get the **Energy-momentum balance law** for this case we notice that the Lagrangian here

$$L = \frac{\rho_0}{2}h_{ij}\phi^i_{,0}\phi^j_{,0} - W(g(X), F^i_I)$$

has the form (43.2) but the expression for the strain tensor contains the material metric g. As a result, participating quantities depend explicitly on the points $X \in B$ of the body but do not depend on the time .

Energy balance law here has the form (25.14) where we take $\Gamma^i_{jk} = 0$:

$$d_\mu T^\mu_0 = -L_{,t} + 2\phi^i_{,0}S^s_{ki}\phi^k_{,\mu}L_{,\phi^s_{,\mu}}. \tag{43.12}$$

Index μ here takes the values $\{0 = t, X^1, X^2, X^3\}$.

Calculate now

$$\begin{cases} T^\mu_0 = L\delta^\mu_0 + y^i_{,0}S^s_{ki}\phi^k_{,\mu}\pi^\mu_s, \\ T^0_0 = L + \phi^i_{,0}L_{,\phi^i_0} = L + \rho_0\|v\|^2_h, \\ T^I_0 = \phi^i_{,0}L_{,\phi^i_I} = -y^i_{,0}P^I_i, \end{cases}$$

because

$$\begin{cases} \phi^k_{,\mu} = \begin{cases} v^k, \ \mu = 0, \\ F^k_I, \ \mu = I \end{cases}, \\ \pi^\mu_s = \begin{cases} L_{,\phi^s_{,0}} = \rho_0 v_s = \rho_0 h_{sl} v^l, \ \mu = 0 \\ -P^I, \ \mu = I. \end{cases} \end{cases}$$

The left side of the energy balance law (43.12) is

$$d_t T^0_0 = d_t(L + \rho_0 \|v\|^2_h) + d_{X^I} T^I_0 = d_t(L + \rho_0 \|v\|^2_h) + d_{X^I}(-v^i P^I_i).$$

The source term in the right side is equal to

$$2\phi^i_{,0} S^s_{ki} \phi^k_{,\mu} \pi^\mu_s = 2\phi^i_{,0} S^s_{ki}(\phi^k_{,0} \rho_0 h_{sl} \phi^l_{,0} - \phi^k_{,I} P^I_s) = 2 S^s_{ki} \phi^i_{,0}(\rho_0 \phi^k_{,0} h_{sl} \phi^l_{,0} - J\sigma^k_s) =$$
$$= 2\rho_0 S^s_{ki} v^i v^k h_{sl} v^l - 2J(\phi) S^s_{ki} v^i \sigma^k_s, \quad (43.13)$$

As a result, the energy balance law has the form

$$d_t(L + \rho_0 \|v\|^2_h) + d_{X^I}(-v^i P^I_i)) = 2\rho_0 S^s_{ki} v^i v^k h_{sl} v^l - 2J(\phi) S^s_{ki} v^i \sigma^k_s. \quad (43.14)$$

44. Dissipative Potentials in Continuum Thermodynamics and the NC-variations.

Systems of balance equations of the continuum thermodynamics can not be obtained in the conventional Lagrangian form due to the presence of dissipative processes of different kind. More then this, not just the energy in such systems dissipates but an additional requirement the **II law of thermodynamics** has to be satisfied. This law is expressed in the form of the requirement that there exists an additional balance equation - *the entropy balance law* that has to be satisfied by all the solutions of the system of balance equations and to have the following property: source term of entropy is the sum of two terms: $\sigma = \sigma^{in} + \sigma^{out}$. First term corresponds to the **entropy production in the system** while the second one corresponds to the entropy exchange with the environment), see ([55, 99]). **II law of Thermodynamics** requires that the **internal entropy production in the system has to be nonnegative.**

Thus, it is interesting to investigate, when such systems of balance equations can be presented in Lagrangian form with the non-commuting variations.

44.1. **Rayleigh dissipative functions in Mechanics and the Dissipative potentials in continuum thermodynamics (see** [140]**).** In this section we first take the base X of the configurational bundle to be of dimension 1 (in a case of finite dimensional dynamical system) or 4-dimensional (in the case of a Field Theory) and use the notation $X = \{R^4, x^0 = t, x^A, A = 1, 2, 3\}$.

We consider here the modification of Lagrangian Field Theory that includes the terms responsible for the dissipative processes.

Such modification was suggested, for mechanical systems (n=1), by Lord Rayleigh, see [116]. More specifically, he suggested adding to the dynamical equations the dissipative forces defined by a function $\mathcal{R}$ now called the **Rayleigh dissipation function** (see [140], Sec.93). Original Rayleigh's dissipative function was quadratic by the velocities $\dot{y}(t)$:

$$\mathcal{D} = \frac{1}{2} \sum (k_x \dot{x}^2 + k_y \dot{y}^2 + k_z \dot{z}^2.) \tag{44.1}$$

Coefficients $k_x, k_y.k_z$ are constant. Euler-Lagrange equations of a mechanical system with kinetic energy T, potential energy U and the Rayleigh dissipation function $-\mathcal{D}$ has the form (see [140], Sec.93)

$$\frac{d}{dt}\left(\frac{\partial T}{\partial y^\mu}\right) - \frac{\partial T}{\partial y^\mu} - \frac{\partial D}{\partial \dot{y}^\mu} = Q_\mu.$$

Here Q is the work of external forces.

44.2. **Dissipative potentials in Field Theory.** In Field Theory, more specifically, in Continuum Thermodynamics corresponding functions were introduced by H.Ziegler, [94, 141, 142] and are now known under the name of **dissipative potentials** - functions $\mathcal{D}(y_{,t}^\mu, \ldots)$ depending on the rates of change $y_{,t}^i = y_{,0}^\mu$ (time derivatives) of dynamical variables y^μ (and, possibly, on other variables), see [94]. Field equations obtained by combining Euler-Lagrange equations with the dissipative potential D have the form

$$\frac{\delta L}{\delta y^\nu} = \frac{\partial \mathcal{D}}{\partial y_{,0}^\nu}, \quad \nu = 1, \ldots, m. \tag{44.2}$$

Here $\frac{\partial \mathcal{D}}{\partial y^\nu_{,0}}$ are derivatives of the function $\mathcal{D}(x^i, y^\mu, y^\mu_i)$ by the rates of change of the fields y^ν, see [93].

It is easy to see that energy balance law for the system (44.2) takes the following form

$$d_i T^i_0 = \dot{y}^\mu \mathcal{D}_{,\dot{y}^\mu}. \tag{44.3}$$

Thus, we get the following

Lemma 9. *If the dissipative potential $\mathcal{D}$ of a system 44.2 is a homogeneous function of the variables y^{mu}_0 of order k, **dissipative potential D is proportional to the rate of (positive or negative) dissipation of energy**.*

Remark 61. Gerard Maugin (see [93]) has noticed that the degree of homogeneity of the dissipative potential by the rates y^μ_0 of dynamical variables y^μ is related to the character of irreversible processes driving by this potential: if **viscosity** or relaxation is involved as an irreversible process in the system described by a dissipative potential (44.2), then the potential $\mathcal{D}$ is *homogeneous of degree two* by the **velocities** y^μ_0 **associates** with these processes, while for a "**plastic**" behavior, dissipative potential is *homogeneous of degree one* by corresponding velocities.

Example 30. Consider a case where $n = m = 1$, $y(t)$ is an unknown function, $L(y) = \frac{1}{2}\dot{y}^2$ and $\mathcal{D} = -\int^{\dot{y}} f(y, s)ds$ where $f(y, s)$ is any function of its variables. Then, the system (44.2) reduces to the equation

$$\ddot{y} + f(y, \dot{y}) = 0,$$

i.e. this is an abstract non-linear oscillator. Thus, any nonlinear oscillator of this form trivially has the form (42.1).

Example 31. Linear hardening([142]). In studying the linear hardening of metals, the basic dynamical quantity - strain tensor field $\epsilon_{ij}(t, x)$ is presented as the sum of elastic strain $\epsilon^e = \frac{1}{2}(u^i_{,x^j} + u^j_{,x^i})$, u^i being the displacement vector and the "plastic strain" ϵ^p:

$$\epsilon_{ij} = \epsilon^e + \epsilon^p.$$

Lagrangian in this case is taken to be $L = K - \Psi$ where K is the kinetic energy density $\frac{\rho}{2}\|\mathbf{v}\|^2$ ($v^i = \dot{u}^i$) and where

$$\Psi = \frac{\lambda}{2}(Tr(\epsilon^e))^2 + \mu I_2(\epsilon^e) + \mu' I_2(\epsilon^p) \tag{44.4}$$

is the **free energy**. Here $I_2(\epsilon^p)$ is the second principal invariant of a tensor ϵ^p which is the second coefficient in the decomposition $Det(\lambda E + \epsilon^p) = \lambda^n + I_1(I)\lambda^{n-1} + I_2(\epsilon^p)\lambda^{n-2} + \ldots$. **Dissipative potential** of linear hardening is taken in the form: $\mathcal{D} = k(2I_2(\dot{\epsilon}^p)^{1/2})$.

Dynamical equations (44.1) for ϵ^e, ϵ^p have the form (here we assume that $\epsilon^p <<I$):

$$\begin{cases} \rho \ddot{u}^j - \partial_{x^k}(\sigma^{jk}) = 0, \\ L_{,\epsilon^p} = \frac{\partial}{\partial \dot{\epsilon}^p} k(2I_2(\dot{\epsilon}^p)^{1/2}) \leftrightarrow 2\mu\epsilon^p = 2kI_2^{-\frac{1}{2}}(\dot{\epsilon}^p) \cdot \dot{\epsilon}^p. \end{cases} \tag{44.5}$$

44.3. Dissipative potentials versus NC-variations. Comparing the Euler-Lagrange system with the non-commuting variations (6.3) with the system (44.2) where a dissipative potential has been used, we may ask two questions:

(1) For which tensors $K^\mu_{\nu i}$ defining the variations (17.1) the force terms in the right side of equations (6.4) has the form $\frac{\partial D}{\partial y^\nu_{,0}}$ for some function D?

(2) For which functions D the system (44.2) can be written in the form (6.3) with some variation rule of the form (17.1)? In the next two subsections we discuss these questions.

Consider first the simplest question where $m = n = 1$ (one independent variable t - time) and one dependent variables $y = y(t)$.). In this case, equating the right sides of EL-system with a NC-variations defined by a vertical connection K (scalar function in this case) and that one generated by a dissipative function $D = D(\dot{y})$ we get the equality

$$\frac{\partial D}{\partial \dot{y}} = -K\pi, \tag{44.6}$$

where $\pi = \frac{\partial K}{\partial y}$. From this relation we immediately get the relations

$$\begin{cases} K = \frac{1}{\pi} \frac{\partial D}{\partial \dot{y}}, \\ D = C - \int^t (K \cdot \pi) d\dot{y}. \end{cases} \tag{44.7}$$

As a result, we see that in the case $m = n = 1$ there is a bijection between the NC-tensors and the dissipative functions defined up to the adding of a function that does not depend on $\dot{y}$.

44.4. EL-systems with NC variations defining the dissipative potential. Consider the relation

$$-K^\mu_{\nu i}\pi^i_\mu = \frac{\partial D}{\partial y^\nu_{,0}}, \quad \nu = 1, \ldots, m \tag{44.8}$$

presenting the dissipative term in right side of Euler-Lagrange equation (6.4) with the NC-variations in the form (17.1). Denote by d_{v0} the "partial vertical" differential $d_{v0}F = F_{,y^\mu_0} dy^\mu_0$ of a function $F \in C^\infty J^1(\pi))$. Then, for a fixed tensor $K^\mu_{\nu i}$, the local condition of solvability of relation (42.8) (existence of a dissipative potential so that $f_\nu = \frac{\partial D}{\partial y^\nu}$) is the closeness condition for the 1-form $K^\mu_{\nu i}\pi^i_\mu dy^\nu_0$) is

$$d_{v0}f_\nu = d_{v0}(K^\mu_{i\nu}\pi^i_\mu dy^\nu_0) = 0 \Leftrightarrow (K^\mu_{i\nu}\pi^i_\mu)_{,y^\sigma_0} = (K^\mu_{i\sigma}\pi^i_\mu)_{,y^\nu_0}, \nu, \sigma = 1, \ldots, m. \tag{44.9}$$

Here $T_{(\nu\sigma)} = \frac{1}{2}(T_{\nu\sigma} + T_{\sigma\nu})$ is the symmetrization by the indices σ, ν.

If that condition is satisfied, the dissipative potential D is defined uniquely up to addition of a function that does not depend on variables y^μ_0.

If we require, additionally, that function D satisfies to the relation

$$y^\mu_0 = 0, \forall\, i \Rightarrow D = 0,$$

then the potential D *is defined uniquely.* Thus, we get

Proposition 14. *The Euler-Lagrange equations with non-commuting variations (4.3) have the form (44.2) if and only if*

$$\left(K^\mu_{i(\nu}\pi^i_\mu\right)_{,y^\sigma)_0} = 0, \ \forall\, \nu, \sigma. \tag{44.10}$$

If this condition is fulfilled, the dissipative potential $\mathcal{D}$ exists and if we require additionally, that the function $\mathcal{D}$ satisfies to the relation

$$y_0^\mu = 0, \forall\, \mu \;\Rightarrow\; \mathcal{D} = 0,$$

, then the Dissipative potential $\mathcal{D}$ is defined uniquely.

Now, let a dissipative potential D is homogeneous of degree **k** by the velocities y_0^μ:

$$D(\alpha y_0^\mu) = \alpha^k D(\alpha y_0^\mu),\ \ \alpha > 0.$$

Recall that the Euler equation for homogeneous functions $f(x^1, \ldots, x^m)$ of degree k:

$$\sum_{i=1}^m x^i \frac{\partial f}{\partial x^i} = kf(x^1, \ldots, x^m). \tag{44.11}$$

Assume that a dissipative potential D is homogeneous of degree k by the variables y_0^μ. Multiply equation (44.8) by y_0^ν and summate by ν. Using now the Euler equation to the right side of the obtained equality we get the **formula for dissipative potential in terms of NC-tensor**

$$D(y_0^1, \ldots, y_0^m) = \frac{1}{k} K_{i\nu}^\mu y_0^\nu \pi_\mu^i. \tag{44.12}$$

44.5. Tensor K of NC-variations defined by a dissipative potential. On the other hand, let us fix, for a given Lagrangian L, a dissipative potential $\mathcal{D}$. Then, by the results of Sec. (9) there always exists a tensor $K_{\mu j}^i$ (canonical tensor K) satisfying the relation (44.2), i.e. the system of equations (44.2) can always be presented as the Euler-Lagrange system with Lagrangian L and the non-commuting variations. Yet, it is useful to know, when there exists a tensor K defining the NC-variations for the system (40.1) in the form (40.5) that is **as simple as possible -** the one that contains minimal number of additional quantities. We study the case where $K_{A\nu}^\mu = 0$, $A = 1, 2, \ldots$, i.e. when only non-zero components of tensor $K_{\nu i}^\mu$ are $K_{\nu 0}^\mu$. Then, (44.8) takes the form

$$-K_{\nu 0}^\mu \pi_\mu^0 = \frac{\partial D}{\partial y_{,0}^\nu}, \ \ \nu = 1, \ldots, m. \tag{23.5'}$$

Thus, dynamical equations of a field system are presented here using the tensor K of the same type as in the case $n = 1$, i.e. in the case of Mechanics.

Next we perform the *partial Legendre transformation* rate replacing variables y_0^μ with the momenta $\pi_\mu^0 = L_{,y_0^\mu}$. This is (locally) possible if the "partial vertical Hessian" $L_{y_0^i y_0^j}$ is nondegenerate. In terms of new "vertical" variables $\pi_\mu^0, \mu = 1, \ldots, m$, the previous relation takes the form

$$K_{\nu 0}^\mu \pi_\mu^0 = -\frac{\partial \tilde{D}}{\partial \pi_\sigma^0} L_{y_0^\sigma y_0^\nu} \Leftrightarrow (L_{y_0^\sigma y_0^\nu})^{-1\,\sigma\nu} K_{\nu 0}^\mu \pi_\mu^0 = -\frac{\partial \tilde{D}}{\partial \pi_\sigma^0}, \ \ \nu = 1, \ldots, m. \tag{44.13}$$

In terms of the tensor field

$$\tilde{K}^{\sigma\mu}(x, y, y_A^\mu, \pi_\mu^0) = -(L_{y_0^\sigma y_0^\nu})^{-1\,kj} K_{\nu 0}^\mu,$$

The previous relation takes the simple form

$$\tilde{K}^{\mu\sigma} \pi_\mu^0 = -\frac{\partial \tilde{D}}{\partial \pi_\sigma^0}, \ \ j = 1, \ldots, m. \tag{44.14}$$

In this equation $\tilde{K}, \tilde{D}$ are functions of $x^i, y^\mu, y^\mu_A, \pi^0_\mu$ instead of y^μ_0.

Present the function $\tilde{D}$ in the form

$$\tilde{D} = \tilde{D}_0 + \tilde{D}^\sigma \pi^0_\sigma + \frac{1}{2}\tilde{D}^{\mu\nu}\pi^0_\mu \pi^0_\nu, \tag{44.15}$$

where $\tilde{D}_0, \tilde{D}^\sigma$ do not depend on variables π^0_σ (but $\tilde{D}^{\mu\nu}$ may depend on these variables).

Substituting $\tilde{D}$ in this form to (6.5) we get it in the form

$$\tilde{K}^{\sigma\mu}\pi^0_\mu = -\tilde{D}^\sigma + \left[\frac{1}{2}\tilde{D}^{\mu\nu}_{,\pi^0_\sigma}\pi^0_\nu + \tilde{D}^{\sigma\mu}\right]\pi^0_\mu.$$

From the last equality we get the condition necessary and locally sufficient for solvability of (44.8) provided K is continuously differentiable by vertical variables

$$\tilde{D}^\sigma = 0, \ \sigma = 1, \ldots m.$$

Provided this condition is fulfilled, the general solution of (44.8) in the form

$$-\tilde{K}^{\sigma\mu} = \frac{1}{2}\tilde{D}^{\mu\nu}_{,\pi^0_\sigma}\pi^0_\nu + \tilde{D}^{\sigma\mu} + \alpha^{\sigma\mu\lambda}\pi^0_\lambda, \tag{44.16}$$

where $\alpha^{\sigma\mu\lambda} = -\alpha^{\sigma\mu\lambda}$ is an arbitrary tensor skew-symmetric by last two indices and independent on the variables π^0_λ.

Multiplying by the matrix inverse to the Hessian matrix $L_{,y^\mu_0 y^\nu_0}$ in the definition of tensor $\tilde{K}$ and returning to the initial vertical variables y^μ_0 we get

$$-K^\mu_{0j} = \frac{1}{2}(\tilde{D}^{\mu\lambda}_{,y^\nu_0})\pi^0_\lambda + L_{,y^\nu_0 y^\sigma_0}\tilde{D}^{\sigma\mu} + L_{,y^\nu_0 y^\sigma_0}\alpha^{\sigma\mu\lambda}\pi^0_\lambda \tag{44.17}$$

as the function of variables (x^i, y^μ, y^μ_i).

Theorem 31. *Let $L, D \in C^\infty(J^1(\pi))$ be two functions on the 1-jet space $J^1\pi$ of the configurational bundle $\pi : Y \to X$. **Euler-Lagrange equations with a dissipative potential** D - (44.2) can be presented in the form (6.4) or, what is the same, in the form (6.6) **with** $K^\mu_{\nu A} = 0, A = 1, 2, 3$ (a natural condition) if and only if the following condition is fulfilled: Make the Legendre transformation of dissipative potential D by the variables y^μ_0 and present the transformed dissipative potential $\tilde{D}$ as the function of conjugated momenta π^0_ν and all other variables*

$$\tilde{D} = \tilde{D}_0 + \tilde{D}^\sigma \pi^0_\sigma + \frac{1}{2}\tilde{D}^{\mu\nu}\pi^0_\mu \pi^0_\nu, \tag{44.18}$$

where $\tilde{D}_0, \tilde{D}^\sigma$ do not depend on variables π^0_σ (but $\tilde{D}^{\mu\nu}$ may depend on these variables). Then,

$$\tilde{D}^\sigma = 0, \ \sigma = 1, \ldots m.$$

If this condition is fulfilled, representation (6.6) with the tensor K such that $K^\mu_{\nu A} = 0, A = 1, 2, 3$ exists, is unique up to adding of an antisymmetric by two indices tensor $\alpha^{\sigma\mu\lambda}$ (this tensor being independent on the momenta - variables π^0_λ) and has the following form

$$-K^\mu_{\nu 0} = \frac{1}{2}(\tilde{D}^{\mu\lambda}_{,y^\nu_0})\pi^0_\lambda + L_{,y^\nu_0 y^\sigma_0}\tilde{D}^{\sigma\mu} + L_{,y^\nu_0 y^\sigma_0}\alpha^{\sigma\mu\lambda}\pi^0_\lambda, \tag{44.19}$$

where $\alpha^{\sigma\mu\lambda} = -\alpha^{\sigma\lambda\mu}$ is an arbitrary tensor skew-symmetric by last two indices and independent on the variables π^0_λ. Returning to the variables y^μ_0 we find the NC-tensor $K^\mu_{\nu 0}$ as the function of original variables (x^i, y^μ, y^μ_0).

Remark 62. It was mentioned above that as G.Maugin had noticed ([94]),that the degree of homogeneity of dissipative potential by the velocities $\dot{y}^\mu$ specifies the type of the dissipative processes going in the physical systems.

If a dissipative potential $\mathcal{D}$ is defined by a non-commuting variations tensor K, homogeneity properties of $\mathcal{D}$ are related to some properties of tensor K.

For instance, if a Lagrangian L is a quadratic function of velocity $\dot{y}^\sigma$ and the tensor K_0^μ is a function of y^σ only (is independent on the velocity $\dot{y}^\sigma$), then, the dissipative potential $\mathcal{D}$ is homogeneous by $\dot{y}^\sigma$ *of the second degree* (see (44.5)). Dual tensor $\tilde{D}$ is homogeneous of second degree by momenta π_σ^0. Then, as G.Maugin has shown, the process related to y^σ exhibits a characteristic (time of relaxation) time and there may appear convex set limiting the values of momenta π_σ. an example of such a situation in the solid state dynamics leading to the visco-plasticity is presented in [93, 94].

On the other hand, if, for a Lagrangian, quadratic by the velocities $\dot{y}^\sigma$, dissipative potential $\mathcal{D}$ is *homogeneous of order one* by the velocities $\dot{y}^\sigma$ (and, as a result, sources f_ν do not depend on $\dot{y}^\sigma$), then tensor K is homogeneous or order -1 by the momenta and the behavior of a material will be of "plastic type" - there is no velocity dependance and there is no characteristic time scale in this model ([93]).

Chapter 6.Material time, NC-Variations and the Material Aging.

45. Introduction.

In this Chapter we present two applications of the method of non-commuting variables in thermodynamics and material science. As the first application, we present the scheme allowing to obtain Entropy Balance as the Euler-Lagrange equation using the scalar field of **thermasy** - a quantity introduced by H.Helmholtz (see [58]) and used later in relativistic thermodynamics (D.van Dantzig, see vL,VD), thermoelasticity (A.Green and Nachdi) and Continuum Thermodynamics (G.Maugin).

As the second application we present the geometrical model of material aging processes, developed by A.Chudnovsky and the author, [15, 16] and the usage of the method of non-commutative variations (more specifically, a form of non-commuting variations related to the material time in solids) to obtain dynamical equations of unconstrained aging, stress relaxation and the material creep - inelastic irreversible deformations abundantly present in the world, [9, 61].

We start with the short review of the **4-dim material space-time** $\mathcal{P}$ fibred over the material body B: $\pi : P \to B$. Manifold P is endowed with the Riemannian metric G (material or inner metric) and the slicing B_t in the space P defined by a deformation history. Detailed presentation of this model and some examples are positioned Later in this Chapter.

The most important feature of this approach is the presence in P of the **material time** τ - "*proper time*" of the material metric G . It is shown in the next section that the use of material time τ instead of the physical (laboratory) time t in the Lagrangian formalism describing heat propagation and related entropy balance law *is equivalent* to the use of the physical (laboratory) time t and the *non-commuting variations* with the NC-tensor K defined by the rate of the material time $S = \frac{\partial \tau}{\partial t}$. It follows that the use of material time τ instead of the conventional physical time t allows us to introduce into the evolutional equations for dynamical fields, the dissipative component naturally defined by the structural fields of the media (here - material metric G and the thermasy γ).

In addition, using the scalar dynamical field of thermasy (see below), we include the Entropy balance law into the system of Euler-Lagrange equations of Variational Theory.

The first three Sections of this Chapter are devoted to the relations of non-commutative variations and the material time). That is why we start with a short introduction of the material space-time endowed with the 4-dim Riemannian metric. More detailed presentation will follow this part in the Chapter.

46. Introduction:4-dim material space-time.

The material body is modeled by the 3-dim manifold B with the boundary ∂B and (local, Lagrange) coordinates $X^A, A = 1, 2, 3$. Manifold B is the base of the bundle $\pi : P \to B$, where P^4 is the **4-dim material space-time** with coordinates $T, X^A,\ A = 1, 2, 3$. P is endowed with the Riemannian **"material metric"** G.

© Springer International Publishing Switzerland 2016
S. Preston, *Non-commuting Variations in Mathematics and Physics*,
Interaction of Mechanics and Mathematics,
DOI 10.1007/978-3-319-28323-4_6

The history of deformation is presented by the **configurations** - diffeomorphic embeddings $\phi = (\phi^0, \phi^1, \phi^2, \phi^3) : P \to \mathbf{E}^4$ of the material space-time P^4 into the classical (Newtonian) space-time $(\mathbf{E}^4, t, x^i, i = 1, 2, 3)$. Newtonian space-time is endowed with the 4-dim Euclidian metric H and the Newtonian slicing by the surfaces of fixed physical time $t = const$, see [90]). Configurations ϕ are assumed to satisfy to the condition: $\frac{\partial \phi^0}{\partial T} > 0$ (we call such configurations **admissible**). This condition guarantees that the material time T and the physical time t flow in the same direction.

It is customary to fix one configuration as the base point for the further development. We will fix such a configuration ϕ_0 and will call it - **the reference configuration.** Reference configuration allows to introduce in the material space-time space P the "physical time" t_p by the condition that for any material point (t_p, X) in P its image in the Euclidian space-time E^4 has the time $t = t_p(t_p, X)$. Simply speaking, we use the reference configuration ϕ_0 to identify time in P with the physical time t in E^4. From now on we will be using the notation t for the time t_p.

Let ϕ be an admissible configuration. Fixing the value of physical time t we define the *slicing* of the space P: $B_{t,\phi} = \phi^{0\ -1}(t)$, where ϕ^0 is the zeroth component of history of deformation ϕ. This slicing defines the ADM-type representation of metric G similar to one used in General Relativity, [97]. In this ADM representation metric G has the form

$$G = g_{AB}(dX^A + N^A d\phi^0) \circ (dX^B + N^B d\phi^0) + Sd\phi^{0\ 2}. \qquad (46.1)$$

Here $\mathbf{N}$ is the shift vector field along the slices $\phi(T, X) = t$ and S is the lapse function, see [15].

Proper time τ - natural parameter along the material world lines, corresponding to the metric G in P - is defined by the relation

$$d\tau = Sd\phi^0 = S(\phi^0_{,0}dT + \phi^0_{,X^A}dX^A).$$

To compare the proper material picture that is obtained in proper coordinates $(\tau, X^A, A = 1, 2, 3)$ with one in physical material coordinates $(t, X^A, A = 1, 2, 3)$ we will use the relation $d\tau = S(t, X^A)dt$. We will be using relations between different time derivatives of the functions in the material space-time P. For an arbitrary scalar function $f(T, X)$ these relations have the form

$$f_{,\tau} = S^{-1}f_{,t} = S^{-1}\phi^{0\ -1}_{,0}f_{,T} \Leftrightarrow \partial_t = S \cdot \partial_\tau. \qquad (46.2)$$

47. Thermasy and the entropy balance as an Euler-Lagrange Equation.

In this section we present an example of a Lagrangian Theory with non-commuting variations for one scalar field -thermasy γ, whose time derivative is proportional to the absolute temperature θ i.e. $\frac{\partial \gamma}{\partial t} = k\theta$. The only Euler-Lagrange equation is the entropy balance (recall that the entropy is - the variable dual to the absolute temperature, [94].

The scalar field γ that will be called **thermasy** was introduced by H. von Helmholtz, see the book of M. von Laue [136], Sec.33 or [92]). The name **"thermasy"**, together with the formula

$$\gamma = \int_0^t k\theta dt, \qquad (47.1)$$

for the thermasy was given to this variable by D.Van Dantzig in his study of thermodynamics of moving matter, see [135]. van Danzig defined thermasy by the integral

$$\gamma = \int_0^t k\theta dt, \tag{47.2}$$

where θ is absolute temperature and k is the Boltzman constant relating energy at the individual particle level with temperature.

Lately, thermasy was used by A.Green and P.Nahdi in thermoelasticity, see [53, 54], by G.Maugin and V.Kalpakides in Continuum Thermodynamics, [92]. P. Podio-Guidugli, in his lecture on the "Progress in the Theory and Numerics of Configurational Mechanics" in Erlangen-Nurnberg, in October 2008 used the term **"thermal displacement"**. This name is probably closer then the term "thermasy" to the ideology of H.Helmholtz. A.Chudnovsky and author have used thermasy for the description of the thermical component of irreversible processes in porous media (in preparation).

Physically, **thermasy** is the measure of "action" of the kinetic energy of micromolecular motion related to it by the relation

$$\gamma \sim k^{-1}\langle \|\mathbf{v}\|^2 \rangle,$$

where $\mathbf{v}$ is the random velocity of micromolecular motion - stochastic component of the velocity field of continuum and k - is the Boltzmann constant. Following [92, 136], **we identify $\dot{\gamma}$ with the absolute temperature ϑ.**

In this Section we take a model Lagrangian L as the function of the variables $\theta = \dot{\gamma}, \nabla\gamma$ characterizing the temperature processes in materials (comp.to [63]). We fix the material metric G. More than this, for simplicity we assume that 3-dim material metric g is fixed (and constant) and that the shift vector field of metric G vanishes. (see previous Section). As a result, the Lagrangian depends only on thermasy and its derivatives as dynamical variables

$$L = L(\dot{\gamma}, \nabla^g\gamma; S, g), \tag{47.3}$$

where L depends on S, g as on parameters.

Lagrangian L is taken to depend on γ only through its derivatives $\dot{\gamma}$ and $\nabla_A^G\gamma$.

Remark 63. If only thermical processes are present in the material, one may enrich this model by taking anzatz with shift vector field N of metric G to be function of temperature gradient $\nabla\theta$.

In applications to the concrete physical situations, Lagrangian will depend on the corresponding dynamical variables and their derivatives as well as to the thermical field γ. More then this, shift field $\mathbf{N}$ may carry information about the dissipative flows in the system at consideration and, therefore, has to be included in the list of dynamical variables.

In thermoelasticity, L will also depend on the Cauchy metric induced by the deformation ϕ from the Euclidian 3-dim metric h in Galilean space-time E^4, on the 3-dim material metric g (see next Section), including, possibly, its curvature and external curvature of 3-dim slices $B_{\phi,t}$.

In this section we take the NC-tensor $K_{\mu i}^\nu$ depending on the 00-component S of 4-dim material metric G and on the zero component ϕ^0 of the configuration history (variables related to the entropy production, see below), i.e. we take

$$K_{\mu i}^\nu = -S_{,t}\delta_i^0 \delta_\mu^\nu. \tag{47.4}$$

This choice of NC-tensor $K^{\nu}_{\mu i}$ is a special case of the NC-tensor delivering the rate type dissipation (see Section 10, Example 7).

In this way, **field variations $\delta\gamma$ do not commute in the sense of (4.4) with the time derivative only** (see Example 7 in Section 10).

Remark 64. Notice that the NC-tensor $K^{\nu}_{\mu i}$ introduced in (47.4) is related to the *"acceleration S of the proper time $\dot{S} = \tau''$"* and, therefore, has a dynamical meaning.

Remark 65. Applying the relations of NC-tensors and the dissipative potentials to our case (see Section 44), where only dynamical variable is the thermasy $\gamma(t = x^0, x^i), i = 1, 2, 3$, we find the dissipative potential (taking into account that $k = S_{,\tau}$) corresponding to the NC-tensor (47.4): $D = S_{,t}L$.

Euler Lagrange Equations for the dynamical fields y^{μ} and the NC-variations defined by the K-tensor (47.4) have the form

$$\frac{\partial L}{\partial y^{\mu}} - d_i\left(\frac{\partial L}{\partial y^{\mu}_i}\right) = -K^{\nu}_{\mu i}\pi^i_{\nu} = S_{,t}\frac{\partial L}{\partial y^{\mu}_0}, \quad \mu = 1, \ldots, m. \tag{47.5}$$

In the case of scalar dynamical field γ, tensor K has the scalar structure with y^1 to be the only dynamical variable and K - to have $i = 0$ as the only space-time index. As the result,

Euler-Lagrange Equation (6.4) for for scalar field of thermacy γ has the form

$$\partial_t\left(\frac{\partial L}{\partial\gamma_{,t}}\right) + \partial_{X^A}\left(\frac{\partial L}{\partial\nabla_A\gamma)}\right) = KL_{,\gamma,t} = -\frac{S_{,t}}{S}L_{,\gamma,t} = -S_{,\tau}L_{,\gamma,t}. \tag{47.6}$$

Consider now the following identifications, standard in Continuum Thermodynamics (see [92, 94]):

(1) $s = \frac{\partial L}{\partial\dot{\gamma}}$ - entropy density,

(2) $S^A = \frac{\partial L}{\partial(\nabla_A\gamma)}$ - entropy flux,

(3) $\sigma = -S_{,\tau}L_{,\gamma,t} = -S_{,\tau}s$ - entropy production.

Then the equation (47.6) **takes the standard form of the entropy balance**

$$\partial_t s + \partial_{X^A}S^A = S_{,\tau}L_{,\gamma,t} = \sigma. \tag{47.7}$$

In these terms, the II law of thermodynamics takes the form of the following condition:

$$S_{,\tau}L_{,\gamma,t} \geqq 0. \tag{47.8}$$

Remark 66. In a case of material aging (see below), we have $S_{,t} < 0$, i.e. $S(t)$ is decreasing in the real processes (or, at least, non-increasing), see ([15, 16]). In this case, condition (47.7) takes the form $L_{,\gamma,t} \leqq 0$.

Identifying the first time derivative $S_{,t}s = S^{-1}S_{,t}s = -\sigma$ we get $Ln(S) = Ln(S(0)) - \int_0^t \frac{\sigma}{s}dt$ or

$$S(t) = S(0)e^{-\int_0^t \frac{\sigma}{s}dt}. \tag{47.9}$$

Comparing this with the expression for the proper time $d\tau = Sdt$ and normalizing $S(t)$ by the condition $S(0) = 1$ we get the basic **relation between the relative entropy production $\frac{\sigma}{s}$** and the **rate of change of the proper material time**

$$d\tau = \left(e^{-\int_0^t \frac{\sigma}{s}du}\right)dt. \tag{47.10}$$

When $\phi^0(T, X) = T$ (internal time T is synchronized with the physical time t) $d\tau = S dt$ and we get the formula for the *proper (material) time rate*

$$S(t) = e^{-\int_0^t \frac{\sigma}{s} dt} S(0). \tag{47.11}$$

Notice that with σ and s being non-negative, S(t) is decreasing (or, at least, non-increasing). Thus, the II law of thermodynamics (47.7) is directly related with the slowing of material time during the irreversible processes (while the entropy is increasing).

Remark 67. Notice that in the relation (47.4) defining the NC-tensor, material time τ is considered as the function of laboratory (physical) time t.

47.1. Internal (Lagrangian) space-time picture and the principle of "material relativity".

One can look at equation (47.7) from a different point of view. To do this, we write down the action principle and the Euler-Lagrange system for the same Lagrangian L *but using the **material** (proper) **time** τ instead of laboratory time t*. We call this Lagrangian $\tilde{L}(\gamma_{,\tau}, \nabla\gamma)$. Now we calculate Euler-Lagrange equation using *the conventional commutativity rule (and the total derivatives by space-time coordinates)*:

$$\frac{\delta\tilde{L}}{\delta\gamma} = 0 \Rightarrow d_\tau\left(\frac{\partial\tilde{L}}{\partial\gamma_{,\tau}}\right) + d_{X^A}\left(\frac{\partial\tilde{L}}{\partial\nabla_A\gamma}\right) = 0. \tag{47.12}$$

Next we rewrite this Euler-Lagrange equation in coordinates (t, X^A), using the relations between the derivatives: $\frac{\partial\tilde{L}}{\partial\gamma_{,\tau}} = S\frac{\partial\tilde{L}}{\partial\gamma_{,t}}$, $\partial_\tau = \frac{\partial t}{\partial\tau}\partial_t = S^{-1}\partial_t$:

$$\frac{\delta\tilde{L}}{\delta\gamma} = S^{-1}\left(S\frac{\partial\tilde{L}}{\partial\gamma_{,t}}\right)_{,t} + \left(\frac{\partial\tilde{L}}{\partial(\nabla_A\gamma)}\right)_{,X^A} = 0 \Leftrightarrow$$

$$\Leftrightarrow \left(\frac{\partial\tilde{L}}{\partial\gamma_{,t}}\right)_{,t} + \left(\frac{\partial\tilde{L}}{\partial(\nabla_A\gamma)}\right)_{,X^A} = \frac{S_{,t}}{S}\frac{\partial\tilde{L}}{\partial\gamma_{,t}} = S_{,\tau}\tilde{L}_{,\dot\gamma} = \sigma. \tag{47.13}$$

In this form the Euler-Lagrange system for the action written in the material (proper) time, and obtained by using *conventional variational rule* coincides with the system (47.7) obtained from the action written in coordinates (t, X^A) *but using the nonstandard commutating rule.*

This result supports the following methodological principle for modeling of material behavior suggested by Prof. A. Chudnovsky (in numerous private communications with author):

"Principle of material relativity" : In the intrinsic material space-time frame of reference, the material appears to be perfectly elastic with no signs of aging such as changes in density, stiffness, creep, plasticity etc.

It also agrees well with the Hindu philosophy position that *"there is no time for the internal being"*.

However, the existence of aging in time is manifested in the evolution of the material space-time metric that is readily detectable **by an external observer**.

Remark 68. Equivalence of using non-commuting variations of derivatives of dynamical fields with the NC-tensor (47.4) with the using conventional variation of derivatives but **in the material time** τ immediately extends to the case described in the Remark 62 where Lagrangian of the first order is the function of the dynamical fields y^α. Euler-Lagrange equations written in laboratory time t coincide. This circumstance is used in the rest of this Chapter where we present several examples of aging processes in solids using the material time or equivalent dissipative potential instead of the non-commutative variations with the NC tensor (47.4).

47.2. **Energy balance law.** Here we consider only γ as the dynamical field with the Lagrangian $L(\gamma, \dot\gamma, \nabla\gamma)$. Scalar field γ enters our *model* Lagrangian through its time and space derivatives. To construct the (Noether) energy balance law (7.7) we calculate the energy density

$$\epsilon = L - \dot\gamma \frac{\partial L}{\partial \dot\gamma} = L - \vartheta s,$$

and the energy flux (Pointing vector field)

$$U^A = \dot\gamma \frac{\partial L}{\partial \gamma_{,A}} = \vartheta S^A.$$

In a case where the entropy flux has the form $\mathbf{S} = \frac{\mathbf{q}}{\theta}$, where q is the heat flux, and where there are no other energy flows, the energy flow coincides with the heat flux (see [94]).

For the second term in the right side of energy balance law (4.5) we are using expression (47.4) for coefficients K and get

$$-y_0^i K_{\mu i}^j L_{z_\mu^j} = \dot\gamma \left(\frac{\partial S}{\partial t}\delta_\mu^0\delta_0^j\right)\frac{\partial L}{\partial z_0^j} = \dot\gamma S_{,t} \cdot s = -\vartheta\sigma.$$

As a result, the energy balance (7.7) takes the form:

$$\partial_t(\mathcal{L} - \vartheta s) + \partial_{X^A}(\vartheta S^A) = -\frac{\partial L}{\partial t}\bigg|_{expl} - \vartheta\sigma, \qquad (47.14)$$

If we recall that the Lagrangian L actually represents the density of *internal energy* and that the difference $\Psi = L - \vartheta s$ is the density of the free energy of Helmholtz, [93, 94], the last relation can be rewritten in the form of the *free energy balance law*

$$\partial_t\Psi + \partial_{X^A}(-\vartheta S^A) = \frac{\partial L}{\partial t}\bigg|_{expl} - \vartheta\sigma. \qquad (47.15)$$

Terms in the RHS of the energy balance correspond to the energy dissipation process.

Heat propagation law.

To get the heat propagation law we assume that $L = L_0(\gamma_{,t}) + \frac{1}{2}\alpha^{AB}\nabla_A\gamma \cdot \nabla_B\gamma$ with a matrix α^{AB} that may depend on the points of the body.

We admit the classical relation $\mathbf{q} = \theta\mathbf{S}$ between the flow of entropy $\mathbf{S}$ and the heat flow q.

Using the expression for the entropy flow used above: $S^A = \frac{\partial L}{\partial \nabla_A\gamma} = \alpha^{AB}\nabla_B\gamma$, we get

$$q^A = \theta\alpha^{AB}\nabla_B\gamma. \qquad (47.16)$$

Differentiating this formula by time and assuming that the time derivative commutes with the covariant derivatives ∇_B we get

$$q_{,t}^A = \theta\alpha^{AB}\nabla_B\gamma = (\theta\alpha^{AB})_{,t}\nabla_B\gamma + \theta\alpha^{AB}(\nabla_B\gamma)_{,t} = (\theta\alpha^{AB})_{,t}(\theta\alpha)_{BC}^{-1}q^C + \theta\alpha^{AB}\nabla_B\theta, \tag{47.17}$$

or

$$q_{,t}^A - \theta\alpha^{AB}\nabla_B\theta = (\theta\alpha^{AB})_{,t}(\theta\alpha)_{BC}^{-1}q^C \tag{47.18}$$

This heat propagation relation has the form of the Cattaneo law

$$\tau\dot{q}^A + k\nabla^A\theta = q^A, \quad A = 1, 2, 3, \tag{47.19}$$

where τ, k are coefficients depending on the properties of the media.

If we take $\theta\alpha^{AB} = -\frac{1}{\tau}k_B^A$ and $(\theta\alpha^{AB})_{,t}(\theta\alpha)_{BC}^{-1} = \frac{1}{\tau}\delta_C^A$ assume that $a > 0, (\theta a)_{,t} < 0$, formula (46.18) takes the form (47.20):

$$\tau q_{,t}^A + k_B^A\nabla^A\theta = q^A, \quad A = 1, 2, 3. \tag{47.20}$$

48. Interlude: History of the "material time".

Here we provide a short historical review of the notion of the "**material time**".

We start with the following warm up example before listing the references on this notion.

Example 32. Harmonic Oscillator in the other time.

Let t be physical (laboratory) time and $\tau = \tau(t)$ - some other parameter which we call the *material time*. Denote $S = \frac{d\tau}{dt}$.

Consider the standard harmonic oscillator *in the material time* with the Lagrangian $L = \frac{1}{2}(u_{,\tau})^2 - \omega^2 u^2$. The corresponding Euler-Lagrange equation has the form

$$u_{,\tau\tau} + \omega^2 u = 0. \tag{48.1}$$

This is a simple conservative system with periodical phase curves. Yet in the laboratory where one uses the (physical) time t, the dynamics of this system look different: using $\partial_\tau = t_{,\tau}\partial_t = -S^{-1}\partial_t$ one rewrites Euler-Lagrange equation 44.1 in the form

$$- S^{-1}\partial_t(-S^{-1}\partial_t u) + \omega^2 u = 0 \quad \Leftrightarrow \quad u_{,tt} - ln(S)_{,t} \cdot u_{,t} + S^2\omega^2 u = 0. \tag{48.2}$$

Thus, observing dynamical behavior of this system in laboratory time we see **oscillator with variable frequency and positive or negative dissipation depending on the material time rate**. So, an *internal observer* sees the simple periodical behavior while an *outside observer* sees the irreversible dissipative evolution of this system (controlled by the material time rate $S(t)$).

The first example of "essentially material time" was the notion of "proper time" in the 4-dim space-time as introduced by A.Einstein in Special Relativity Theory in 1905 and in General Relativity, in 1916). Since then is was widely used by physicists studying relativistic processes in physical systems, [80]. Of course, this is still the time in "empty" space but essential step to the second, different from the "laboratory" time was done.

In Biology, an "internal (to the live organism) time(s)" measured by the "internal clocks" associated with different processes in the living organisms is the well recognized notion.

Works by the well known etnographic/antropological business specialist E.T.Hall ([56]) showed that the notion of time **"internal to a specific human society"** is a **natural and very useful notion** in the ethnography, sociology (and, as the experience of Hall shows, in the international business relations).

First known to author, the introduction of the notion of **"material time"** and its use (for studying of the growth of crystals, evolution of plasma and of biological structures) was done by A.A.Vlasov in 1966 in his (rarely cited) monograph "Distribution Functions,[129] (in Russian), see Ch.10 of this monograph.

We also want to refer readers to the "INNER TIME" introduced by Ilia Prigogine in the book "From the Being to Becoming". Published by W.H. Freeman and Co. in 1980.

In the 1967 A.Chudnovsky discussed this notion (in the directions mentioned above in this section) (unpublished) with A. Lurie. A couple years later, A. Vakulenko published the paper where he used "thermodynamical time", [124, 126].

After this, the notion of a media with the "internal" or "material" time appeared from time to time - in works of K. Valanis (under the name "endoscopic systems") ([127, 128]), in the reviews of Kadashevich ([62]), Fedorovksy ([32]) and in works of A.Chudnovsky and author, [15, 16]. Finally,we mention the work of J.Kijowski, A.Smolski and A. Gornicka, [63] where material time is used for the description of the thermal properties of the perfect fluid in the Hamiltonian formulation of self-gravitating fluid.

49. Introduction to the material space-time.

The content of the rest of this Appendix IV represents a part of the work of A.Chudnovsky and the author on the geometrical modeling of aging phenomena in solids.

We seek to develop a model of inelastic processes in the aging materials by employing a 4-D inner material metric tensor G as the aging (damage) parameter of a material continuum.

Aging here implies any variation in the chemical make-up, i.e., chemical degradation, phase transformation, phase coarsening, nucleation, and growth of micro-defects such as dislocations and voids, shear bands, crazes, micro-cracks, etc. Material engineering and failure analysis indicate that, in addition to the stress and strain tensors, a parameter of state (the "aging" parameter) is needed to represent on a continuum level the sub-micro and micro-structural changes of a material. A kinetic equation for the evolution of the aging parameter will represent the aging process of a material. The equations of evolution for the material metric G are the Euler-Lagrange equations resulting from a Variational Principle. The conjugate force of the evolution of metric G (and of the related quantities characterizing the properties of the material) is the Energy-Momentum Tensor of Elasticity introduced by J.Eshelby (see [29, 30] or Sec.57 below).

A 3-D material metric g has long been employed as an internal variable in continuum mechanics. For example, it was used for studying the duality of material and physical relations of the Doyle-Erikson type in article [91], the thermodynamics of a continuum in [12], and in [28] where the curvature of material metric g defined by a uniformity mapping of a uniform material was employed as the driving force of the material evolution. We use the 4-D material metric G as the dynamical field that characterize the aging processes. G is introduce with the largest covariance group allowed by the condition that a small vicinity of each point of the material preserves its topology during the aging process (see Sec.48 below). We consider the 3-D material metric g on the slices B_t of constant physical time as one of the main dynamical variables following the ADM (Arnovitt, Deser, Misner) presentation of G (see [97, 2] or Sec.48.3 below). In that respect, we follow the tradition of the cited works. What is new in our work is that the smaller (in comparison to the General Relativity) covariance group of the Lagrangian allows us to use the lapse function S and the shift vector $\vec{N}$ as independent dynamical variables reflecting the proper material time scale and the intrinsic material flows respectively.

This aging parameter is justified by the observation of shrinkage associated with aging and the subsequent material density variation as well as a change of the resonance atomic frequencies and characteristic relaxation times measured in macroscopical studies. In other words, the internal length and time scales change with aging when compared to the corresponding absolute (physical) scales. The most sensitive indicator of aging is a variation of an intrinsic material time scale. The measurement of time in the laboratory as well as in material (intrinsic time) can be accomplished by several methods, the most common of which is the use of oscillating processes such as those found in clocks with a pendulum or crystal-based timepieces. Another way of measuring time is the use of a unidirectional evolution of state. In medieval Europe, for example, time was measured by burning a candle which had numbered and colored beeswax strips. Still another method is associated with relaxation processes which require an external input to enable a fading response. Electronic relaxation generators employing the discharge of a capacitor and the fading luminescence of phosphorus are both examples of relaxation processes, which are well suited for measuring intrinsic time scale changes because they reflect atomic or interatomic events.

Consider an external deformation of a material which responds with a specific change in its state. The decay or fading of the response constitutes the relaxation process. The decay can be described by an exponential function (within certain limits) e^{t/τ_0} where t is time and τ_0 is the time constant characterizing the rate of relaxation. Usually τ_0 becomes smaller with an increase in temperature or decrease in pressure. Phosphorous fades more slowly at colder temperatures, for example.

In the next section we discuss the kinematics of a media with a variable Riemannian metric G in a 4-D material space-time P, embedded into 4-D Absolute (Newton's) space-time M^4 with the Euclidean metric H. Thus, the 3-D "ground state" metric tensor is introduced together with the "proper time lapse function" and the material shift vector field. We consider mass conservation law in section 49. Elastic and inelastic strain tensors $E^{el}; E^{in}$ are introduced in section 50 as a measure of deformation and the "unstrained state" respectively. The Lagrangian describing inelastic and elastic processes in media is discussed in section 51. A variational formulation of aging theory and the Euler-Lagrange equations (equations of elasticity coupled with the aging equations) are considered in section 52. We present the combined system of elasticity and aging equations in section 53 and discuss special cases of the aging equations in section 54. In Section 55 we present two examples:Necking phenomena in polymers and the chemical degradation in the polymer pipes. Corresponding to the Material and laboratory symmetries, and the material and spacial balance laws corresponding to these symmetries are discussed in section 55. In section 56 the Energy-Momentum balance Law is presented in material form - using the Eshelby energy-momentum tensor (see [29, 30]). We write down the decomposition of the Energy-Momentum tensor into components, corresponding to the factors of the aging processes.

In section 58 we explore the application of this model to the basic inelastic processes- unconstrained aging, stress relaxation, and creep of a homogeneous rod. In appendix A to this Chapter we realize the strain energy as the perturbation of the "ground state energy" corresponding to the background nonelastic (usually slow) irreversible processes. In Appendix B we present, in the form of table (Table 3) the form of variations of the Action we use by the dynamical variables defined by the material metric G and defining inelastic (-aging) processes - g_{IJ}, S, N (see Section 48).

50. 4D kinematics of media with an inner (material) metric.

In this section we introduce the basic elements of the kinematics of a continuum with a variable metric, including material space-time P, 4D material metric G, 4D deformations ϕ, slicing of the material space-time by the surfaces of constant physical time $B_{\phi,t}$, and total, elastic and irreversible strain tensors.

50.1. **Physical and Material Space-Time.** Let us consider the 4-D Euclidean vector space $M = \mathbb{R} \times \mathbb{R}^3$ (**physical space-time**) with the standard Euclidean metric H. There exists the volume form d^4v corresponding to this metric.

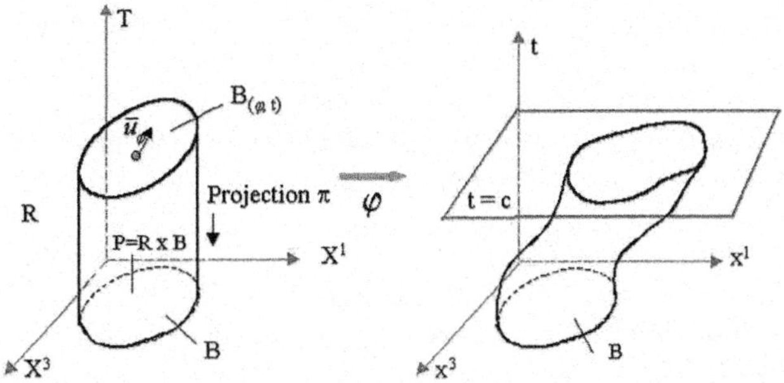

FIGURE 2. 4D Deformation History and the Physical time Foliation.

We select global coordinates $x^i, i = 1, 2, 3$, in the physical space $\mathbb{R}^3$ and $x^0 = t$ on the time axes $\mathbb{R}$. We have $H = dt^2 + h = dt^2 + \sum_i dx^{i\,2}$.

Hyperplanes $t = c$ are endowed with the 3D Euclidean metric h induced by H. We extend 3D tensor h to the degenerate (0,2)-tensor $\hat{h}$ in M, taking $\hat{h}_{0i} = \hat{h}_{i0} = \hat{h}_{00} = 0$.

A solid is considered here, in a conventional way as a 3D manifold B with the boundary ∂B, i.e. a set of "idealized" material points. We will use local coordinates $X^I, I = 1, 2, 3$ which, incidently, may be global coordinates induced by a **reference configuration** i.e., a diffeomorphic embedding $\phi_{0\,3} : B \longmapsto \mathbb{R}^3$ ([90]).

Cylinder $P = \mathbb{R} \times \bar{B}$ (with the coordinates $(X^0 = T, X^I, I = 1, 2, 3)$) is equipped with the 4D Riemannian metric G (material metric) with the components G_{IJ} relative to the coordinates $X^0 = T$, $X^I, I = 1, 2, 3$. Space (P, G) is further referred to as the **material "space-time"**.

Metric G defines the 4D volume form $dV = \sqrt{|G|}d^4X$, where $|G|$ is the determinant of the matrix (G_{IJ}).

An example of such a material metric G can be constructed as follows. Extend the "reference configuration" $\phi_{0\,3}$ to the diffeomorphic embedding $\phi_0 : P \longmapsto M$, $\phi_0(T, X^I) = (T, x^i = \phi^i_{0\,3}(X^I))$. Let G_0 be the metric $\phi_0^*(H)$ (here and below we denote by $\phi^* Q$ the pullback of a covariant tensor Q by the differentiable mapping ϕ). In the coordinates (X, T) the matrix of the metric G_0 is $\begin{pmatrix} 1 & 0 \\ 0 & h_{IJ} \end{pmatrix}$. Denote by $dV_0 = \sqrt{|G_0|}d^4X = \phi_0^*(dv^4)$ the 4D-volume element defined by the metric G_0.

Projection $\pi : P \longmapsto B$ along T-axes plays the same role in the construction below as in the relativistic elasticity theory ([11],[64]). In particular, we require invariance of Lagrangian theory with respect to the automorphisms of the bundle (P, π, B) (diffeomorphisms of material space-time P onto itself, projecting to B, so that material points retain their identity during the material evolution) preserving the direction of the flow of the "intrinsic" time (see below), but not with regard to the whole group of diffeomorphisms of P as in Gravity Theory.

50.2. Deformation History.

The **history of the deformation** of the body B is represented by a **diffeomorphic embedding** $\phi : P \longmapsto M$ of the material space-time P into the physical (Newtonian) space-time M (see Fig.1).

A deformation history ϕ for which $t = \phi^0(X) = T$ will be called **"synchronized"**. The synchronization can practically performed for relatively slow deformation processes (in comparison to sound wave velocity).

Using the deformation ϕ, we introduce the slicing of the material space-time P by the level surfaces of the zeroth component of ϕ

$$B_\phi = \{B_{\phi,t} = \phi^{0\;-1}(t)\} = \{(T,X) \in P | \phi^0(T,X) = t\}. \tag{50.1}$$

For a synchronized deformation $B_{\phi,t} = \{(T,X) \in P | T = t\}$.

There is a **time flow vector field** u_ϕ in P, associated with the slicing $B_{\phi,t}$ of the space-time P ([97], [11]). This (future directed) vector field represent the flow of "intrinsic" (proper in Relativity Theory) time in the material. Lifting the index in the 1-form $d\phi^0$ with the help of the metric G and normalizing obtained vector field, we define the time flow vector field as

$$u_\phi = \frac{(d\phi^0)^\#}{\|d\phi^0\|_G}. \tag{50.2}$$

The norm of the 1-form $d\phi^0$ is defined as $\|d\phi^0\|^2 = (G^{AB}\phi^0_{,A}\phi^0_{,B})^{1/2}$ (summation agreement by repeating indices is used here). Thus, u_ϕ is the unit vector G-orthogonal to the slices $B_{\phi,t}$. In the local coordinates X^I,

$$u_\phi = \frac{G^{AB}\phi^0_{,B}}{\sqrt{G^{AB}\phi^0_{,A}\phi^0_{,B}}}\frac{\partial}{\partial X^A}. \tag{50.3}$$

For the synchronized deformations does not depend on ϕ:

$$u_\phi = u_G = \frac{G^{I0}}{\sqrt{G^{00}}}\frac{\partial}{\partial X^I}. \tag{50.4}$$

Additionally, if the metric G has the block-diagonal form in the coordinates $(X^0 = T, X^I)$ (shortly, BD - metric), we have $u_G = [G_{00}]^{-\frac{1}{2}}\frac{\partial}{\partial T}$.

Let $u_0 = \frac{\partial}{\partial T}$ be the flow vector associated with the metric G_0 and the corresponding 3D slicing B_0.

We require fulfillment of the following condition ensuring the irreversibility of the flow of time:

$$< \langle u_\phi, u_0 >_G \rangle > 0. \tag{50.5}$$

Deformation history ϕ for which the condition (48.5) is satisfied is called **admissible**.

In coordinates (X^I) this condition reduces to the following simple inequality

$$\phi^0_{,0} > 0 \tag{50.6}$$

and, therefore is a restriction on the deformation history only.

For a synchronized deformation history ϕ, this condition is trivially satisfied.

Time component ϕ^0 of the deformation history may be excluded from the list of dynamical variables by an appropriate "gauging". Namely, we use the invariance of Lagrangian under the automorphisms of the bundle (P, π, B) to make the deformation history synchronized.

An automorphism $F : P \longmapsto P, X^I = F^I(Y^A)$ of the bundle (P, π, B) determines the diffeomorphism of the base B that can be considered as a change of variables $X^I = F^I(Y^A)$.

In the new variables, the condition (48.5) takes the form $\frac{\partial \phi^0}{\partial Y^0} = \phi^0_{,I} F^I_{,0} = \phi^0_{,0} F^0_{,0} > 0$ (since F^I, $I = 1, 2, 3$ do not depend on Y^0). Thus, the class of admissible deformation histories ϕ is stable under the action of the subgroup $Aut^+(P)$ of all automorphisms of P with $F^0_{,0} > 0$.

The group $Aut^+(P)$ of automorphisms of the bundle contains two subgroups. One is the subgroup TC of the "time change" proper gauge diffeomorphisms ($X^0 = T; X^I, I = 1, 2, 3) \longmapsto (F(T, ; X^J, J = 1, 2, 3), X^I, I = 1, 2, 3)$ for an arbitrary smooth function $F(X^I)$ with $F_{,0} > 0$. The other one (denoted $D(B)$) consists of the lifts to the slices $B_{\phi,t}$ of the manifold P of the orientation preserving diffeomorphisms of the base B (group of such transformations of B is denoted $Diff^+(B)$). To lift a diffeomorphism we use (diffeomorphic) projections $\pi_{\phi,t} = \pi|_{B_{\phi,t}} : B_{\phi,t} \longmapsto B$. If ϕ is synchronized, lifted diffeomorphisms do not depend on T.

Any automorphism $\phi \in Aut^+(P)$ of the bundle $\pi : p \to B$ generates the time independent diffeomorphism ϕ_B of the base B, which is an element of $Diff^+(B)$. Lifting this element to the element of $D(B)$ we represent the group $Aut^+(P)$ as the semi-direct product of the normal subgroup $TC(P)$ and the subgroup $D(B)$. Thus, we have proved the first of two following statements

(1) Automorphism group $Aut^+(P)$ is the semidirect product of the subgroup $D(B) \sim Diff^+(B)$ of orientation preserving diffeomorphisms of the base B and the subgroup TC of proper gauge transformations ξ_F of the fibers

$$\xi_F : ((X^0 = T; X^I, I = 1, 2, 3) \longmapsto (F(T, ; X^J, J = 1, 2, 3), X^I, I = 1, 2, 3))$$

with $F_{,0} > 0$.

(2) For any admissible history of deformations ϕ one can choose a transformation $\xi_F \in TC$ such that the history of deformation $\phi \circ \xi_F$ is synchronized.

To prove the second statement let ϕ be an admissible history of deformation. Define the element $F \in TC(P)$ as follows: $F : (T, X^I, I = 1, 2, 3) \longmapsto (\phi^0(T, X^J, J = 1, 2, 3), X^I, I = 1, 2, 3)$. Then, $\phi = \phi_1 \circ F$ where ϕ_1 is another admissible history of deformation with the same components $\phi^i, i = 1, 2, 3$ and the identity component $\phi^0_1(T, X^I, I = 1, 2, 3) = T$. Transformation F is admissible since $F^0_{,0} = \phi^0_{,0} > 0$, thus $F \in TC(P)$. Apparently, the deformation history ϕ_1 is synchronized.

Therefore we may restrict our consideration to the synchronized histories of deformation keeping in mind that the covariance group of the theory reduces from the group $Aut^+(P)$ to the group $D(B)$ of time-independent orientation preserving diffeomorphisms of base B.

50.3. ADM-decomposition of Material Metric, Lapse and Shift.

Slicing $B_{\phi,t}$, generates the (3,1)-decomposition of the material metric G employed (for a Lorentz type metric) in General Relativity ([97],[2]). Specifically, the sandwich structure of the part of the manifold P bounded by the surfaces $B_{\phi,t}$ and $B_{\phi,t+\Delta t}$ allows one to introduce the time-dependent lapse function N and the shift vector field $\vec{N}$ tangent to the slices $B_{\phi,t}$ such that the metric is block-diagonalized in the moving coframe $(dT, dX^A + N^A dT)$:

$$ds^2 = G_{IJ} dX^I dX^J = g_{IJ}(dX^I + N^I dT)(dX^J + N^J dT) + S^2 dT^2. \qquad (50.7)$$

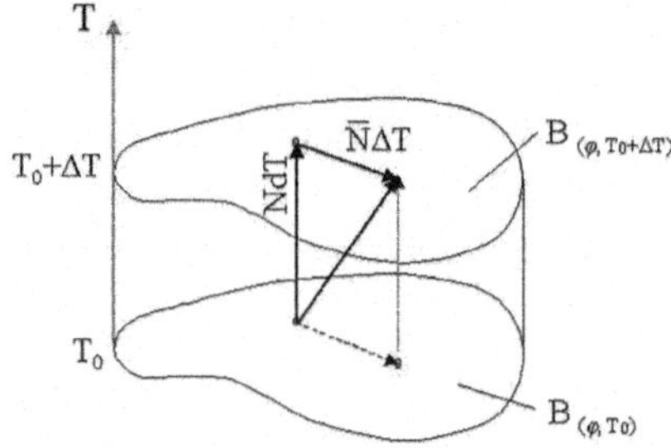

FIGURE 3. Lapse Function S and Shift Vector $\vec{N}$.

The matrix representation of the material metric tensor G and inverse tensor G^{-1} have in these notations, the forms

$$\begin{pmatrix} G_{00} & G_{0J} \\ G_{I0} & G_{IJ} \end{pmatrix} = \begin{pmatrix} N_A N^A + S^2 & N_J \\ N_I & g_{IJ} \end{pmatrix},$$

$$\begin{pmatrix} G^{00} & G^{0J} \\ G^{I0} & G^{IJ} \end{pmatrix} = \begin{pmatrix} \frac{1}{S^2} & -\frac{N^J}{S^2} \\ -\frac{N^I}{S^2} & g^{IJ} + \frac{N^I N^J}{S^2} \end{pmatrix}, \quad (50.8)$$

where g is the 3D-metric induced by G on the slices $B_{\phi,t}$ and g^{-1} is the corresponding inverse tensor. In this notation $\sqrt{|G|} = S\sqrt{|g|}$.

In what follows we assume that the 4D-deformation history ϕ is synchronized. Thus, the slices $B_{\phi,t}$ have the form $T = t = const$.

The flow vector u_G and the corresponding 1-form have, in these notations, the form

$$u_G = \frac{1}{S}\partial_T - \frac{N^A}{S}\partial_{X^A}, \quad u_G^\flat = SdT. \quad (50.9)$$

The last formula gives the "material time differential" $d\tau = SdT$ for the material metric G. In our context, the coordinate X dependence of lapse function S(T,X) accounts for heterogeneity of material aging in different points of the solid. On the Fig. 3 the local observer at different points of body at different moments of time T sees the different rate of the local time in comparison with the laboratory clocks.

Moreover, the lapse function S can be considered as an intrinsic measure of material age, associated with its cohesiveness. It can be normalized to be equal 1 in the reference state of the solid. As a result of energy dissipation in various inelastic processes, material loses its cohesiveness with aging. In the formalism presented here it is manifested in the slowing down of the material (intrinsic) time, i.e. increasing of $S(X,T)$.

Here we do not consider thermodynamics. However, monotonic increase of $S(X,T)$ resembles and can be linked to the principle of non-negative entropy production of the thermodynamics of irreversible processes: $\frac{dS_{in}}{dt} > 0$.

The requirement $S > 1$ leads to the strong constraints on the form of the "ground state term" of the material Lagrangian L_m, (see Sec. 51).

In this context, the shift vector field $\vec{N}$ in the metric G reflects a propagation of the phase transition or chemical transformation boundary through the material as reflected, for instance in the mass conservation law (see below).

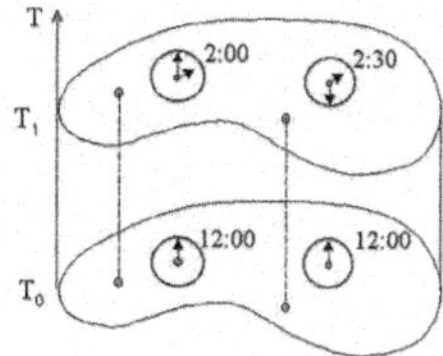

FIGURE 4. Illustration of Time Rate S (Lapse Function) by Variation of Time Interval $(\tau - \tau_0) = S(T_1 - T_0)$.

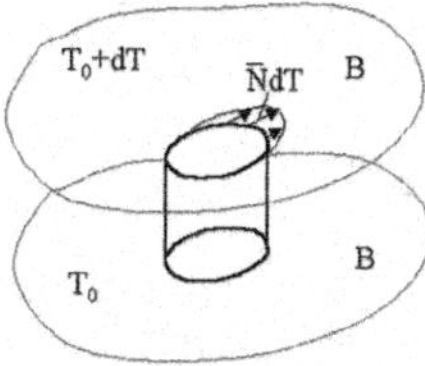

FIGURE 5. Propagation of Phase Boundary Represented by Shift Vector $\vec{N}$

The separation of the evolution of 3D material metric g, material transformation process in B_t, characterized by $\vec{N}$ and the inhomogeneity of the local time $d\tau = SdT$, is the main reason for introducing (3+1) ADT-representation of 4D-metric in Gravity Theory ([97]). In addition, an adoption of this view leads to a very clear separation of the "physical" degrees of freedom in the canonical formalism and to the explicit hyperbolic formulations of Einstein Equations ([1, 34]).

51. MASS CONSERVATION LAW

The **mass form** $dM = \rho_0 dV$ defined in P is introduced here, in addition to the volume form dV_G of metric G. The reference mass density ρ_0, defined by this representation, satisfies the **mass conservation law** ([90],[11])

$$\mathcal{L}_{u_\phi} dM = d(i_{u_\phi} dM) = 0. \tag{51.1}$$

Here $\mathcal{L}_{u_\phi}$ is the Lie (substantial) derivative of the exterior 4-form dM in the direction of the vector field u_ϕ. Recall that the Lie derivative of a differential form ω along a vector filed u is defined as $\mathcal{L}_u \omega = \frac{d}{dt}\phi_t^* \omega|_{t=0}$, where $\phi_t^* \omega$ is the pullback of the form ω by the flow $t \longrightarrow \phi_t = exp(tu)$ of the vector field u ([90]).

Equation (49.1) is equivalent to the condition $div_G(\rho_0 u_\phi) = 0$, where divergence is taken with respect to the volume form dV. In local coordinates the Mass Conservation Law has the form

$$\left(\frac{G^{IB}\phi^0_{,B}}{\|d\phi^0\|} \rho_0 \sqrt{|G|} \right)_{,I} = 0. \tag{51.2}$$

Due to the properties of the metric G and the deformation ϕ, the material space-time manifold P is foliated by the phase curves of the flow vector field u_ϕ and thus the value of the reference mass density $\rho_0(0, X)$ at $T = 0$ uniquely defines its values for all $T > 0$.

In the synchronized case $\phi^0 = T$, $\|d\phi^0\| = \sqrt{G^{00}} = S^{-1}$, $G^{M0} = -\frac{N^M}{S^2}$, $G^{00} = S^{-2}$ and (49.2) take the form of the following balance law

$$(\rho_0\sqrt{|g|})_{,0} = \sum_{A=1}^{3}\left(N^A\rho_0\sqrt{|g|}\right)_{,A}. \tag{51.3}$$

From (3.3) we note that the shift vector field $\vec{N}$ can describe the **matter (density) flow due to the some internal processes** such as the phase or chemical transformations.

If, in addition to being synchronized, the material metric G is also in the BD-form ($\vec{N} = 0$), the flow term in (49.3) disappears and the mass conservation law is equivalent to the following representation of the reference mass density in terms of its initial value $\rho_0(0, X) = \rho_0$:

$$\rho_0(T, X^I) = \rho_0(0, X)\sqrt{\frac{G_{00}}{|G|}} = \frac{\rho_0(0, X)}{\sqrt{|g(T, X)|}}, \tag{51.4}$$

where $G(0, X)$ is assumed to be Euclidean metric. If metric G is not changing with time T, we get the classical local mass conservation law $\frac{\partial \rho_0}{\partial T} = 0$.

52. Elastic, Inelastic and Total Strain Tensors

In this section we introduce the principal quantities characterizing both elastic and inelastic deformation processes. Total deformation is presented as a composition of elastic and inelastic ones and is integrable. Its elastic and inelastic "components" are non-integrable, in general, but might be such in special situations (see Sec.11). We recall that the presentation of total deformation as a composition of this type was studied in different forms in many works ([73, 82, 95], to name a few). What is new here is the 4D-approach to this decomposition and direct definition of elastic, inelastic and total strain tensors in terms of material metric g_t as an independent dynamical variable, reference (undeformed) Cauchy metric g_0 and the current Cauchy metric $C_3(\phi)$ rather then using the "deformation gradients" (integrable or not) of elastic and inelastic (plastic) deformations.

Slicing $B_{\phi,t}$ of P defines the covariant tensor $\gamma = G - u_\phi \otimes u_\phi =^s \begin{pmatrix} N_A N^A & N_J \\ N_I & g_{IJ} \end{pmatrix}$

([11],[97]). Here and later the sign s over $=$ means that this equality is true in synchronized case. Tensor γ induces the time dependent 3D-metric g_t on the slices $B_{\phi,t}$ (see, for example, [34]).

To obtain the expression for g_t in material coordinates X^I, we notice that the tangent vectors

$$\xi_I = -\frac{\phi^0_{,I}}{\phi^0_{,0}}\partial_{X^0} + \partial_{X^I}, \quad I = 1, 2, 3 \tag{52.1}$$

form the basis of the tangent spaces to the slices $B_{\phi,t}$. In this basis, g_t is given by

$$g_{AB} = g(\xi_A, \xi_B) = G_{AB} - G_{0B}\frac{\phi^0_{,A}}{\phi^0_{,0}} - G_{A0}\frac{\phi^0_{,B}}{\phi^0_{,0}} + G_{00}\frac{\phi^0_{,A}\phi^0_{,B}}{\phi^0_{,0}{}^2}, \quad A, B = 1, 2, 3. \tag{52.2}$$

For a synchronized deformation $\gamma = G - (G^{00})^{-1}dT \otimes dT$ and g_t is simply the restriction of 4D-metric G to the slices $B_{T=c}$, i.e. $g_{t\,IJ} = G_{IJ}|_{T=t}$, see (48.8).

Denote by h_t the 3D metric on the leaves $B_{\phi,t}$ induced by the metric G_0 (that is by the tensor $\gamma_0 = G_0 - u_0 \otimes u_0$).

Associated with the tensor γ there is the projector $((1,1)$-tensor$)$

$$\Pi = G^{-1}\gamma = I - u_\phi^* \otimes u_\phi =^s I - \frac{G^{I0}}{G^{00}}\partial_{X^I} \otimes dT = \begin{pmatrix} 0 & 0 \\ N^I & I_3 \end{pmatrix}. \qquad (52.3)$$

on the tangent spaces to the slices $B_{\phi,t}$, with the last equality being true for synchronized deformations ϕ.

Let us consider the pullback $C_4(\phi) = \phi^*\hat{h}$ of the degenerate tensor $\hat{h}$ by the 4D-deformation mapping ϕ. Tensor $C_4(\phi)$ is degenerate in P, its kernel is generated by the vector $\phi_*^{-1}(\frac{\partial}{\partial t})$. In coordinates (X^I) we have

$$C_4(\phi)_{IJ} = \begin{pmatrix} h_{ij}\phi_{,0}^i\phi_{,0}^j = \|\mathbf{V}\|_h^2 & h_{ij}\phi_{,0}^i\phi_{,J}^j \\ h_{ij}\phi_{,I}^i\phi_{,0}^j & \sum_{i,j=1}^3 h_{ij}\phi_{,I}^i\phi_J^j \end{pmatrix}. \qquad (52.4)$$

The spatial part of this tensor is the conventional Cauchy-Green strain tensor $C_3(\phi)$ of the Elasticity Theory. Components of this tensor with indices $(0J)$ and $(I0)$, $I,J = 1,2,3$ have the form *velocity* $\times$ *deformation covector* (see [90]). (00)-component of $C_4(\phi)$ is the square of the material velocity $\mathbf{V} = \phi_{3*}(\frac{\partial}{\partial T}) =^s \frac{\partial\phi^i}{\partial t}\frac{\partial}{\partial x^i}$.

52.1. Elastic Strain Tensor.

Here we *define* the 4D *(1,1)-elastic strain tensor* in P. We will do it first in linear approximation and then, using logarithm of a $(1,1)$-tensor function, in another way, more suitable for large deformations.

We start with the following, conventional definition:

$$\hat{E}_{\cdot}^{el} = \frac{1}{2}G^{-1}(C_4(\phi) - \gamma) =^s$$

$$= \frac{1}{2}\begin{pmatrix} S^{-2}\|\mathbf{V}\|_h^2 - S^{-2}N^I\langle\mathbf{V},\phi_{,I}\rangle_h & S^{-2}\langle\mathbf{V},\phi_{,J}\rangle_h - S^{-2}N^K\langle\phi_{,K}\phi_{,J}\rangle_h \\ -S^{-2}N^I\|\mathbf{V}\|_h^2 - N^I+ & -S^{-2}N^I\langle\mathbf{V},\phi_{,J}\rangle_h + g^{IK}C_3(\phi)_{KJ}- \\ +(S^{-2}N^IN^K + g^{IK})\langle\phi_{,K},\mathbf{V}\rangle_h & -\delta_J^I - S^{-2}N^IN^K\langle\phi_{,K}\phi_{,J}\rangle_h \end{pmatrix}. \qquad (52.5)$$

This tensor contains the square of material velocity and the shift vector field. Having in mind the general, dynamical situation it is more appropriate to use the following tensor as the proper Elastic Strain Tensor

$$E_{\cdot}^{el}(\phi) = \Pi\hat{E}^{el}\Pi = \frac{1}{2}\Pi G^{-1}(C_4(\phi)-\gamma))\Pi =^s \frac{1}{2}\begin{pmatrix} 0 & 0 \\ g^{IK}\langle\phi_{,K},\phi_{,0}\rangle_h - N^I & g^{IK}C_3(\phi)_{KJ} - \delta_J^I \end{pmatrix}, \qquad (52.6)$$

Here Π is the projector on the slices $B_{\phi,t}$ defined in (48.1). The last equality is valid in the synchronized case. Notice that the basic invariants $Tr(A^k)$ for the tensor (50.6) are the same as for the tensor $C_3(\phi) - g$.

For simplicity we use the same symbol E^{el} for the restriction of this tensor to the slices $B_{\phi,t}$.

Tensor $E(\phi)^{el}$ is a measure of the deviation of the Cauchy metric $C_3(\phi)$ of the actual state from the "ground state" G. For a synchronized deformation ϕ and a material metric G with the zero shift vector, $E^{el} = \frac{1}{2}(C(\phi) - g_t)$ has the form of the conventional elastic strain tensor.

Remark 69. The deformation ϕ is **essentially 3-dimensional** in the sense that only the spatial Euclidean metric $g_0 = h$ in B is deformed. The 4D-tensor $C_4(\phi) = \phi^*h$ defines the metric in the material space-time P. Spacial part of this metric is the conventional Cauchy metric of elasticity. It is instructive to compare $C_4(\phi)$ with the (degenerate) tensor $\gamma = G - u_\phi \otimes u_\phi$. The elastic strain tensor E^{el} measures the deviation of $C_4(\phi)$ from γ on the slices $B_{\phi,t}$. Thus, the scheme presented here is essentially different from relativistic elasticity theory ([11],[64]) as well as from the 4D version of conventional elasticity theory.

We see from (50.6) that $E^{el} = 0$ if and only if the following two conditions are fulfilled:

$$\begin{cases} 1)\ g_{ij} = C_3(\phi)_{IJ} = \phi_3^*(h)_{IJ}, \\ 2)\ N^I = g^{IK}\langle \phi_{;0}, \phi_{;K}\rangle. \end{cases} \tag{52.7}$$

In particular, metric g coincides with the Cauchy metric induced by deformation ϕ and is flat.

. If $E^{el} = 0$ then $\hat{E}^{el} = 0$ if and only if, in addition to the conditions (50.7), the following condition is fulfilled

$$3)\ \|V\|_h^2 = g^{IJ}\langle \phi_{;0}, \phi_{;I}\rangle\langle \phi_{;0}, \phi_{;J}\rangle. \tag{52.8}$$

52.2. Inelastic Strain Tensor.

Now we introduce the *inelastic strain tensor* in linear approximation

$$\hat{E}^{in} = \frac{1}{2}G^{-1}(\gamma - \gamma_0) =^s \frac{1}{2}\begin{pmatrix} 0 & S^{-2}N^K h_{KJ} \\ N^I & \delta_J^I - g^{IK}h_{KJ} - S^{-2}N^I N^K h_{KJ} \end{pmatrix}, \tag{52.9}$$

(The last equality being true in the synchronized case) and the *total strain tensor* $\hat{E}^{tot}$ of the body at each given moment T to characterize the deviation of the deformed Euclidean metric $\phi^* h|_{B_{\phi,t}}$ from the initial (Euclidean) 3D-metric h (h being the restriction of G_0 to the slices $B_{\phi,t}$)

$$\hat{E}^{tot} = \frac{1}{2}G^{-1}(C_4(\phi) - \gamma_0). \tag{52.10}$$

Tensor $\hat{E}^{tot}$ can be represented as the sum of the elastic strain tensor $\hat{E}^{el}$ and of inelastic strain tensor $\hat{E}^{in}$:

$$\hat{E}^{tot} = \hat{E}^{el} + \hat{E}^{in}. \tag{52.11}$$

To obtain the corresponding decomposition for the 3D strain tensors we apply projector Π to the total and inelastic strain tensors. In particular we introduce

$$E^{in} = \Pi\hat{E}^{in}\Pi =^s \frac{1}{2}\begin{pmatrix} 0 & 0 \\ N^I - g^{IK}h_{KB}N^B & \delta_J^I - g^{IK}h_{KJ} \end{pmatrix}, \tag{52.12}$$

As a result, we get from (50.11) the corresponding decomposition of the 3-dim total strain tensor"

$$\Pi E^{tot}\Pi = E^{el} + E^{in}. \tag{52.13}$$

Restriction of these tensors on the 3D slices $B_{\phi,t}$ leads to the more conventional (t-dependent) version of this decomposition.

For a synchronized deformation history, restriction of E^{in} to the slices B_t takes the form

$$E^{in}|_{B_t} = \frac{1}{2}g^{-1}(g - g_0) = \frac{1}{2}(I - g^{-1}g_0), \tag{52.14}$$

that describes the decline of 3D material metric g from its initial (reference) value $g_0 = \phi_0^* h$.

Another way to define strain tensors, more suitable for description of large deformation is to take

$$\hat{E}^{el} = \frac{1}{2}ln(G^{-1}C_4(\phi))), \hat{E}^{in} = \frac{1}{2}ln(G_0^{-1}G)), \hat{E}^{tot} = \frac{1}{2}ln(G_0^{-1}C_4(\phi))). \qquad (52.15)$$

We can define E^{el}, E^{in}, E^{tot} correspondingly, using projector Π. Strain tensors, defined in such a way will, in some simple cases, enjoy the same additive relations as (50.11),(50.13). In the other case, if elastic deformation happens in the directions different from the principal axes of inelastic deformation, the relation between these deformations becomes more complex.

The relationship between these definitions and those of the linear approximation above is established by using the fact that for a couple A, B of (0,2)-tensors such that A is invertible, $ln(A^{-1}B) \approx A^{-1}(A-B)$ provided $A-B$ is small enough. Thus, when linear approximation is allowable, the first definition is a good approximation of the second. For instance

$$ln(g^{-1}C_3(\phi)) = ln(I + g^{-1}(C_3(\phi) - g)) \approx g^{-1}(C_3(\phi) - g), \qquad (52.16)$$

provided $C_3(\phi) - g$ is small.

52.3. **Strain Rate Tensor.** One can also define the material elastic strain rate tensor as follows

$$\dot{\hat{E}}^{el} = \mathcal{L}_{u_\phi}\hat{E}^{el}, \qquad (52.17)$$

as well as inelastic strain rate tensor

$$\dot{\hat{E}}^{in} = \mathcal{L}_{u_\phi}\hat{E}^{in}. \qquad (52.18)$$

In the case where $G = G_0$ and $\phi^0 = T$, elastic strain rate tensor defined in (50.17) has, the same spatial components as the conventional strain rate tensor ([90]).

Denote by $\dot{G}$ the the Lie derivative $\dot{G} = \mathcal{L}_{u_\phi}G$ of the metric tensor G with respect to the flow vector $\vec{u}_\phi$. Then the calculation of the Lie derivative in (50.17) results in the following relation

$$\dot{\hat{E}}^{el} = \mathcal{L}_{u_\phi}\hat{E}^{el} = -G^{-1}\dot{G}\hat{E}^{el} + \frac{1}{2}G^{-1}(C_4(\phi) - K), \qquad (52.19)$$

where $K = \mathcal{L}_{u_\phi}\gamma$ is the **extrinsic curvature tensor** of the slices $B_{\phi,t}$.

Remark 70. Here we are using material coordinates and tensors only. In order to obtain the corresponding "laboratory" quantities (seen by an external observer), one defines the laboratory (Euler) Elastic Strain Tensor

$$\epsilon^{el\ i}_j = \phi^i_{,A}\phi^{-1\ B}_j E^{el\ A}_B \qquad (52.20)$$

and recalculate all the other quantities accordingly.

Figure 5 presents the above decomposition of total deformation into the inelastic and elastic deformations.

The actual state under the load at any given moment T results from both elastic (with the variable elastic moduli) and inelastic (irreversible) deformations. The "ground state" of the body is characterized by the 3D-metric g_t. This state is the background to which the elastic deformation is added to reach the actual state.

The transition from the reference state to the "ground state" that manifests in the evolution of the (initial) Euclidean metric h to the metric g_t cannot be described, in general, by any point transformation. Transition from the "ground state" to the

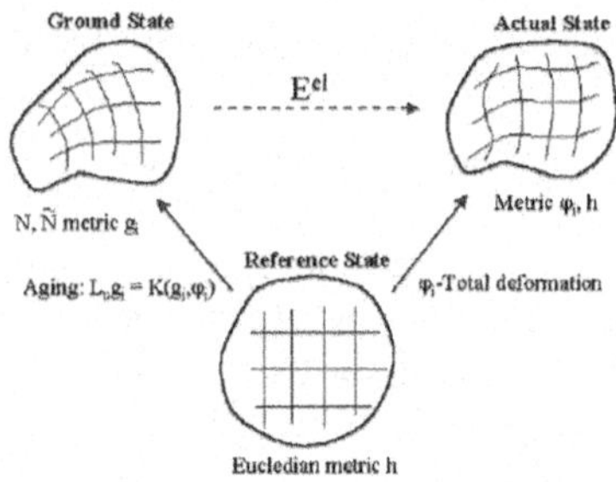

FIGURE 6. Decomposition of Deformation History.

actual state at the moment t is also not compatible in general. Yet the transition from the reference state to the actual state is represented by a diffeomorphism ϕ_t.

Here we are considering the **material 4D-metric** G and the **deformation** ϕ (or elastic strain tensor $E^{el}(\phi)$) to be the dynamical variables of the field theory. The reference mass density ρ_0 is found (for the synchronized deformation ϕ and the BD-metric G) by the formula (49.4) if its initial value $\rho_0(T = 0)$ is known. In this study we consider mainly the quasi-static version of the theory, i.e. inertia forces and kinetic energy are assumed to be negligible.

53. Parameters of Material Evolution, metric Lagrangian.

In examining the processes of deformation and aging of a solid with a synchronized deformation history ϕ we use both general (G_{IJ}) and ADM $(S, \vec{N}, g)$ notation for the 4D material metric G.

At this step one has to choose the "dissipation driver" the factor responsible for the evolution of material metric and therefore, that of inelastic strain to use in the Lagrangian formalism.

There are several alternatives here as presented in the Chapters 2, 4 and 5.

Specifically, one can use the non-commuting variations defined by a vertical connection in the bundle $J^k\pi \to Y$ with the appropriate choice of the NC-tensor K. Another way is to use a special case of the NC-method, - to introduce a dissipative potential D (see Sec.42).

Finally, one can use **material time** τ instead of physical time T (see below, Sec.55).

Remark 71. Using material time τ instead of the conventional (laboratory) time t and considering the material time rate S as the dynamical variable has the effect similar to the use of non-commutative variations. It was clearly demonstrated in Section (45).

Following the framework of Classical Field Theory, see [10, 80], we take a Lagrangian density $\mathcal{L}(G, R(G); \phi)$ referred to the **volume form** dV as a function depending on 4-dimensional material metric G and the invariants of metric G (with respect to the group of $Diff^+(B)$ of the orientation preserving diffeomorphisms of the base B) - Extrinsic curvature $\mathcal{K}$, scalar curvature, inelastic strain tensor and the elastic strain tensor: $L(G, E^{el})$.

Remark 72. In cases where some thermal and other dissipative processes may be present, the Lagrangian can contain corresponding terms, see Sections 44=45.

The Lagrangian $L = L(G, E^{el})dV$ is taken to be a sum of the two parts: the **"ground state"** term $L_m(G)$ depending, in general, on the external curvature $\mathcal{K}$ and the scalar curvature $R(G)$ and, as the *perturbation of the ground state*, the **elastic part** $L_e(G, E^{el})$ associated with elastic deformations

$$L = L_m(G) + L_e(E^{el}, G). \tag{53.1}$$

Lagrangian $L_m(G)$ in (51.1) is introduced to account for the "cohesive energy" or strength of the solid state, the strain energy of "residual strain" and the energy of the change associated with a evolution of material properties in time, for instance material aging processes of phase transitions.

Lagrangian $L_m(G)$ is the sum of several terms with the coefficients that may depend on the 3D volume factor $|g_t|$ (actually $|g_t|/|g_0|$) and the lapse function S (see below (49.4)). These volume factors are associated with the solid state ability to retain its intrinsic topology in contrast to the fluid and gaseous states.

The second (kinetic) term in L_m (see (51.4) below) is the function of invariants of the tensor $K = \mathcal{L}_{u_G}\gamma$ of extrinsic curvature of the slices B_T in the material space-time P ([34, 97]).

In the ADM notations, the (1,1)-tensor K has the following form

$$(K^I_J) = \begin{pmatrix} 0 & 0 \\ (*) & S^{-1}g^{IS}(\xi \cdot g)_{SJ} \end{pmatrix}, \tag{53.2}$$

where $(*)$ represents the terms in the (2x1) cell of matrix (48.8). These terms do not enter in the invariants of K and $\xi \cdot g = \frac{\partial}{\partial T}g - \mathcal{L}_{\vec{N}}g$. Lie derivative $\mathcal{L}_{\vec{N}}g$ of the metric tensor g with respect to the vector field $\vec{N}$ is calculated on each 3D slice $B_{\phi,t}$ for a fixed t.

In the case of a block-diagonal metric G (no shift: $\bar{N} = 0$) and tensor γ has the form $\gamma = \begin{pmatrix} 0 & 0 \\ 0 & g_t \end{pmatrix}$; therefore, K is, essentially, the time derivative of the metric g_t:

$$K^I_J = \begin{pmatrix} 0 & 0 \\ 0 & S^{-1}g^{IA}\frac{\partial}{\partial T}g_{AJ} \end{pmatrix}. \tag{53.3}$$

As a result, tensor K represents **the rate of change of intrinsic length scales that reflects the aging processes.** It shall be noted that K is also related to the elastic strain rate (see (50.19)).

Lagrangian $L_m(G)$ may also depend on the shift vector field $\vec{N}$ through its norm $\|\vec{N}\|^2_g = N_A N^A$ (entering the "ground state energy" F) and, possibly, divergence $div_g(\vec{N})$ and the "proper time derivative" $\mathcal{L}_u\vec{N}$.

We also include the term reflecting the residual strain energy (incompatibility) of E^{tot} which is accounted for by the scalar curvature $R(g)$ of 3D material metric g.

Summarizing the above assumptions we construct the metric Lagrangian L_m as the scalar function of the parameters listed above:

$$L_m = F(S, \|\vec{N}\|^2_g, E^{in}) + \chi(\mathcal{K}) + \alpha \cdot div_g(\vec{N})^2 + \beta R(g_t). \tag{53.4}$$

The first term of L_m is the "ground state energy" $F(E^{in}, S, \|\vec{N}\|^2_g)$ (shortly GS) - initial ("cohesive") energy (per unit volume).

Dissipative potential χ - a function of invariants of tensor K^I_J is the the moving factor of inelastic processes in the material [b].

Coefficients α, β may depend on S and $|g|$.

In the case of a homogeneous media or in the 1D case, the scalar curvature $R(g_t)$ of the metric g_t is zero and the corresponding term in (51.4) vanishes.

Given the diversity of the material properties and the different conditions (of loading, boundary, forces, heat,etc.) of inelastic processes affecting the material it is especially important to choose the material Lagrangian of different materials appropriately. It appears as if the different conditions activate different "layers" of structural changes for a given material and, correspondingly, turn on different terms in the "ground state energy" and in the "dissipative potential" that are responsible for the given type of aging. For example, the slow process of unconstrained aging in a homogeneous rod (see Sec.55.2 below or [16]) is overcome by the scale processes of a stress relaxation or creep each of which begins in a different loading situation after the strain energy (density) reaches a (different) activation level. For these two processes both ground energy F and the dissipative potential $\chi(\mathcal{K})$ have the same form different from those for unconstrained aging.

Remark 73. The scalar curvature $R(G)$ of the 4D metric G can be expressed, by the Gauss equation, as the combination of scalar curvature of 3D metric g and of invariants of its **extrinsic curvature**: $R(G) = -(tr(\mathcal{K}^2) - (tr\mathcal{K})^2) + R(g_t)$, up to a divergence term ([34]). As a result, the above form of Lagrangian for an aging media (5.4) is a generalization of the Hilbert-Einstein Lagrangian $R(G)\sqrt{|G|}$ of General Relativity ([97]). By breaking the invariance group of general relativity to the smaller group of automorphisms of the bundle $P \to B$ we can use a more general form of metric Lagrangian.

The perturbation of Lagrangian due to elastic deformation is taken in the form of the Lagrangian of Classical elasticity ([90], Sec.5.4)

$$L_e(E^{el}, G) = \frac{\rho_0}{2}\|V\|^2_h - \rho_0 f(E^{el}, G) - \rho_0 U \circ \phi, \tag{53.5}$$

where $\rho_0\|V\|^2_h = \rho_0 \sum_{ij} h_{ij}\phi^i_{,0}\phi^j_{,0}$ is the density of kinetic energy, f is the strain energy per unit of mass, and U is the potential of the body forces. Strain energy f is assumed to be a function of two first invariants of the (1,1)-strain tensor E^{el}. Strain energy may depend on the metric G through the invariants of $g_0^{-1}g_t$, S, vector field $\vec{N}$, scalar curvature $R(g)$ etc.

Because we are considering a quasistatic synchronized theory here we ignore the inertia effects and, therefore, omit the kinetic energy term in (5.5).

The strain energy density in linear elasticity is conventionally presented as follows

$$f(E^{el}) = \frac{\mu}{2}Tr(E^{el\ 2}) + \frac{\lambda}{2}(Tr(E^{el}))^2,$$

where μ, λ are the initial values of the Lame constants ([80]).

---

[b]Initially ([16]) we've considered function χ to be a quadratic function of invariants $Tr(\mathcal{K})$, $Tr(\mathcal{K}^2)$ but as the examples of stress relaxation and creep in a rod demonstrate this function should be chosen differently, corresponding to the material studied. In particular, if the Dorn relation $\dot{\eta} = D(exp^{\beta\sigma} - 1)$ between the stress σ and the strain rate $\dot{\eta}$ ($\eta(t)$ is the volume preserving part of inelastic strain) has to be achieved, one should take $\chi(x) = cx + \frac{x}{\beta D}ln(\frac{x}{D}) - \frac{1}{\beta}(1 + \frac{x}{D})ln(1 + \frac{x}{D})$.

We assume that the strain Energy f and the "ground state" term F are independent of each other. Yet, below we introduce a scheme where elastic deformation (presented by the elastic strain tensor E^{el}) is considered as (small) perturbation of (large) inelastic deformation (presented by inelastic strain tensor E^{in}). Therefore, strain energy $f(E^{el})$ is obtained by decomposition of the function $F(S, E^{tot})$ into the "Taylor series" by the parameter E^{el}. This leads to the expression of elastic moduli of a media in terms of the invariants of material metric g and the lapse function S.

54. ACTION, BOUNDARY TERM, HOOKE'S LAW.

The Action functional is the integral of the Lagrangian density $\mathcal{L}(G, \phi)$ over a 4D domain $U = [0, T_0] \times V$. Here $(V, \partial V)$ is an arbitrary subdomain of B with the boundary ∂V, combined with the 3D boundary integral that accounts for the work W of surface traction ([90]).

$$A_U(G, \phi) = \int_U (L_m(G) + L_e(E^{el}, G))dV + \int_{[0, T_0] \times \partial V_1} V_\tau(\phi, G)d^3\Sigma. \qquad (54.1)$$

Here $d^3\Sigma$ is the area element on the 3D boundary ∂U of the cylinder U ([90]).

The second term on the right represents boundary conditions put on the deformation history ϕ. Typically the boundary ∂V of the domain $V \subset B$ is divided into two parts $V = \partial V_1 \cup \partial V_2$. The deformation is prescribed on the part ∂V_2: $\phi_{\partial V_2} = \psi(t, X)$, while along the part ∂U_1 of the boundary the traction τ is prescribed. Function V_τ depends (conventionally) on the velocity $\vec{V}$ of deformation ϕ and on the traction 1-form τ, which is chosen in such a way as to have $-\nabla_\phi V_\tau = \tau$ - traction ([90]). In Euclidean space with the dead load one takes $V_\tau = -\tau \cdot \phi$.

Deformation $\phi(0, X)$ and the velocity $\vec{V} = \phi_*(\frac{\partial}{\partial t})$ of material points are assumed to be given at the moment $t = 0$. This determines initial conditions for the deformation history.

The boundary conditions for the metric G (including initial conditions for 3D material metric g) require a special attention. Initial values of $S, g, \vec{N}$ are known - prescribed by the material manufacturing process and by the previous history of the material deformation.

On the part ∂V_1 of lateral surface ∂V where deformation $\phi|_{\partial V_1}$ is prescribed (for instance when this part of surface is not moving at all, see [16] for examples) we can find $\phi_*(g|_{\partial V_1})$ by measuring distances between the material points on the boundary of the body in the physical space at moment t and recalculating them back to B by the tangent to the (prescribed) mapping ϕ_t.

If a part of the surface is free from load, one can use the natural (Neumann type) condition that the mean curvature (with respect to the metric induced by g_t) of this part of the surface is zero.

Along the part V_2 of the surface where the load τ is applied one may use for g an analog of the Laplace-Young condition for liquid surfaces relating difference of pressure with the surface tension and the mean curvature. The formulation of corresponding boundary conditions are the subject of another work.

From the requirement that the variation of the action near the lateral sides of cylinder U are zero we come to the **natural boundary condition of the form**

$$P \cdot N = \tau, \ or \ \sum_{IJ} g_{IJ} P_j^I N^J = \tau_j$$

in terms of the first Piola-Kirchoff stress tensor P_j^I defined by the equation (material form of the Hooke's law, see [90]):

$$P_j^I = -\frac{\partial L_e}{\partial \phi_{,I}^j} = \frac{\partial f}{\partial \phi_{,I}^j}. \tag{54.2}$$

Notice that if the kinetic energy is included in L_e, Piola-Kirchoff Tensor has the density of the linear momentum vector as its P_i^0 components ([31, 95]).

We will be using the second (material) Piola-Kirchoff tensor $S_J^I = P_i^I \phi_{,J}^i$.

It is useful to recall that the (laboratory) Cauchy stress tensor σ_{ij} is related to the first Piola-Kirchoff tensor by the following formula $\sigma_{ij} = J^{-1}(\phi) h_{ik} \phi_{,I}^k P_j^I$, $J(\phi)$ being the Jacobian of the deformation ϕ.

The zero condition for the variation at the top $(T = T_0)$ and the bottom $(T = 0)$ of the cylinder leads to the relation between linear momentum and kinetic energy in the classical case (Legendre transformation). In the scheme presented here these variations also include terms related to the aging processes.

55. Euler-Lagrange Equations.

The variation principle of the extreme action $\delta A = 0$ taken with respect to the dynamic variables ϕ and G results in a system of Euler-Lagrange equations that represent the coupled Elasticity and "Aging" equations

$$\frac{\partial}{\partial T}(\rho_0 \phi_{,0}^m) + \frac{\partial \mathcal{L}_e}{\partial \phi^m} - \sum_{I=1}^{I=3} \frac{\partial}{\partial X^I}\left(\frac{\partial \mathcal{L}_e}{\partial \phi_{,I}^m}\right) - \rho_0 \sqrt{|G|}(\nabla B)_m = 0, \ m = 1, 2, 3. \tag{55.1}$$

$$\frac{\delta \mathcal{L}_m}{\delta G_{IJ}} = -\frac{\delta \mathcal{L}_e}{\delta G_{IJ}} = \sqrt{|G|} T^{IJ}, \ I, J = 0, 1, 2, 3. \tag{55.2}$$

The Elasticity Equations (53.1) are obtained by taking the variation δA with respect to the components ϕ^i within the domain U. In the case of a BD metric G and the synchronized deformation ϕ, these equations coincide with the conventional dynamical equations of Elasticity Theory. However their special features are associated with the different form of the elastic strain tensor E^{el} and with the dependence of the elastic parameters on time through the invariants of the metric G. The evolution of these parameters is defined by the equations (53.2) (referred to as Aging equations).

The Aging Equations (53.2) resulting from the variation of action with respect to the metric tensor G describe the evolution of the material metric G for a given initial and boundary conditions.

The right side of the equations (53.2) represents the (symmetrical) **"Canonical Energy-Momentum Tensor"** $\sqrt{|G|} T^{IJ} = -\frac{\delta \mathcal{L}_e}{\delta G_{IJ}}$ ([2]). In our situation this tensor is closely related to the **Eshelby EM Tensor** b_{IJ}.

In his celebrated works J.Eshelby ([29, 30]), introduced the 3D and then 4D dynamical energy-momentum tensor (**Eshelby EM Tensor**) b (denoted P_{lj}^* in

[30]).

$$b^I_J = f\delta^I_J - \sum_{i=1}^{i=3} \frac{\partial f}{\partial \phi^i_{,I}} \phi^i_{,J} = f\delta^I_J - S^I_J, \tag{55.3}$$

f being the elastic energy per unit volume.

The tensor b includes the 3D-Eshelby stress tensor ([29, 30]) , the 1-form of *quasi-momentum (pseudomomentum)* $\mathcal{P} = b^0_J$, $J = 1,2,3$, (see [30, 95]), strain energy density $b^0_0 = -L_e = f$ (plus kinetic energy, if the last one is present) and the energy flow vector $s = b^I_0 = -P^I_i \phi^i_{,0}$, $I = 1,2,3$. In the quasi-static case $b^0_J = 0$ for $J = 1,2,3$. In the case of a BD metric $(\vec{N} = 0)$ G we have $\mathcal{P}_B = b_{0B} = 0$, $B = 1,2,3$.

Tensor b_{IJ} is, in general, not symmetric (although its 3x3 space part is symmetric with respect to the Cauchy metric $C_3(\phi)$, see [95].

It was proved in [15] that if metric G is block diagonal (i.e. if $\vec{N} = 0$) and the body forces are zero, then

$$T^{IJ} = \frac{1}{2} b^{(IJ)} + \left(\frac{\delta f(S, g, E^{el})}{\delta g_{IJ}} \right)_{exp}, \quad I, J = 1,2,3, \tag{55.4}$$

where $b^{(IJ)}$ is the symmetrical part of the 4D Eshelby tensor and the symbol $_{exp}$ refers to the derivative of L_e by the **explicit dependence of** G (not through E^{el}).

For the Lagrangian $L = L_m + L_e$ defined in Section 51, the Aging Equations (53.2) can be rewritten in the more convenient ADM notations.

The aging equations (53.2) take the form of the system of PDE for the lapse function S, shift vector field $\vec{N}$ and the 3D material metric g. The explicit form of the above equations can be readily obtained for the Lagrangian in a form (5.4-5). In order to achieve this the variational derivatives of components of Lagrangian with respect to the variables $S, \vec{N}, g_{IJ}$ need to be calculated. In Appendix B to this Chapter we calculate the variations of some of these terms and present them in tabular form.

Variation by S (assuming that f does not depend on S):

$$\frac{\delta \mathcal{L}}{\delta S} = 0 \Leftrightarrow (F + S\frac{\partial F}{\partial S}) + (\chi(K) - \frac{\partial \chi}{\partial K} : K) + \alpha div_g(\vec{N})^2 + \beta R(g) = f + S\frac{\partial f}{\partial S}. \tag{55.5}$$

Variation by N^I:

$$\frac{\delta \mathcal{L}_g}{\delta N^I} = 2\frac{\partial F}{\partial(\|\vec{N}\|^2)} N_I - \frac{\partial}{\partial X^I}(2\alpha \cdot ln(\rho_0 S) \cdot div_g(\vec{N})) + [-S^{-1}\frac{\partial \chi}{\partial K^A_J} g^{AS} \partial_{X^I} g_{SJ} +$$

$$+ \frac{1}{\rho_0 S\sqrt{|g|}} \partial_{X^S}(\rho_0 \sqrt{|g|} \frac{\partial \chi}{\partial K^A_J} g^{AS} g_{IJ}) + \frac{1}{\rho_0 S\sqrt{|g|}} \partial_{X^J}(\rho_0 \sqrt{|g|} \frac{\partial \chi}{\partial K^I_J})] = 0. \tag{55.6}$$

Variation by g_{IJ} (for simplicity,in this equation we omit the terms coming from $div_g(\vec{N})^2$ in Lagrangian (51.4-5), for the corresponding term in the equation see

Appendix B):

$$[-\beta\mathcal{E}^{AB} + \frac{1}{\rho_0 S}(\Delta_g(\rho_0\beta S)g^{AB} + Hess^{AB}(\rho_0\beta S))]+$$

$$+[-\frac{\partial\chi}{\partial K_J^I}S^{-1}g^{IA}\frac{\partial N^B}{\partial X^J} - \frac{\partial\chi}{\partial K_B^I}S^{-1}g^{IS}\frac{\partial N^A}{\partial X^S} - \frac{\partial\chi}{\partial K_J^I}g^{IA}K_J^B - \frac{1}{\rho_0 S\sqrt{|g|}}\partial_t\left(\rho_0\sqrt{|g|}\frac{\partial\chi}{\partial K_B^I}g^{IA}\right)+$$

$$+\frac{1}{\rho_0 S\sqrt{|g|}}\partial_{X^K}\left(\rho_0\sqrt{|g|}\frac{\partial\chi}{\partial K_B^I}g^{IA}N^K\right)] + \frac{1}{2}L_m g^{AB} + \frac{\partial F}{\partial g_{AB}}+$$

$$+\frac{\partial F}{\partial\|\vec{N}\|_g^2}\cdot[N^A N^B + \frac{1}{2}\|\vec{N}\|_g^2 g^{AB}] = \frac{1}{2}(f+U)g^{AB} - \frac{1}{2}S^{(AB)} + \frac{\partial f}{\partial g_{AB}}\exp = \frac{1}{2}b^{(AB)} + \frac{\partial f}{\partial g_{AB}}\exp + \frac{1}{2}U g^{AB}.$$

$$(55.7)$$

Here $\mathcal{E}^{AB} = Ric(g)^{AB} - \frac{R(g)}{2}g^{AB}$ is the Einstein tensor of metric g. In the first line of equations (53.7) (left side) Δ_g is the 3D Laplace operator is defined by the metric g, $Hess(f) = f_{;M;N}$ stands for the Hessian of the function f (double covariant derivative tensor of f).

On the right side of (53.7) remains the symmetrized Second Piola-Kirchoff Stress Tensor S (here and thereof $S_{(AB)} = \frac{1}{2}(S_{AB} + S_{BA})$) or the Eshelby stress tensor since $S^{(AB)} = -b^{(AB)} + \mathcal{L}_e g^{AB}$. The Eshelby EM Tensor b is thus the driving force of the evolution of material metric g (comp. [30]).

Equations (53.5-7) together with equation (49.1-3) for the reference density, form a closed system of equations for dynamic variables (G_{IJ}, ϕ^i, ρ_0). Complemented with the initial and boundary conditions, these equations provide a closed non-linear boundary value problem for the deformation of solid and evolution of the material properties.

In general, the system (53.1-2) seems rather complex, especially if L_e depends on the metric G and its (differential) invariants explicitly. Nevertheless, leaving a detailed analysis of this system for future studies, we make some brief remarks about special cases where system (53.2) is effectively simplified.

56. SPECIAL CASES AND EXAMPLES.

56.1. Block-diagonal metric G. In a case of a BD-metric, $\vec{N} = 0$ (no shift). Therefore, the metric Lagrangian has the form $L_m = F(S, E^{in}) + \chi(K) + \beta R(g)$ that includes time derivatives of 3D metric g (in $\chi(K)$) and the space derivatives of g in the curvature term $R(g_t)$.

No derivatives of the lapse function S appear anywhere in the Lagrangian. In particular, equation obtained by variation of S **is not a dynamical equation but rather a constraint**, similar to the "energy constraint" in the Einstein equations ([34]).

In the case when the elastic coefficients do not depend on S, equation (53.5) takes the form

$$(F + S\frac{\partial F}{\partial S}) + (\chi(K) - \frac{\partial\chi}{\partial K} : K) + \beta R(g) = f, \qquad (56.1)$$

where $\rho_0 f\sqrt{|g|}S$ is the density of strain energy (per unit of unperturbed volume).

This relation represents an equilibrium between the strain energy in the material (residual stresses presented by the scalar curvature of g) and the internal material constituents (the "ground state term" and the terms defined by the kinetic of material processes). In the case of a homogeneous tensile rod ([16]) this relation determines the domain of admissible evolution in the phase space and the "stopping surface" where evolution of the material under the fixed conditions stops (see Sec.58 below).

As $f \to 0$ and the kinetic processes are stopped, the system tends to the "natural" limit state which determines the relation between the "ground state energy" $F(S, E^{in})$ and the residual stresses (see Sec.58.3 below).

56.2. Spacial subsystem. The spatial part (53.7) of aging equations represents the system of PDE for the metric g_{IJ} having the form

$$-\beta \mathcal{E}^{AB}(g) - S^{-1}S^{IA}Q_{IM}^{BN}(\xi_p^2 g)_N^M + W = S^{(AB)}. \tag{56.2}$$

Here $Q_{IM}^{BN} = \frac{\partial^2 \chi}{\partial K_B^I \partial K_N^M}$ and $\xi_p = \partial_t - \vec{N}$ is the principal part of the 1st order linear operator $\xi = \partial_t - \mathcal{L}_{\vec{N}}$. The term W in the left side depends on the metric coefficients, function S and their first derivatives. Einstein tensor $\mathcal{E}(g)$ is linear by the second-order space derivatives of g. Thus, this system is quasilinear evolutional second order system for metric g. It can be easily transformed to the normal form under simple conditions on the dissipative potential χ.

56.3. Statical case. Consider the case where $\vec{N} = 0, U = 0, f$ does not depend on G, S explicitly, S, g are time-independent, and $\beta = const$. Then the system of aging equations is reduced to the following form (here and below $\tilde{F} = SF$)

$$\begin{cases} \frac{\partial \tilde{F}}{\partial S} + \beta R(g) = f(E^{el}), \\ (S)^{-1}\frac{\partial \tilde{F}}{\partial g_{AB}} - \beta \mathcal{E}^{AB} + (\rho_0 S)^{-1}[\Delta_g(\rho_0 S)g^{AB} + Hess_g^{AB}(\rho_0 S)] + \frac{1}{2}(F + \beta R(g))g^{AB} \\ \quad = \frac{1}{2}(fg^{AB} - S^{(AB)}). \end{cases}$$
$$\tag{56.3}$$

In the absence of the strain energy, i.e. when $f(E^{el}) = 0,\ b^{(AB)} = 0$, system (54.3) has the trivial solution $S = const, g = g_0$.

Calculate S^{AB} through the Cauchy stress tensor using (6.3) as follows: $S^{AB} = P_j^A \phi_C^j g^{CB} = (J(\phi)\phi_s^{-1}\, {}^A h^{is}\sigma_{ij})\phi_C^j g^{CB} = J(\phi)g^{BC}\phi_C^j\phi_s^{-1}\, {}^A h^{is}\sigma_{ij} = \frac{\sqrt{|h|}}{\sqrt{|g|}}g^{BC}\sigma_C^A$.

Multiplying the first equation in (8.3) by $\frac{1}{2}g^{AB}$ and subtracting from the second we get

$$\left[\frac{\partial F}{\partial g_{AB}} - \frac{1}{2}\frac{\partial \tilde{F}}{\partial S}g^{AB}\right]\sqrt{|g|} + (\rho_0 S)^{-1}\sqrt{|g|}[\Delta_g(\rho_0 S)g^{AB} + Hess_g^{AB}(\rho_0 S)] - \beta \mathcal{E}^{AB}\sqrt{|g|}$$

$$= -\frac{1}{2}\sqrt{|h|}g^{(B|C}\sigma_C^{A)}. \tag{56.4}$$

This is the balance equation between the metric characteristics (Einstein tensor, "ground state energy", lapse function S) and the stresses in the body. It is especially simple in the case where $S \equiv 1$ is absent from F:

$$\frac{\partial F}{\partial g_{AB}}\sqrt{|g|} - \beta \mathcal{E}^{AB}\sqrt{|g|} + \rho_0^{-1}\sqrt{|g|}[(\Delta_g \rho_0)g^{AB} + Hess_g^{AB}\rho_0] = -\frac{1}{2}\sqrt{|h|}g^{(B|C}\sigma_C^{A)}. \tag{56.5}$$

Here we can see how the curvature of the material metric and the density of non-homogeneities may be a source of the stresses in the body in the absence of elastic deformation, i.e. when the conventional strain tensor $E^{el\ con} = \frac{1}{2}ln(g_0^{-1}C_3(\phi))$ is zero. Namely, in the case, where the conventional strain tensor is zero, decline of the Cauchy metric $C_3(\phi)$ from the material metric g is not zero. Subsequently stress tensor S is not zero. Equation (54.5) thus describes the self-equilibrated stress resulting from the curvature of the metric g and is related to the incompatibility of embedding of the solid into the physical space. The first term on the left in (54.5) is related to the deviation of the total energy from its stationary value.

One example of this situation a nonhomogeneous chemical transformation (oxidation) of material, which results in the variation of material density and an incompatibility with the reference configuration. A more specific example of stress induced chemical transformation is discussed below in section 54.5.

56.4. **Almost flat case.** Here we use essentially that the dimension of B is 3. In the case, where $Ric(g_t) \approx 0$, a good approximation of the general system (53.1-2) can be proposed. If the total deformation ϕ is approximated by the *"ground deformation"* $\bar{\phi}$ (i.e. deformation $\bar{\phi}(X,T)$ such that $\bar{\phi}^*h = g_T$, recall that this is the synchronous case!) in the evaluation of the EMT T^{IJ} on the right side of aging equations (53.2), the latter becomes decoupled from the equilibrium equations (53.1). This allows us to study the aging equations separately from the elasticity equations and, after obtaining solution for G, substitute them into the elastic equilibrium equation (53.1) and solve it as the conventional elasticity equation **with variable elastic moduli**.

56.5. **Homogeneous media.** In the case of a homogeneous material ([34]) metric G depends on T only, and Einstein tensor $\tilde{\mathcal{E}}_{IJ}(g)$ is identically zero. As a result, (7.2) becomes a system of quasi-linear **ordinary** differential equations of the second order for the lapse function N and the material 3D metric g_{IJ}. The Cauchy problem for this system is correct under some mild conditions to the dissipative potential χ.

The linearized version of aging equations of 1D homogeneous rod was discussed in ([17]). In Sec.58 we shall briefly present the study of some aging problems for a homogeneous rod. More detailed presentation will be published elsewhere.

We conclude this section with two model examples that show the type of material behavior that can be studied using presented approach.

57. Two examples.

57.1. **Modeling of Necking Phenomena in Polymers.** Delayed Necking, observed in various engineering thermoplastics, is a pictorial illustration of traveling wave solution. Necking in general is a localized large deformation (drawing) of a polymer with a distinct boundary between the drawn and undrawn material domains (see [61, 72]).

Delayed necking takes place in a rod in uniaxial tension, i.e., under constant applied load when the initial Piola-Kirchoff stress S_{11} (see [90]) is less then the yield stress. At first a uniform creep takes place, i.e., a uniform stretching with a draw ratio $\lambda = l/l_0$, where l stands for an actual (current) length scale. After certain time interval when the increasing stress S_{11} reaches the yield stress value, necking, also called "cold drawing" with a natural draw ratio $\lambda = l_1/l_0$ starts, i.e., strain localization is formed and propagates along the rod with a constant speed N^1. The observed elongation results exclusively from a transformation of the original material adjacent to the neck boundary into the drawn (oriented) state and propagation of the boundary along the rod, as depicted in Fig. 6.

We consider here a 1D model of a rod, with the lapse function $S = 1$ and the shift vector field $\vec{N} = N^1 \frac{\partial}{\partial X}$ being constant (see [16] for a 3D model of the necking process). Denote by $g = g_{11}(t, X) = \lambda^2 g_0$ the only component of material metric. Take the "ground state" energy in the form

$$F(\lambda) = (\lambda - \lambda_0)^2 (a + b(\lambda - \lambda_1)^2)$$

where the elongation of the rod $\lambda(t, X) = \frac{1}{2}\frac{g}{g_0}$ is the drawing variable, $\lambda_0 = 1, \lambda_1$ are two states (to compare with example of the creep in Sec.11 put $\lambda = e^\eta \approx 1 + \eta$). This is the simplest function that admits two different stable states (metrics) with equal chances when true stress reaches a critical value.

The metric Lagrangian is reduced to $L_m = F(\lambda) + \chi(K)$, where $K = (g^{-1} u_G \cdot g) = 2\partial_\tau \eta$, $\eta = ln(\lambda)$ where $\partial_\tau = u_G = \partial_t - N^1 \partial_X$.

Experimental data suggest that in the necking the material density variation is negligible, thus we take $\rho_0(t, X)\lambda(t, X) = \rho_0(0, X)$. As a result, the action takes the form

$$A(\lambda(t, X)) = \int_{[0, t_1] \times [0, L]} [F(\lambda) + \chi(2\partial_\tau \ln(\lambda))] dt \wedge dX =$$

$$\int_{[0, t_1] \times [0, L]} [(\lambda - \lambda_0)^2 (a + b(\lambda - \lambda_1)^2) - \frac{\mathcal{F}}{A_0}(\lambda - \lambda_0)^2 + \chi(2\partial_\tau ln(\lambda))] dt \wedge dX,$$

$$(57.1)$$

where A_0 is the initial cross section of the rod and $\mathcal{F}$ is the force acting on the right end pulling in X-direction. The second term here represents the work of the load on non-elastic deformation. Consider the case where $\mathcal{F}$ is large enough to change sign of the quadratic part of the "ground energy" F. For simplicity we take $\frac{\mathcal{F}}{A_0} = a$.

Variation by λ leads to the aging equation in the form

$$\chi'' \lambda_{\tau\tau} - \frac{\lambda_\tau}{2}[\chi' + (2\lambda^{-1}\lambda_\tau)\chi''] - \frac{\lambda^2}{4}\bar{F}'(\lambda) = 0,$$

where $\bar{F}(\lambda) = b(\lambda - \lambda_0)^2 (\lambda - \lambda_1)^2$.

The 2D dynamical system corresponding to this equation has equilibria points $(\lambda_i, 0)$, $i = 0, 1, 2$ at the roots $\lambda_0, \lambda_2 = \frac{1}{2}(\lambda_0 + \lambda_1), \lambda_1)$ of the polynomial $F'(\lambda)$. If the dissipative potential $\chi(u)$ satisfies the conditions $\chi'(0) = 0, \chi''(0) > 0$, root λ_2 is the center while other two are saddles whose separatrix loop enclose the elliptic region.

For a given $F(g)$, when the stress reaches the initiation level and from the trivial solution for g there bifurcates the separatrix solution, then, we get the "traveling wave solutions" in the form of "kink" ([16]), propagating with the the speed N^1 along the rod, for the metric $g(X, t)$ (and, by the mass conservation law, for the density $\rho_0(X, t)$).

57.2. Example: Variation of Material Metric g due to the Chemical Degradation. Here the evolution of the uniform material metric to the piecewise constant metric with the jump along the interface between a layer of chemically

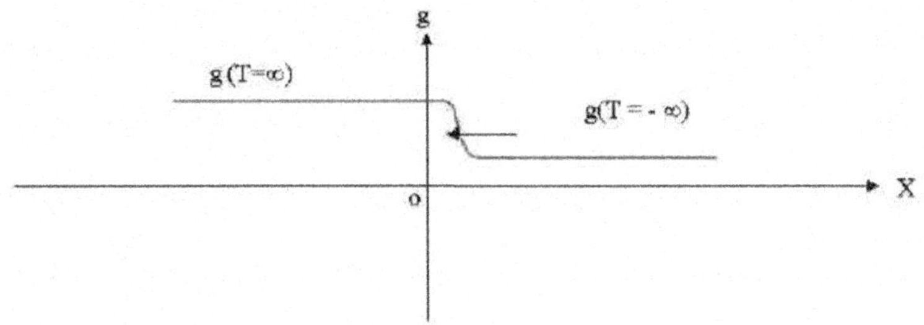

FIGURE 7. Necking of 1D Rod.

degraded material and the original material follows the kinetics of chemical degradation (see [18]).

Consider a thin-walled thermoplastic tubing employed for transport of chemically aggressive fluid. In time, the inner surface layer of material undergoes chemical degradation due to interaction with aggressive fluid flow.

Chemical degradation is manifested in an increase of the material density ρ_0, significant reduction in toughness (resistance to cracking) and a subtle change in yield strength, Young's modulus and other thermo-mechanical properties.

Assuming the homogeneity of degraded layer we see that the original euclidian material reference metric in degraded ring evolves (see the mass conservation law (3.4)) which generates a jump on the interface with the outer layer of unchanged material. Continuity of normal stresses on the interface allows us to describe the final state of the system by elementary methods presented below.

Consider a thin ring (see Figure 8) which represent the 2D cross-section of the tubing. The wall thickness $t = R_o - R_i$ is small in comparison to the outer radius R_o: $t/R_o \ll 1$. R_d in Fig ** stands for the radius of interface between the layer of degraded material and unchanged layer. The depth of degradation $t_d = R_d - R_i$ is relatively small: $t_d/t \ll 1$.

Select the polar coordinate system (r, θ). 2D material metrics of the initial (g^0) and degraded (g') states are

$$g^0 = \begin{pmatrix} 1 & 0 \\ 0 & r(0)^2 \end{pmatrix}, \quad |g^0| = r^2(0); \quad g' = \begin{pmatrix} 1+\epsilon & 0 \\ 0 & r'^2 \end{pmatrix}, \quad |g'| = (1+\epsilon)r'^2,$$

where $r' = (1 + \epsilon)r$ and ϵ is a small variation of scale in the radial direction.

Mass conservation law $\rho_0 \sqrt{|g^0|} = \rho_0' \sqrt{|g'|}$ relates density variation $\rho_0' = \rho_0 + \Delta\rho_0$, $\frac{\Delta\rho_0}{\rho_0} \sim 10^{-3}$ with the change in material metric

$$\frac{\rho_0}{\rho_0 + \Delta\rho_0} = \sqrt{\frac{(1+\epsilon)r'^2}{r(0)^2}} = (1+\epsilon)^{3/2} \implies 1 - \frac{\Delta\rho_0}{\rho_0} \approx 1 + \frac{3}{2}\epsilon + O(\epsilon^2). \quad (57.2)$$

Therefore $\epsilon \approx -\frac{2}{3}\frac{\Delta\rho_0}{\rho_0} + O(\epsilon^2)$.

Thus, the densification (i.e. $\Delta\rho_0 > 0$) leads to the shrinkage of the thin ring of degraded material. If we remove the constrains on shrinkage applied by the outer ring of original material the gap

$$w = R_d^0 - (1 - \frac{2}{3}\frac{\Delta\rho_0}{\rho_0})R_d^0 = \frac{2}{3}\frac{\Delta\rho_0}{\rho_0}R_d^0 \quad (57.3)$$

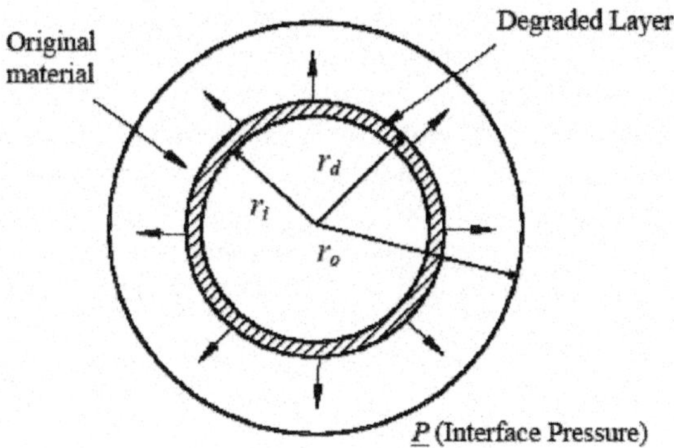

FIGURE 8. A sketch of polymer tubing cross section with inner degraded layer

appears.

As a result of such constrains, the degraded material should be elastically stretched to close the gap w. This elastic deformation has the form

$$\phi(r,\theta) = \begin{cases} r''(r) = (1 + \frac{2}{3}\frac{\Delta\rho_0}{\rho_0})r', \\ \theta'' = \theta' = \theta. \end{cases}$$

Under such a deformation elastic strain tensor $E^{el} = \frac{1}{2}g^{-1}(C_3(\phi) - g), C_3(\phi) = h_{ij}\phi^i_{,I}\phi^j_{,J}$ has the form

$$E^{el} \approx \frac{2}{3}\frac{\Delta\rho_0}{\rho_0}\begin{pmatrix} 1 & 0 \\ 0 & 1 \end{pmatrix}. \tag{57.4}$$

The tensile strain (50.9) is directly translated into the tensile radial stress via Hooke's law

$$\sigma_{rr} = \frac{Y}{1-\nu}\frac{2}{3}\frac{\Delta\rho_0}{\rho_0}.$$

Although hoop stresses $\sigma_{r\theta}$ may be discontinuous, the equilibrium conditions requires continuity of radial stress across the interface,

$$\sigma_{rr}n_r|_{r=R_d-0} = \sigma_{rr}n_r|_{r=R_d+0}.$$

This implies that the outer ring of original material experiences compressive stresses while the inner degraded layer is under tension. The elastic strains E^{el} jointly close the gap w and restore the compatibility of the Cauchy metric g_{final} in the whole domain:

$$g_{final} = g' + E^{el}.$$

Therefore while the material metric g' has the jump leading to the nonzero singular curvature along the interface, the final metric is continuous and flat.

58. PHYSICAL AND MATERIAL BALANCE LAWS

As it is typical for a Lagrangian Field Theory, the action of any one-parameter group of transformations of the space $P \times M$, commuting with the projection to P, leads to the corresponding balance law (See [90]). In particular, translations in the "physical space-time" M lead to the dynamical equations (53.1-2), rotations in

M lead to the angular momentum balance law (conservation law in the absence of applied torque). Respectively, translations in the "material space-time" P lead to the **energy balance law** (translations along the time T axis) and to the **material momentum balance law** ("pseudomomentum" balance, [36, 57, 95], rotations in the material space B lead to the "material angular momentum" balance law ([95]).

In the table below we present basic balance laws together with the transformations generating them. It is instructive to compare the space and material balance laws as it has been considered previously by several authors ([36], [57]).

TABLE 2. Space and Material Balance Laws

Symmetry	Physical space-time (Material independent)	Material space-time (Space independent)
Homogeneity of 3D-space	Linear momentum balance law (Equilibrium equations) $div(\sigma) = f$	Material momentum (pseudo-(-momentum) balance law **div(b)=fmat**
Time homogeneity	Energy balance law: $\partial_t \mathcal{E}^{tot} = div(\mathcal{P}^{tot})$	Energy balance law
Isotropy of 3D-space	Angular momentum balance law $\equiv$ h-symmetry of Cauchy stress tensor σ: $\quad \mathbf{I} : \sigma = \sigma : \mathbf{I}$	Material angular momentum balance law $\equiv C$-symmetry of Eshelby stress tensor $\mathbf{b}$ $\mathbf{b} : \mathbf{C} = \mathbf{C} : \mathbf{b}$

Space and Material balance (conservation) laws are related via the deformation gradient $d\phi$. Restricting ourselves to the *synchronized case* (see Sec.48) and writing the material balance laws in the form $\eta_I = 0$ and their "physical" counterparts in the form $\nu_i = 0$ we get the relationsship between these families of balance laws

$$\begin{pmatrix} \eta_0 \\ \cdots \\ \eta_4 \end{pmatrix} = \begin{pmatrix} 1 & \phi^1_{,0} & \cdots & \phi^3_{,0} \\ 1 & \phi^1_{,1} & \cdots & \phi^3_{,1} \\ \cdots & \cdots & \cdots & \cdots \\ 0 & \cdots & \cdots & \phi^3_{,3} \end{pmatrix} \cdot \begin{pmatrix} \nu_0 \\ \cdots \\ \nu_4 \end{pmatrix} \tag{58.1}$$

Similar to the case of relativistic elasticity ([64]), the system of material balance laws $\eta_I = 0$, $I = 1, 2, 3$ is equivalent to the elasticity equations $\nu_i = 0$, $i = 1, 2, 3$, while the energy balance law $\eta_0 = 0$ (which here is the **material** conservation law as well as the **physical one**: in the case of synchronized history of deformation ϕ material time T and physical time t coincide). As a result, the energy conservation law is the consequence of the time translation invariance in both senses and follows from any of these two systems: $\eta_0 = \sum_{i=1}^{i=3} \phi^i_{,0} \nu_i$. This reflects the fact that the deformation we consider here are not truly 4-dimensional.

Balance laws (with the source terms) can be transformed into conservation laws by adding new dynamical variables. In the theory of uniform materials ([26, 95, ?]) it is a zero curvature connection in the frame bundle over M that is added to the list of conventional dynamical variables, in our scheme - it is the 4D material metric G.

59. Energy-Momentum Balance Law and the Eshelby Tensor.

In this section we consider the Energy-Momentum balance law resulting from the Least Action Principle and the space-time symmetries.

Consider local rigid translations in the material space-time $X^J \longmapsto X^J + \delta X^J$. They generate a variation of components ϕ^i of the deformations, components G_{IJ} of material metric and their derivatives (we follow the arguments of J.Eshelby ([29, 30]).

Taking the variation of the Lagrangian density $\mathcal{L} = \sqrt{|G|}L = \sqrt{|G|}(L_g(G) + L_e(G, E^{el}))$ with respect to the material coordinates X^J, one obtains

$$\frac{\delta \mathcal{L}}{\delta X^J} \;=\; \sum_{i=0}^{i=3} \frac{\delta \mathcal{L}}{\delta \phi^i}\phi^i_{,J} + \frac{\partial}{\partial X^I}\Big(\sum_{i=1}^{i=3}\frac{\partial \mathcal{L}}{\partial \phi^i_{,I}}\phi^i_{,J}\Big) + \frac{\delta \mathcal{L}}{\delta G^{AB}}G^{AB}_{,J} +$$

$$\frac{\partial}{\partial X^I}\left(\mathcal{E}(G)^I_J = \frac{\partial \mathcal{L}}{\partial G^{AB}_{,I}}G^{AB}_{,J} + \frac{\partial \mathcal{L}}{\partial G^{AB}_{,IK}}G^{AB}_{,KJ} - \frac{\partial}{\partial X^K}\left(\frac{\partial \mathcal{L}}{\partial G^{AB}_{,IK}}\right)G^{AB}_{,J}\right) \quad (59.1)$$

The last term in the right side of (57.1) includes a definition of the (1,1)-tensor density $\mathcal{E}(G)$. Employing the Euler-Lagrange equations (53.1-2), we obtain for the Total Energy-Momentum Tensor (density)

$$\mathcal{E}^{tot} = -\mathcal{L}\delta^I_J - S^I_J + \mathcal{E}(G)^I_J, \qquad (59.2)$$

the conservation law

$$div_{G_0}(\mathcal{E}^{tot}) = \frac{\partial}{\partial X^I}\mathcal{E}^{tot\,I}_J = 0, \;\; J = 0,1,2,3. \qquad (59.3)$$

Divergence here is taken with respect to the 4D "reference" metric G_0. Since L_m does not depend on deformation ϕ and the body forces potential U does not depend on its derivatives while $L_e = -\rho_0 f - \rho_0 U$,

$$S^I_J = -\sum_{i=1}^{i=3}\frac{\partial \mathcal{L}}{\partial \phi^i_{,I}}\phi^i_{,J} = \sum_{i=1}^{i=3}P^I_i\phi^i_{,J}S\sqrt{|g|}, \qquad (59.4)$$

which is the 4D-version of the (density of) Second Piola-Kirchoff Stress Tensor.

Rewrite $\mathcal{E}^{tot}$ in the form: $\mathcal{E}^{tot} = \mathcal{B} + U\delta^I_J\sqrt{\|G\|} + \mathcal{L}_m\delta^I_J + \mathcal{E}(G)$ where we denoted by $\mathcal{B}$ the tensor density $\mathcal{B} = b\sqrt{|G|}$ of the Eshelby EM Tensor. Then the equation (57.3.3) takes the form

$$div_{G_0}\mathcal{B} = \mathcal{B}^I_{J,I} = -div_{G_0}(U\delta^I_J\sqrt{\|G\|} + \mathcal{L}_m\delta^I_J + \mathcal{E}(G)), \qquad (59.5)$$

where in the right side only metrical quantities and the potential U of the body forces are left.

The second term and the metrical part of the third term in the right side of (57.5) are related to the ground state of the Lagrangian density i.e. to the inhomogeneity of "cohesive energy" and the "material flows". The elastic part of the third term on the right is related to a variation of elastic moduli if these moduli depend on the derivatives of the metric G. Equality (57.5) can be easily rewritten in terms of covariant derivatives with respect to the metric G.

Taking $J = 0$ in (57.6) we arrive at the energy conservation law in ADM notations (using t instead of X^0)

$$\frac{\partial}{\partial t}((f + U + L_m)S\sqrt{|g|} + \mathcal{E}^0_0) = \sum_{I=1}^{I=3}\frac{\partial}{\partial X^I}\left(\sum_{i=1}^{i=3}P^I_i\phi^i_{,0}S\sqrt{|g|} - \mathcal{E}(G)^I_0.\right) \quad (59.6)$$

This equation has the form

$$\frac{\partial(TotalEnergyDensity)}{\partial T} = TotalFlowDensity,$$

with the total (inner) energy density given by

$$\mathcal{E}_0^{tot\ 0} = (f + U + L_m)S\sqrt{|g|} + \mathcal{E}(G)_0^0. \tag{59.7}$$

The total energy is the sum of the following parts:

(1) **Elastic (strain) energy** f,

(2) **Potential energy of the volume forces** U ,

(3) **Cohesive "ground state" energy"** - the term $F(E^{in}, S)$ in L_m,

(4) **Kinetic "material" energy** depending on the curvature $R(g)$
of material metric G and the external curvature $\chi(K)$ of the foliation B_ϕ
(see Section 48).

The sum on the right side of (57.5) consists of the flow of the Piola-Kirchoff
stress tensor density $\sum_{I=1}^{I=3}[P_i^I \phi_{,0}^i S\sqrt{|g|}]_{,I}$ and the flows related to the change of
the material metric: internal material flows, flows of inhomogeneities (coming from
the curvature $R(g)$), etc.

If the metric G does not depend on time (i.e. $\bar{N} = 0, K = 0,\ G = G_0$) and
if $Ric(g_t) = 0$, one obtains the conventional energy conservation law of Elasticity
Theory ([90], Chapter 5, Sec.5):

$$\frac{\partial(f + U)}{\partial T} = -\sum_{I=1}^{I=3} \frac{\partial}{\partial X^I}(P_i^I \phi_{,0}^i).$$

Example 33 (Block diagonal metric G, synchronous deformation and homogeneous
media). In the case presented here, we have $Ric(g_t) = 0,\ \tilde{N} = 0, g = g(t), S = S(t)$,
the extrinsic curvature has the form $K_J^I = \begin{pmatrix} 0 & 0 \\ 0 & S^{-1}g^{IK}g_{KJ,0} \end{pmatrix}$ and $\chi(K)$ is the
only term in the Lagrangian containing time derivatives.

In addition to this, no flow terms except the usual Piola-Kirchoff flow appear on
the right side in the energy balance law which takes the form

$$\frac{\partial}{\partial t}\ \left[(f + U + L_m)S\sqrt{|g|} + S^{-1}g_{AB,0}\ \left(\frac{\partial\chi}{K_B^M}g^{MA} + \frac{\partial\chi}{K_A^M}g^{MB} \right) S\sqrt{|g|} \right] = +\sum_{I=1}^{I=3}(P_i^I \phi_{,0}^i S\sqrt{|g|})_{,I}.$$

$$\tag{59.8}$$

This equation describes how the energy supplied by the boundary load spreads,
not just to the increase of the strain energy, but also to the change of its "cohesive
energy" of the material ($F(S, |g|)$) and to the acceleration of the aging processes.

60. Aging of a homogeneous rod.

Here we consider three types of inelastic processes in a tensile homogeneous rod: unconstrained aging, stress relaxation and the Creep (see [9, 61, 72] for more detailed exposition). To simplify the situation, we assume that $\vec{N} = 0, R(g) = 0$, $S(t = 0) = 1$ and that the material time rate $S(t)$ - dynamical variable characterizing the rate of aging processes is increasing to a certain level depending on the initial state of the body and the process, going in the solid.

Remark 74. We will be using the material time τ or,equivalently, dissipative potential, in order to formulate dynamical system of aging material. Using the dissipative potential is equivalent to the using of the appropriate non-commuting variations defined by the NC- tensor field K (see Section 44,Chapter 5) or [94].

In the first example we are using dissipative potential to model the aging behavior of solid (Comp. to Remark 58). In the examples of stress relaxation and creep we introduce corresponding term controlling the inelastic evolution directly into the Lagrangian.

60.1. **Deformation, strain tensors and tensor K.** We introduce material cylindrical coordinates (R, Θ, Z) in the reference state of a rod B. Spacial cylindrical coordinates (r, θ, z) are introduced in the physical space R^3. In addition we normalize the initial state of the 3-dim material metric $g(0)$ taking $g(0) = g_0$.

We consider the class of time dependent (total) deformations ϕ of the from

$$\phi_t : (R, \Theta, Z) \to (r = \mu(t, Z)R, \theta = \Theta, Z = k(Z, t)), \tag{60.1}$$

with the parameters μ, k representing amount of "stretch" in radial and axial directions respectively.

Material metric g_t is flat (homogeneous case!) and is generated by a global deformation ϕ_m of the same type as in Section 51, with μ_m, k_m the same as above: $g_t = \phi_m(t, \cdot)^* h$.

As a result

$$g = \phi_m^* h = \begin{pmatrix} \mu_m^2 & 0 & R\mu_m\mu_{m,Z} \\ 0 & R^2\mu_m^2 & 0 \\ R\mu_m\mu_{m,Z} & 0 & \lambda^2 + R^2\mu_{m,Z}^2 \end{pmatrix}, \quad C(\phi) = \begin{pmatrix} \mu^2 & 0 & R\mu\mu_{,Z} \\ 0 & R^2\mu^2 & 0 \\ R\mu\mu_{,Z} & 0 & \lambda^2 + R^2\mu_{,Z}^2 \end{pmatrix},$$

$$\tag{60.2}$$

where $\lambda = k_{,Z}, \lambda_m = k_{m,Z}$. For a homogeneous rod $k(Z, t) = \lambda(t)Z$, $k_m(Z, t) = \lambda_m(t)Z$.

In the elasticity theory (see [52]) it is customary to present deformation (total and inelastic as well) as the composition of a uniform dilation with the axial expansion factor $\lambda_v(t)$ and of the volume preserving normal expansion with the factor λ_d: $\lambda = \lambda_v\lambda_d$. We will obtain $\mu = \lambda_v\lambda_d^{-1/2}$ and $\sqrt{|g|} = \lambda_v^3 R$ (and the same for $\lambda_{mv}, \lambda_{md}$).

As a result, the inelastic strain tensor can be written in the following form

$$E^{in} = \frac{1}{2}(g_0^{-1}g) = \begin{pmatrix} ln(\mu) & 0 & 0 \\ 0 & ln(\mu) & 0 \\ 0 & 0 & ln(\lambda) \end{pmatrix} = \begin{pmatrix} \xi - \frac{1}{2}\eta & 0 & 0 \\ 0 & \xi - \frac{1}{2}\eta & 0 \\ 0 & 0 & \xi + \eta \end{pmatrix} \tag{60.3}$$

Here we introduced the dynamical variables $\xi = ln(\lambda_{m\,v}), \eta = ln(\lambda_{m\,d})$. As a result, calculating basic invariants of these tensors we see that the "ground state" energy F is the function of 3 scalar dynamical variables S, ξ, η^2.

Decomposing the total deformation as the composition of inelastic and elastic one and assuming that elastic deformation is small compared to 1, we write:

$$\lambda_v = \lambda_{v\ m}(1 + \epsilon_v), \ \ \lambda_d = \lambda_{d\ m}(1 + \epsilon_d).$$

In these notations elastic strain tensor takes the conventional diagonal form

$$E^{el} = \frac{1}{2}ln(g^{-1}C(\phi^{tot})) \approx diag(\epsilon(v) - \frac{1}{2}\epsilon_d, \epsilon_v - \frac{1}{2}\epsilon_d, \epsilon_v + \epsilon_d).$$

Strain energy for our (homogeneous) rod will now take the form

$$f = \frac{K}{2}\epsilon_v^2 + \frac{3\mu}{2}\epsilon_d^2 \tag{60.4}$$

with the bulk coefficient K and the Lame coefficient μ.

Mass conservation law (49.1) takes the form $\rho_0(t) = \lambda_v^3\rho_0(0)$.

The spacial part of the tensor K for a homogeneous rod takes the diagonal form

$$K = S^{-1} \cdot \begin{pmatrix} 2\xi_t - \eta_t & 0 & 0 \\ 0 & 2\xi_t + 2\eta_t & 0 \\ 0 & 0 & 2\xi_t - \eta_t \end{pmatrix}. \tag{60.5}$$

Thus, $Tr(K) = 6S^{-1}\xi_t$, $Tr(K - \frac{1}{3}Tr(K)I)^2 = 6S^{-2}\eta_t^2$ and the dissipative potential $\chi(K)$ is the function of arguments $S^{-1}\xi_t, S^{-1}\eta_t$.

Remark 75. In general, the aging equation for the described situation has the form of a 3D degenerate Lagrangian system (we refer to [16], or [112] for more details). In the cases of the processes studied below this system reduces to the 2D degenerate dynamical system. In all three cases one can trivially solve elasticity equations, exclude elastic variables ϵ_v, ϵ_d from aging equations and, therefore, close the system of aging equations.

60.2. **Unconstrained aging.** Unconstrained aging (shortly UA) is the simplest example of a material evolution. A sample of material (rod) is prepared and then left without any constraints or load applied to it. Usually the process of aging is manifested in a variation of material density, or a specific volume change up to a saturation point, when the observable evolution stops. In many polymers the aging is accompanied by shrinkage up to a few percent of initial volume. This diminishing in volume (2-5%) is called **unconstrained aging**. We discuss here a model for UA in terms of variables (S, ξ) (dilatational deformation ν plays negligible role in UA). No strain energy is present; stress is zero.

We take the "ground state energy" to be

$$F_{UA}(S, \xi) = (c_1 + c_2 S + (p\xi S^{-1} + k\xi^2)$$

with $k > 0, p < 0, c_1 < 0, c_2 > 0$ and the dissipative potential

$$\chi_{UA}(K) = \alpha(S^{-1}\xi_t)^2.$$

Remark 76. Notice that $S^{-1}\xi_{,t} = \xi_{,\tau}$, where τ is the material time defined by the metric G and the (2-dim in the case of rod) variable configuration $4\phi(\tau)$ (see above, Sec.46). As a result, dissipative potential has the form

$$\chi_{UA}(K) = \alpha(S^{-1}\xi_t)^2 = \alpha(\xi_\tau)^2) \tag{60.6}$$

of the "material kinetic energy".

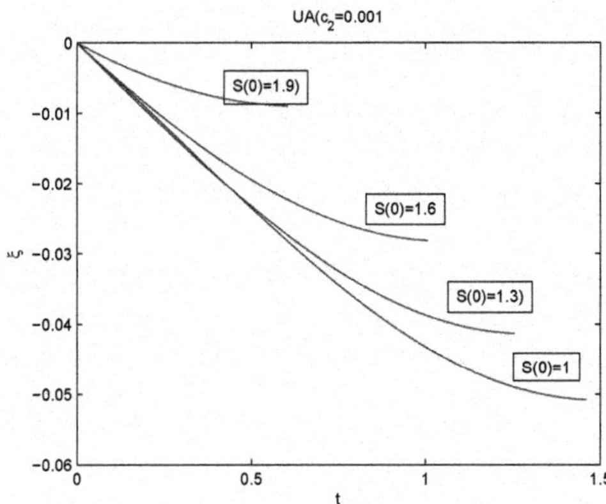

FIGURE 9. Aging curves, UA-case.

Integrating over the volume of the rod we get the action in the form

$$A(\xi, S) = V \int_0^T ((c_1 S + c_2 S^2 + (p\xi + k\xi^2 S) + \alpha S(S^{-1}\xi_t)^2)dt. \qquad (60.7)$$

Euler-Lagrange Equations of UA can be reduced to the following dynamical system

$$\begin{cases} \xi_t &= -S\left(\frac{c_1 + 2c_2 S + k\xi^2}{\alpha}\right)^{\frac{1}{2}}, \\ S_t &= -\frac{p}{2c_2}\left(\frac{c_1 + 2c_2 S + k\xi^2}{\alpha}\right)^{\frac{1}{2}}. \end{cases} \qquad (60.8)$$

We have here $\xi_t \leqq 0, S_t \geqq 0$.

Equation (59.6.5) takes here the form $\alpha\xi_t^2 = S^2 \frac{\partial(\tilde{F})}{\partial S}$, where $\tilde{F} = SF$.

Take $\alpha = -1$. Then, the domain of admissible dynamics defined by the positivity of expression under the square root in the first equationis

$$S \leqq -\frac{1}{2c_2}(c_1 + k\xi^2), \qquad (60.9)$$

and the curve where evolution stops when the phase trajectory reaches the final state is $S = -\frac{1}{2c_2}(c_1 + k\xi^2)$.

The "ground state energy" F is negative at initial moment and is increasing during the evolution.

Dynamical system (58.8) has the first integral $J = c_2 S^2 - p\xi$.

Choosing an initial point $(S(0), \xi(0) = 0)$ of a trajectory Γ in the domain of admissible dynamics we may draw the phase picture of this system: Figure 9 shows a family of shrinkage curves corresponding to the various values of $S(0) = 1, 1.3, 1.6, 1.9$ (which represent the initial aging of the material). Apparently, the higher the initial age, the less shrinkage is observed.

When a load applied to the rod reaches a certain level, new processes may start. These new processes (going on the background of the UA) initiate an action of a new part of the "ground state" $F(S, \xi, \eta)$ and activates the new kinetic potential $\chi_2(K)$. For the description of stress relaxation and creep we choose the dissipative potential $\chi_2(K) = \frac{K}{\beta D}ln(\frac{K}{D}) - \frac{1}{\beta}(1 + \frac{K}{D})ln(1 + \frac{K}{D})$ corresponding to the phenomenological

Dorn relation between the stress and the strain rate $\dot{\eta}$ (see [9], Sec.2.3 and the footnote in Sec.51 above). Unconstrained aging is much slower and leads to smaller changes then both stress relaxation and the creep. That is why we may, with good accuracy, disregard the UA while describing two other processes.

60.3. Stress relaxation. In the case of a **stress relaxation** (SR) we fix the rod of initial length L at the left end and then quickly pull (or compress) it uniaxially and fast (elastically) until it reaches certain length L^*. Then we fix the right end as well, leaving the lateral surface of the rod free. In this configuration the only component of Cauchy stress that is nonzero is σ_{zz}. For the SR the volume change is negligible and we have $\lambda_{m\ v} = 1, \lambda_m = \lambda_{m\ d}$

Initially all the stretching is due to elastic process and $\lambda_d^* = L^*/L = (1 + \epsilon_z(0))$. Then the inelastic deformation starts to increase in expense of the elastic one maintaining the total strain constant. The reduction of elastic strain is directly translated into the reduction of stresses via Hooke's law. The total elongation at moment t can be decomposed as follows

$$\lambda_d^* = (1 + \epsilon_z(t))\lambda_{m,d}(t) = (1 + \epsilon_z(t))e^{\eta(t)} \tag{60.10}$$

and therefore $\epsilon_z(t) = \eta^* - \eta(t)$, where $\eta^* = ln(\lambda^*)$.

From Hooke's law $\sigma_{zz} = Y\epsilon_z$, where Y is the Young module. Thus for the strain energy expression, we obtain

$$f(\eta) = \frac{1}{2}\sigma_{zz}\epsilon_z = \frac{1}{2}Y\epsilon_z^2 = \frac{Y}{2}(\eta^* - \eta(t))^2. \tag{60.11}$$

For pure stress relaxation (without background UA),

$$F = F_{SR}(S,\eta) = (q_1 + q_2 S) + \eta(b_0 S^{-1} + b_1 + a_1\eta)| \quad q_1 < 0, q_2 < 0, b_0 < 0, a_1 < 0,$$

with the coefficients different from those of the slow UA.

Action $A(S(t), \eta(t))$ now takes the form

$$A(S,\eta) = \int_0^T \left[\tilde{F}_{SR}(S,\eta) + S\chi(S^{-1}\eta_t)\right] dt =$$

$$= \int_0^T \left[(q_1 + q_2 S) + \eta(b_0 S^{-1} + b_1 + a_1\eta) + \frac{Y}{2}(\eta^* - \eta(t))^2 + S\chi(S^{-1}\eta_t)\right] S dt \tag{60.12}$$

where

$$\tilde{F}_{SR}(S,\eta) = SF_{SR} = q_2 S^2 + p_2(\eta)S = (q_1 S + q_2 S^2) + \eta(b_0 + b_1 S + a_1\eta S) + S\frac{Y}{2}(\eta^* - \eta(t))^2.$$
$$, \tag{60.13}$$

Here $p_2(\eta) = (q_1 + \frac{Y}{2}\eta^{*\,2}) + (b_1 - Y\eta^*)\eta + (a_1 + \frac{Y}{2})\eta^2$.

Domain of admissible motion is defined by the conditions

$$D_{ad} = \{(\eta, S)|\eta > 0, S \geq 1, S < -\frac{p_2(\eta)}{2q_2}\},$$

while the **stoping curve** has the form $S = -\frac{p_2(\eta)}{2q_2}$.

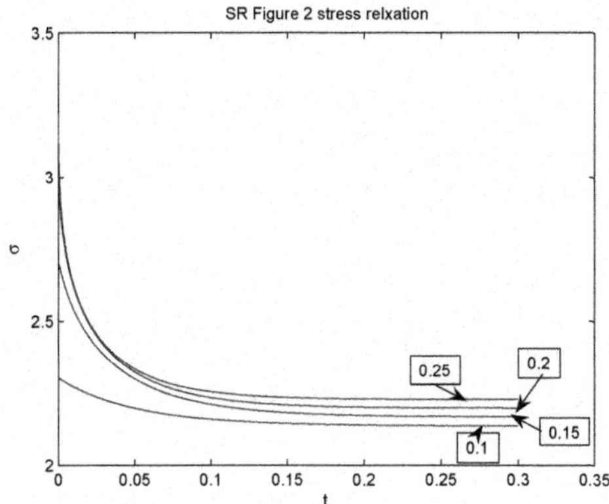

FIGURE 10. Stress Relaxation: Stress $\sigma_{zz}(t)$. for $\eta^* = 0.1, 0.15, 0.2, 0.25$.

euler Lagrange equations (for dynamical variables S, η) (see Section 54) takes now the form

$$\begin{cases} \eta_t = S\psi^{-1}(2q_2 S + p_2(\eta)) = S\psi^{-1}(2q_2 S + [(q_1 + \tfrac{Y}{2}\eta^{*\,2}) + (b_1 - Y\eta^*)\eta + (a_1 + \tfrac{Y}{2})\eta^2]), \\ S_t = \tfrac{b_0}{2q_2}\psi^{-1}(2q_2 S + p_2(\eta)) = S\psi^{-1}(2q_2 S + [(q_1 + \tfrac{Y}{2}\eta^{*\,2}) + (b_1 - Y\eta^*)\eta + (a_1 + \tfrac{Y}{2})\eta^2]). \end{cases}$$
$$(60.14)$$

This system has the first integral $J = b_0\eta - q_2 S^2$, and the phase trajectory corresponding to the initial value $S = S(0), \eta(0) = 0$ has the form $S = \sqrt{S(0)^2 + \tfrac{b_0}{q_2}\eta}$. Using this one can find analytic solutions $\eta(t)$ in terms of elliptic functions ([AS01]).

In the Figure 9 we present results of calculations of the stress relaxation $\sigma_{zz}(t)$ for several values of initial stretching $L = 0.1, 0.15, 0.2, 0.5$ and for realistic values of parameters of the problem. Values of $\sigma_{zz}(t)$ are found by solving system (58.14) numerically for $\eta(t)$, calculating elastic strain $\epsilon_z(t)$ and using the Hooke's law. Apparently, the higher the value of initial stretching is, the sharper is the stress relaxation curve and the higher is the asymptotic value of stress.

60.4. **Creep.** In the case of the **creep** we fix the left end of the rod and apply force $\mathcal{F}$ in the Z-direction to its right end. If this force is large enough (i.e. if the concentration of elastic energy f in the rod is larger then an activation threshold), the creep starts: inelastic deformation that goes on for some time until the rod brakes. Thus, at the moment when the inelastic strain $\eta(t)$ starts growing from zero, there should be a supply of strain energy obtained from the work of the stress $\sigma_{zz} = \tfrac{\mathcal{F}}{A(t)}$ on **elastic deformation**. Denote this strain energy (of initiation) by f_{in}.

During the creep the homogeneous component σ_{zz} of stress is equal to

$$\sigma_{zz}(t) = \frac{\mathcal{F}}{A(t)} = \frac{\mathcal{F}}{\lambda_v(t)^2 \lambda_d^{-1}(t) A_0} = \frac{\lambda_d(t)\mathcal{F}}{A_0} = \frac{e^\eta \mathcal{F}}{A_0}. \qquad (60.15)$$

The last equality is true provided the assumption (natural for a conventional creep) that **inelastic volume change is negligible**, i.e. $\lambda_v = 1$, $\xi \approx 0$ for a constant force $\mathcal{F}$ and variable cross-section area $A(t)$ *holds true*.

Using Hooke's law one can show that $\epsilon_z = \frac{\sigma_{zz}}{Y}$ and that the strain energy is equal to $f = \frac{\mathcal{F}^2 e^{2\eta}}{2Y A(0)^2}$.

Calculating the work of the load $\mathcal{F}$ on the total way $L(e^{\eta(t)} - 1)$ of the right end of the rod we get the additional term in the Lagrangian (work of load on the inelastic deformation) equal to $\frac{\mathcal{F}}{A_0}(e^\eta - 1)$.

Overall Lagrangian for the creep takes the form $(\xi(t) = 0)$

$$\mathcal{L}(S(t), \eta(t)) = [F_{CR}(S, \eta) + \frac{\mathcal{F}}{A_0}(e^\eta - 1) + \phi(S^{-1}\eta_t) + \Lambda \mathcal{F}^2 e^{2\eta}]S dt, \qquad (60.16)$$

where $F(S, \eta)$ is the same as for stress relaxation.

Acting as in the previous two examples, we get the following dynamical system for parameters η, S:

$$\begin{cases} \eta_t &= S\psi^{-1}(\frac{\mathcal{F}}{A_0}(e^\eta - 1) + \Lambda \mathcal{F}^2 e^{2\eta} + (q_1 + 2q_2 S + b_1\eta + a_1\eta^2)), \\ S_t &= \frac{b_0}{2q_2}\psi^{-1}(\frac{\mathcal{F}}{A_0}(e^\eta - 1) + \Lambda \mathcal{F}^2 e^{2\eta} + (q_1 + 2q_2 S + b_1\eta + a_1\eta^2)). \end{cases} \qquad (60.17)$$

where $\psi(x) = (e^{Dx} - 1)_+$, as in the case of stress relaxation.

During the creep evolution, we have to have in this dynamical system $\eta_t \geqq 0, S_t \geqq 0$ for the admissible initial values $S(0), \eta(0)$.

For the creep to start, the argument of the function $\psi(x)$ in the system should be positive at initial moment. Since $\sigma_{zz} = \frac{\mathcal{F}e^\eta}{A_0}$, this condition takes the form

$$\frac{\sigma(0)^2}{2Y} + q_1 + 2q_2 S(0) + a_1\eta(0)^2 > 0. \qquad (60.18)$$

Since q_1, q_2, a_1 are negative parameters, the inequality (58.18) defines the stress (or strain energy - f_{in}) threshold for the initiation of the creep processes (see [72], p.7).

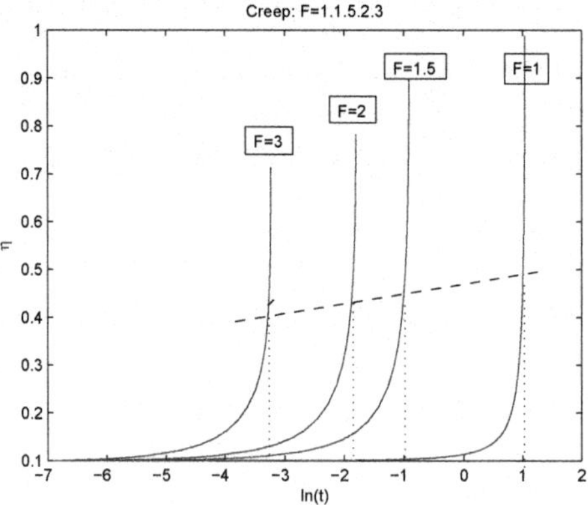

FIGURE 11. Creep: inelastic strain $\eta(t)$ for $F = 1, 1.5, 2$ and 3.

In Figure 11 the graphs of $\eta(t)$ for the creep are presented for different values of the force $\frac{F}{A_0} = 1, 1.5, 2$ $(A_0 = 1)$.

There is a point on each trajectory, corresponding to the instability of creep deformation where the cross-section of the rod diminishes to zero and the rod fails in so called ductile manner. As it can be seen from the results of calculations, the higher the force, the faster creep deformation develops and the time to the ductile failure becomes significantly shorter.

For example, three times increase in force results in more then 3 orders of magnitude in time to failure.

Comparing these graphs with the experimental data ([72, 9]) we see the good qualitative (and, for some materials, quantitative) agreement.

Remark 77. In all three examples presented above, one can exclude one of the variables and reduce the system to the nonlinear second order ODE. These equations can be explicitly solved in elliptic functions in the cases of Unconstrained aging and stress relaxation and in theta functions in the case of creep.

61. Appendix A. Strain energy as a perturbation of the "ground state energy".

In this section we discuss the perturbation scheme of a ground state Lagrangian $F(E^{in}, S, \bar{N})$, $E^{in} = \frac{1}{2}\ln(g_0^{-1}g)$ by elastic deformation, assuming that the elastic strain tensor is small in compare to the inelastic one.

In the pure inelastic mode of behavior (free aging which is the special load that produces $C(\phi) = g$) only metric quantities $g, S, \bar{N}$ enter the total Lagrangian L. Under the general load, the material metric g is deformed into $C(\phi)$ and the total deformation $E^{tot} = \frac{1}{2}\ln(g_0^{-1}C(\phi))$ takes the place of E^{in}.

Assuming that $g^{-1}C(\phi) \approx I + small$, so that $E^{el} \ll E^{in}$ we decompose $g_0^{-1}C(\phi) = (g_0^{-1}g) \cdot (g^{-1}C(\phi))$.

Taking the logarithm, we obtain for the total energy following expression:

$$E^{tot} = \frac{1}{2}\ln(exp(2E^{in}) \cdot exp(2E^{el})) \simeq \frac{1}{2}\ln(exp(2E^{in}) \cdot (I + 2E^{el}))$$

Here we've used the linear approximation $exp(2E^{el}) \simeq I + 2E^{el}$. Using the Campbell-Hausdorff-Dynkin formula we get

$$E^{tot} \simeq \frac{1}{2}[\ln(exp(2E^{in})) + Ad(exp(2E^{in}))2E^{el}] = E^{in} + Ad(exp(2E^{in}))E^{el}. \quad (61.1)$$

We consider the "total ground state" Lagrangian F depending on E^{tot} through its invariants $I_k(E^{tot})$, $k = 1, 2, 3$. Then we decompose it into Taylor series by considering E^{el} small compared to E^{in}. The approximate expression for E^{tot} above gives us up to the second order terms

$$I_k(E^{tot}) \simeq I_k(E^{in}) + dI_k(E^{in})(E^{el}\,g_0^{-1}g) + d^2I_k(E^{in})(E^{el}\,g_0^{-1}g, E^{el}\,g_0^{-1}g) + h.o.t.. \quad (61.2)$$

Substituting this into the metric term $F(I_k(E^{tot})$ (for this discussion we suppress in F all other arguments it depends on) we get, after recombining its terms, its decomposition up to the second order

$$F(E^{tot}) \simeq F(E^{in}) + \sum_k F_{,I_k}(E^{in})\langle dI_k(E^{in}), E^{el}\,g_0^{-1}g \rangle +$$

$$+ \{\sum_k F_{,I_k}(E^{in})\langle d^2I_k(E^{in})(E^{el}\,g_0^{-1}g, E^{el}\,g_0^{-1}g)\rangle +$$

$$+ \sum_{ij} F_{I_iI_j}(E^{in})\langle dI_i(E^{in}), E^{el}\,g_0^{-1}g >< dI_j(E^{in}), E^{el}\,g_0^{-1}g \rangle\}. \quad (61.3)$$

In this formula first term on the right is the basic metric ("ground state") energy describing, in particular, equilibrium values for the metric g (see example below).

The second term, linear by E^{el} describes the interaction of the elastic and inelastic processes. In the absence of such interactions, or other material processes, g takes the value delivering a minimum to the basic energy $F(E^{in})$. Therefore, its differential takes a zero value for the corresponding value of the argument E^{in} and the linear term vanishes.

Finally, the quadratic form in formula (60.3) is the conventional elastic (strain) energy with variable, and possible inhomogeneous, elasticity tensor.

During the active processes of configurational changes in the material (aging, in the zone of phase transition) this elasticity tensor, as well as the basic energy plays an active role in the evolution. But when such processes stops (no aging happens or wave of phase transition passed) and the material metric g is locked in some stable state (local minimum of F?), the value of this tensor is also locked at the corresponding value (see below).

In order to calculate the elasticity tensor e we have to calculate differentials of invariants I_k of the inelastic strain tensor E^{in}. We choose momenta $Tr(E^k)$ as the basic invariants of a (1,1)-tensors ([?]). In our, 3D case we have $I_1(E) = Tr(E), I_2(E) = Tr(E^2), I_3(E) = det(E)$. Thus, we have for its first differentials ([?])

$$\frac{\partial I_1(E)}{E_J^I} = \delta_J^J;\ \frac{\partial I_2(E)}{E_J^I} = 2E_I^J;\ \frac{\partial I_3(E)}{E_J^I} = 3E_K^J E_I^K = 3E_I^{2\ J}.$$

and $\langle dI_1(E), P \rangle = Tr(P);\ \langle dI_2(E), P \rangle = 2Tr(EP);\ \langle dI_3(E), P >= 3Tr(E^2 P))$. Here we are using multiplication of (1,1)-tensors. The second differentials of momenta have the form $d^2 I_1(E) = 0;\ d^2 I_2(E)(B, B) = 2Tr(B^2);\ d^2 I_3(E)(B, B) = 6Tr(EB^2)$.

Before using these expressions for differentials we notice that since $g_0^{-1}g = exp(2E^{in})$, $(E^{in})^{g_0^{-1}g} = E^{in}$ and due to the properties of Tr for all natural powers a, b one has $Tr(E^{in\ a}((E^{el})^{g_0^{-1}g})^b) = Tr(E^{in\ a} E^{el\ b})$. Thus, conjugation by $g_0^{-1}g$ disappear from the formulas for Elastic energy.

Substituting the expression for differentials in (12.3) we get the expression for $F(E^{tot})$ as the sum of "constant", linear by E^{el}, and quadratic by E^{el} terms

$$F(E^{tot}) \simeq F(E^{in}) + Tr(CE^{el}) + Tr(\mathbf{e} : E^{el} : E^{el}) \tag{61.4}$$

Here C is the (1,1)-tensor

$$C = F_{,I_1}(E^{in})I + 2F_{,I_2}(E^{in})E^{in} + 3F_{,I_3}(E^{in})(E^{in})^2, \tag{61.5}$$

and $\mathbf{e}$ the elasticity tensor

$$\mathbf{e}_{AC}^{BD} = 2F_{,I_2}(E^{in})\delta_C^B \delta_A^D + 6F_{,I_3}(E^{in})_C^B \delta_A^D + 2F_{,I_1 I_1}(E^{in})\delta_A^B \delta_C^D + 4F_{,I_2 I_2}(E^{in})_A^B (E^{in})_C^D +$$
$$9F_{,I_3 I_3}(E^{in})_A^{2\ B}(E^{in})_C^{2\ D} + 4F_{,I_1 I_2}\delta_A^B (E^{in})_C^D + 6F_{,I_1 I_3}\delta_A^B (E^{in})_C^{2\ D} + 12F_{,I_2 I_3}(E^{in})_A^B (E^{in})_C^{2\ D}.$$
$$\tag{61.6}$$

If the ground state function F is given as a function of variables $g, S, \bar{N}$, these formulas determine values of the elastic moduli in a material which depends on the point X and on time t through the metric variables $g, S, \bar{N}$. If these variables take stationary values, we get isotropic, but nonhomogeneous, material. If they are constant, we return to the conventional linear elasticity.

Example 34. Isotropic material. For isotropic material of linear elasticity, elastic tensor (in its (1,1)-version) has the form ([90])

$$e_{ik}^{jl} = 2\mu \delta_i^l \delta_k^j + \lambda \delta_i^j \delta_k^l. \tag{61.7}$$

Comparing with (59.6) we see immediately that there are two simple cases when (15.3) determines an isotropic material.

Case 1 - generic. Take $E_B^{in\ A} = h\delta_B^A$ where h is a scalar function of $g, S, \bar{N}$. In this case

$$e_{AC}^{BD} = [2F_{,I_2} + 6F_{,I_3} h]\delta_C^B \delta_A^D +$$
$$+ [2F_{,I_1 I_1} + 4F_{,I_1 I_2} h^2 + 9F_{,I_3 I_3} h^4 + 4F_{,I_1 I_2} h + 6F_{,I_1 I_3} h^2 + 12F_{,I_2 I_3} h^3]\delta_A^B \delta_C^D. \tag{61.8}$$

The first bracket gives an expression for 2μ while the second one gives an expression for λ.

Case 2 - simple elasticity. In this case we have no aging, $E^{in} = 0$. Then we get material with

$$e_{AC}^{BD} = 2F_{,I_2}(0)\delta_C^B \delta_A^D + 2F_{I_1 I_2}(0)\delta_A^B \delta_C^D. \tag{61.9}$$

Thus, in this, restricted case $2\mu = 2F_{,I_2}(0)$, $\lambda = 2F_{I_1 I_2}(0)$.

Example 35. Consider the model 1D case with one component of strain tensors E^{el}, E^{in}, trivial decomposition $E^{tot} = E^{el} + E^{in}$, and simple Taylor decomposition of the basic energy function $F(E^{tot})$:

$$F(E^{tot} = F(E^{in}) + F'(E^{in})E^{el} + \frac{1}{2}F''(E^{in})(E^{el})^2 + h.o.t. \tag{61.10}$$

As a result, strain energy here has the form

$$U(E^{el}) = \frac{1}{2}F''(E^{in})(E^{el})^2, \tag{61.11}$$

and the Young's module (or compressional stiffness, in a case of an elastic bar) is

$$Y = \frac{1}{2}F''(E^{in}).$$

Consider two special cases:

1. Classical elasticity:
In this case we take $F(E) = F_0 + cE^2$. further, there is one equilibrium - (minimum $E = 0$) that corresponds, for $E = E^{in} = \frac{1}{2}\ln(g_0^{-1}g)$ to the value $g(X,T) = g_0(-constant)$. We have $Y = c$.

2. Two-phase material (material that can exist in two stable phases). In this example function $F(E)$ has two (locally) stable states $g = g_0, g_1$, or $E = 0$, $E = Q = \frac{1}{2}\ln(g_0^{-1}g_1)$ and

$$F(E) = F_0 + \frac{1}{4}c_4 E^2 (E + Q^{-1})^2 + \frac{1}{2}c_2 E^2. \tag{61.12}$$

Then, for $E = E^{in}$,

$$F(E) = F_0 + \frac{1}{4}c_4(\frac{1}{2}\ln(g_0^{-1}g))^2(\frac{1}{2}\ln(g_1^{-1}g))^2 + \frac{1}{2}c_2(\frac{1}{2}\ln(g_0^{-1}g))^2. \tag{61.13}$$

We have

$$F''(E) = c_2 + \frac{1}{2}c_4((E + Q^{-1})^2 + 4E(E + Q^{-1}) + E^2).$$

The Young module in the state $g = g_0$ is equal to

$$Y_0 = c_2 + \frac{1}{2}c_4 Q^{-2}.$$

While in the second stable state $g = g_1$ and

$$Y_1 = c_2 + \frac{1}{2}c_4((Q + Q^{-1})^2 + 4Q(Q + Q^{-1}) + Q^2) = Y_0 + 3c_4(1 + Q^2).$$

Thus, in a case of a wave of phase transition going along the bar, the Young module changes by the amount $Y_1 - Y_0 = 3c_4(1 + Q^2)$.

62. APPENDIX B. VARIATIONS.

Variations of some expressions for the Lagrangian (51.4) are calculated here and presented in a table. All terms in Lagrangian density $L(G, \phi)$ will be refereed to by the mass form $dM = \rho_0 d_G V = \rho_0 S \sqrt{|g|} d^4 X$. In other terms we calculate variation $\int f(A) dM$ by A. The result of variation has the form $\mathcal{V} dM : \delta f(A) dM = \mathcal{V} dM$. In the calculations we repeatedly use the following standard relation $\delta \sqrt{|g|} = \frac{\sqrt{|g|}}{2} g^{IJ} \delta g_{IJ}$ (see, for instance, [2]). In the table below we present tensors $\mathcal{V}$ for different f.

As an example of such a calculation we provide calculation of $\delta \left[div_g(\vec{N}) dM \right]$ where $div_g(N) = \frac{1}{\sqrt{|g|}} \frac{\partial}{\partial X^I}(\sqrt{|g|} N^I)$:

$$\delta div_g(\vec{N}) dM = div_g(\vec{N}) \rho_0 (\sqrt{|g|} \delta S + S \frac{\sqrt{|g|}}{2} g^{IJ} \delta g_{IJ}) + \delta(\frac{1}{\sqrt{|g|}} \frac{\partial}{\partial X^I}(\sqrt{|g|} N^I)) dM =$$

$$= (div_g(\vec{N}) S^{-1} \delta S - \frac{\partial}{\partial X^I} ln(\rho_0 S) \delta N^I + \ div_g(\vec{N}) \frac{1}{2} - \frac{1}{2}(div_g(\vec{N}) - \frac{1}{2} N^I \frac{\partial}{\partial X^I} ln(\rho_0 S))$$

$$g^{AB} \delta g_{AB}) dM = (div_g(\vec{N}) S^{-1} \delta S - \frac{\partial}{\partial X^I} ln(\rho_0 S) \delta N^I + \ -\frac{1}{2} N^I \frac{\partial}{\partial X^I} \ ln(\rho_0 S)) \ g^{AB} \delta g_{AB}) dM$$

$$\tag{62.1}$$

since

$$\delta(\frac{1}{\sqrt{|g|}} \frac{\partial}{\partial X^I}(\sqrt{|g|} N^I)) dM = -(\frac{1}{\sqrt{|g|}^2} \delta \sqrt{|g|} \frac{\partial}{\partial X^I}(\sqrt{|g|} N^I)) dM - \frac{\partial}{\partial X^I}(\rho_0 S) \delta(\sqrt{|g|} N^I) d^4 X =$$

$$= [-\frac{1}{2}(div_g(\vec{N}) - \frac{1}{2} N^I \frac{\partial}{\partial X^I} ln(\rho_0 S)) g^{AB} \delta g_{AB} - \frac{\partial}{\partial X^I} ln(\rho_0 S) \delta N^I] dM \tag{62.2}$$

The formula of variation of $hR(g)\sqrt{|g|}$ in the 5th row of the table is taken from [34], Prop.3.2.

To find variation of the strain energy density $f(G, E^{in}) \rho_0 S \sqrt{|g|} d^4 X$ we first take the variation of $dM = \rho_0 S \sqrt{|g|} d^4 X$ to get the first two terms in the last row of the Table, then - explicit variation by g if the strain energy function f depends on g not just through E^{el}. Finally for variation by g through the strain tensor $E^{el}{}^I_J = \frac{1}{2} g^{IK}(C(\phi)_{KJ} - g_{KJ})$ we have

Term	Variation by S, N^I, g_{IJ}
$f(\lvert g\rvert)$	$S^{-1}f(\lvert g\rvert)\delta S + [\tfrac{1}{2}f(\lvert g\rvert)g^{IJ} + f'(\lvert g\rvert)\lvert g\rvert g^{IJ}]\delta g_{IJ}$
$f(S)$	$[f'(S) + S^{-1}f(S)]\delta S + \tfrac{1}{2}f(S)g^{IJ}\delta g_{IJ}$
$\lVert\vec{N}\rVert_g^2$	$\lVert\vec{N}\rVert_g^2 S^{-1}\delta S + 2N_I\delta N^I + [N^I N^J + \tfrac{1}{2}\lVert\vec{N}\rVert_g^2 g^{IJ}]\delta g_{IJ}$
$\chi(K_J^I)$	$S^{-1}K_J^I\frac{\partial\chi(K)}{\partial K_J^I}\delta S + [-S^{-1}\frac{\partial\chi}{\partial K_J^A}g^{AS}\partial_{X^I}g_{SJ} + \frac{1}{\rho_0 S\sqrt{\lvert g\rvert}}\partial_{X^S}(\rho_0\sqrt{\lvert g\rvert}\frac{\partial\chi}{\partial K_J^A}g^{AS}g_{IJ}) +$ $+\frac{1}{\rho_0 S\sqrt{\lvert g\rvert}}\partial_{X^J}(\rho_0\sqrt{\lvert g\rvert}\frac{\partial\chi}{\partial K_J^I})]\delta N^I \;+\; [-\frac{\partial\chi}{\partial K_J^I}S^{-1}g^{IA}\frac{\partial N^B}{\partial X^J} - \frac{\partial\chi}{\partial K_B^I}S^{-1}g^{IS}\frac{\partial N^A}{\partial X^S} -$ $\frac{\partial\chi}{\partial K_J^I}g^{IA}K_J^B - \frac{1}{\rho_0 S\sqrt{\lvert g\rvert}}\partial_t\;\rho_0\lvert g\rvert\frac{\partial\chi}{\partial K_B^I}g^{IA} + \frac{1}{\rho_0 S\sqrt{\lvert g\rvert}}\partial_{X^K}\partial_t\;\rho_0\lvert g\rvert\frac{\partial\chi}{\partial K_B^I}g^{IA}N^K\;]\delta g_{AB}$
$hR(g)$	$S^{-1}hR(g)\delta S + [-h(Ric(g) - \tfrac{1}{2}R(g)g) + \frac{1}{\rho_0 S}[\Delta_g(\rho_0 hS)g^{AB} + Hess(\rho_0 hS)]]\delta g_{AB}$
$div_g(\vec{N})$	$div_g(\vec{N})S^{-1}\delta S - \frac{\partial}{\partial X^K}ln(\rho_0 S)\delta N^K + [-\tfrac{1}{2}N^K\frac{\partial}{\partial X^K}ln(\rho_0 S))]\,g^{IJ}\delta g_{IJ}$
$f(G, E^{el})$	$S^{-1}f\delta S + [\tfrac{1}{2}fg^{AB} + \frac{\partial f}{\partial g_{AB}}\exp - \tfrac{1}{2}S^{(AB)}]\delta g_{AB}$

TABLE 3. Table of Variations.

$$\delta f(E^{el}) = \frac{\partial f}{\partial E^{el}{}_J^I}\delta E^{el}{}_J^I = \frac{1}{2}\frac{\partial f}{\partial E^{el}{}_J^I}\delta g^{IK}C(\phi)_{KJ} = \frac{1}{2}\frac{\partial f}{\partial E^{el}{}_J^I}[-g^{IA}g^{KB}\delta g_{AB}]C(\phi)_{KJ} =$$

$$-\frac{1}{4}\frac{\partial f}{\partial E^{el}{}_J^I}[(g^{IA}g^{KB}+g^{IB}g^{KA})\delta g_{AB}]C(\phi)_{KJ}\delta g_{AB} = -\frac{1}{4}\frac{\partial f}{\partial E^{el}{}_{MN}}[(\delta_M^J g_{IN} + \delta_N^J g_{IM}) \times$$

$$\times ((g^{IA}g^{KB} + g^{IB}g^{KA})\delta g_{AB})]C(\phi)_{KJ}\delta g_{AB} = -\frac{1}{4}(\frac{\partial f}{\partial E^{el}{}_{JA}}g^{KB} + \frac{\partial f}{\partial E^{el}{}_{JB}}g^{KA} +$$

$$+ \frac{\partial f}{\partial E^{el}{}_{AJ}}g^{KB} + \frac{\partial f}{\partial E^{el}{}_{BJ}}g^{KA})C(\phi)_{KJ}\delta g_{AB} = -\frac{1}{2}S^{(AB)}\delta g_{AB}, \qquad (62.3)$$

where $S^{(AB)}$ is the symmetrization of the second Piola-Kirchoff Tensor S (see [90]), the last equality is proved in [15].

63. Conclusion.

In this Chapter, we considered the intrinsic material metric tensor to be an additional parameter of state, i.e., an internal variable that characterizes material degradation and aging. The material metric tensor G is conjugate (with respect to a particular Lagrangian) to the canonical Energy-Momentum Tensor (or to the Eshelby energy-stress tensor to some degree).

Equations of metric evolution, (i.e., the aging equations), are derived as the Euler-Lagrange equation of a corresponding variational problem. The canonical energy-momentum tensor (or Eshelby Tensor) plays a role of the source of metric evolution. This represents an alternative approach to numerous phenomenological damage models, which usually have more adjustable parameters than practical testing is able to determine. Thus it is difficult to validate the models since they can almost always be adjusted to reach an agreement with the experiment. In contrast, a variational approach prescribes a functional form of the aging equations, limits the number of constants (adjustable parameters) employed in the Lagrangian, provides a simple physical interpretation of the constants, and admits an essential experimental examination of the validity of the basic assumptions of the model. Particular examples (aging homogeneous rod, see Sec. 11 or [20], cold drawing (necking) [18], residual stress and others) can be analyzed theoretically and unambiguously tested in the experiments as a natural continuation of the present work.

APPENDIX . Fibre bundles,jet bundles and the Noether balance laws.

In this **Appendix** we sketch the basic notions concerning the differentiable manifolds and their mappings, fibre bundles (Appendix 1), jet bundles and the basic geometrical structures on these objects used above (metrics, connections of different type, etc. (Appendix II). Finally, in Appendix III we discuss the symmetries of Lagrangian problems and differential equations and and present the Noether formalism in the form suited to the Euler-Lagrange systems with forces (sources in the case of Field Theory). For more detailed exposition of these topics we refer to the sources [?, 60, 70, 69, ?, 122, 138] (Differentible manifolds, fiber bundles, metrics, connections), to [47, 71, 75, 106, 119] (Variational Calculus,Geometrical theory of differential equations, jet bundles).

APPENDIX I. Fibre bundles and their geometrical structures.

64. Differentialbe manifolds.

In this section we introduce the Differentiable manifolds, their mappings and present some examples of these objects. For more detailed presentation of the theory of Differentiable manifold we refer to [69, 138].

A (real) **differentiable manifold** M^k of dimension k is the space (set) endowed with the structures allowing the development of Calculus on M and the formulating and study differential equations as well as the systems of such equations.

This *differential structure* is defined on the "topological background" in the space M. Topological background ensures that, locally, space M looks like an open subset in R^k: any point $x \in M$ has a neighbourhood U_x endowed with the continuous one to one mapping $\phi_x : U_x \to \mathbb{R}^k$ onto its image - open subset of R^k. These mappings, called **topological charts**, are in agreement: if domains of two neighborhoods $(U_x, \phi_x), (U_y, \phi_y)$ have nonempty intersection: $U_x \bigcap U_y \neq \varnothing$, then the one-to one mapping $\phi_\beta \circ \phi_\alpha^{-1}$ from the image $\phi_x(U_x \bigcap U_y)$ onto the image $\phi_y(U_x \bigcap U_y)$ is continuous together with its inverse. This structure defines M as a *topological manifold*.

A collection of charts $\{U_\alpha, \alpha \in I\}$ compatible in the sense introduced above, whose domains together cover the whole space M: $M = \bigcup U_\alpha$ is called the *continuous atlas*, or *covering* of M. This "topology" defined on M allows us to define R-valued *continuous functions* as functions on the domains of M whose compositions with the inverse mappings ϕ_x^{-1} are continuous functions defined in the domain $\phi_x(U_x) \subset R^k$. We refer the reader to the monograph [138] for further development of the properties of the topological manifolds.

Next we define the *infinitelydifferentiable* or $C^\infty - manifolds$. All smooth manifolds in this text are assumed to be $C^\infty - manifolds$. We refer to [138] for definitions and properties of k-differentible manifolds, where $k \geqq 1$.

Definition 32. *An **infinitely differentiable manifold** (called also **infinitely smooth** or $C^\infty - manifold$) is a topological manifold endowed with a continuous atlas $\{U_\alpha, \phi_\alpha\}$ having the following property: if two charts in M with the domains and coordinate mappings $U_\alpha, \phi_\alpha : U_\alpha \to R^k$ and $U_\beta, \phi_\beta : U_\beta \to R^k$ respectively have nonzero intersection: $U_\alpha \cap U_\beta \neq \emptyset$, then the composition $\phi_\beta \circ \phi_\alpha^{-1}$ of chart mappings defined in the corresponding domains of two copies of $\mathbb{R}^k$ is differentiable (of the type C^∞) together with its inverse mapping $\phi_\alpha \circ \phi_\beta^{-1}$. An atlas whose charts have this property is called a **differentiable atlas**. A topological manifold with a given differentiable atlas is called a differentiable (smooth or C^∞-smooth) manifold of dimension k.*

Differential structure on a manifold M allows us to define differentiable functions on M, differentiable mappings between differentiable manifolds, smooth surfaces, smooth submanifolds, to define differential equations and study these equations.

© Springer International Publishing Switzerland 2016
S. Preston, *Non-commuting Variations in Mathematics and Physics*,
Interaction of Mechanics and Mathematics,
DOI 10.1007/978-3-319-28323-4_7

We recall the definition of a class of smooth manifolds which is especially convenient for the introduction and investigation of geometrical and analytical structures in a manifold.

To do this we use the notion of a **covering**: a covering of a smooth manifold M is a collection I of open subsets U_α, $\alpha \in I$ of manifold M such that $M = \cup U_\alpha$. Let $(U_\alpha, \alpha \in I)$ and $(W_\beta, \beta \in J)$ be two covering of the manifold M. We say that the covering $U_\alpha, \alpha \in I$ is *inwritten* to the covering $W = \{W_\beta, \ \beta \in J\}$ (or that the cover W is the *refinement* of the cover U) if for any index $\alpha \in I$ of open sets in the first covering there exist the set W_β of second covering such that $W_\beta \subset U_\alpha$.

Let M be a differentiable manifold and $M = \bigcup_\alpha N_\alpha$ where N_α, $\alpha \in I$ is a covering of manifold M by open subsets. This covering is called **locally finite** if any point $m \in M$ has an open neighborhood that has nonempty intersections only with a finite number of sets N_α.

Definition 33. *A differentiable manifold M is called **paracompact** if it is a Hausdorff space (is separable) and if for any covering $M = \bigcup_\alpha U_\alpha$ of M by open sets U_α, there exists a locally finite covering of M inwritten to the covering $\{U_\alpha\}$.*

One of the most specific properties of paracompact manifolds is: such manifolds have a partition of unit subordinate to any covering, see [138], Sec.1.7 for more details.

Example 36. Simple examples of k-dim manifolds are: vector space $\mathbb{R}^k$ and its open subsets, m-dim sphere S^m in R^k and the l-dim tori T^l in R^k, k-dimensional surfaces without singular points in the space $\mathbb{R}^n$ ($k < n$),etc.

64.1. **Differentible mappings of manifolds.**

Definition 34. *Let M^m, N^n be two differentiable manifolds of dimension m and n respectively. Let $f : M \to N$ be a mapping of manifold M to the manifold N. Mapping f is called C^∞-differentiable or **smooth** if for any couple of charts: $(U, \phi : U \to R^m)$ for M and $(V, \psi : V \to R^n)$ for N, composition $\psi \circ f \circ \phi^{-1}$ defining the mapping from the image $\phi(U) \subset R^m$ of coordinate mapping ϕ to the image $\psi(V) \subset R^m$ of chart mapping ψ is $C^\infty - differentiable$ (as the mapping from the open subset in R^m to the space R^n.)*

Example 37. For example, a real valued function $M \to R$ on a manifold M^m is smooth if for any chart (U, ϕ) in M, composition $f \circ \phi^{-1} : \phi(U) \to R$ is the smooth (infinitely differentiable) function in the domain $\phi(U) \subset R^m$.

65. **Fibre bundles.**

In this section we introduce **fibre bundles** - the basic notion of the Geometrical Theory of Differential Equations and, in particular, of Variational Calculus. For more detailed descriptions of the fibre bundles, their properties and related structures, see [60] or [70] .

Simplest fiber bundles are the **trivial bundles (called also product bundles)**. A trivial fiber bundle is the triple of manifolds X^n, Y^{n+m}, F^m and the projection $\pi : X \times F \to X$ to the factor X. Manifold X is called the **base space** and Y is the **total space** of the trivial bundle, F is the **standard fiber** of the bundle and mapping π - is the **projection of the trivial bundle**.

General fiber bundle is the construction glued from the trivial bundles by the smooth transition mappings. More specifically,

Definition 35. *A **fibre bundle** (Y^{n+m}, π, X^n) consists of the smooth manifolds Y, X, F of dimensions $(n+m), n, m$ respectively, a smooth epimorphism (smooth surjective mapping) of manifolds $\pi : Y^{n+m} \to X^n$ such that any point $x \in X$ has an open neighborhood $U \subset X$ with the property that the restriction of the bundle to the subset $\pi^{-1}(U) \subset Y$ (i.e. the bundle $\pi : \pi^{-1}U \to \pi(pi^{-1}U))$ is diffeomorphic to the **trivial bundle** (see above) $U \times F \to U$ via a **fibre preserving diffeomorphism** $\simeq \psi : u \times F \simeq \pi^{-1}U$ as in the commutative diagram*

$$
\begin{array}{ccccc}
\psi : U \times F & \xrightarrow{\;\simeq\;} & \pi^{-1}U & \xrightarrow{\;\hookrightarrow\;} & Y \\
\pi \downarrow & & \pi \downarrow & & \pi \downarrow \; . \\
U & \xrightarrow{\;\simeq\;} & U & \xrightarrow{\;\hookrightarrow\;} & X
\end{array}
\qquad (65.1)
$$

As in the case of trivial bundle, manifold X is called the **base space**, Y is the **total space** of the fiber bundle π, F is the **standard fiber** of the bundle and mapping π - is the **projection of the fiber bundle**.

The couple (U, ψ) where $U \subset X$ is an open subset of X and $\psi : U \times F \to \pi^{-1}U)$ is the diffeomorphism , making the left part of the diagram above **commutative**, is called a **fibred chart over U or a local trivialization** of the bundle π over the open set $U \subset X$.

Remark 78. Notice that **every bundle defined above is locally trivial**.

Let $\pi : Y \to X$ be a fiber bundle with the m-dimensional fibres Y_x, $x \in X$ and n-dim base X. Let $\phi : W \to R^{n+m}$ be a diffeomorphism of an open domain $W \subset Y$ onto the open subset $\phi(W) \subset R^{n+m}$. Mapping ϕ defines the coordinates (x^i, y^μ) in W borrowing them from R^{n+m}. We say that this coordinate system **is adopted** (to the bundle structure) or that these coordinates are **fibred coordinates** if for all point couples $a, b \in W$ such that $\pi(a) = \pi(b)$, one has

$$
pr_1(\phi(a)) = pr_1(\phi(b))
$$

where $p_1 : R^n \times R^n \to R^n$ is the first projection in the product bundle $R^{n+m} \to R^n$.

Example 38. If U is a domain of a coordinate system (U, x^i) in X and if $(V \subset F, y^\mu)$ is a local chart in F, then the coordinates (x^i, y^μ) in the open subset $(\pi^{-1}(U)) \subset Y$ form the **fibred chart**.

Remark 79. If (Y^{n+m}, π, X^n) is a fibre bundle with each fibre $\pi^{-1}(x)$, $x \in X$ being diffeomorphic to F. Any fibre chart (U, ψ) with $x \in U$ realizes this diffeomorphism for the fibers Y_x with F over the domain U.

Remark 80. Notice that the epimorphic mapping $\pi : Y \to X$ of a fibre bundle is the *smooth mapping of constant rank $n = dim(X)$*.

Definition 36. *An **atlas of a fibre bundle** $\pi : Y \to X$ is the collection of fibred charts (W_α, ψ_α) such that the domains of these charts cover the whole space Y: $Y = \bigcup_\alpha W_\alpha$ and the **transition mappings** $\psi^\beta \circ \psi_\alpha^{-1} : \bar{W}^\alpha \to \bar{W}^\beta$ on the intersections of the domains $\bar{W}_\alpha \bigcap \bar{W}_\beta$ are diffeomorphisms.*

A natural way fibre bundles appears in mathematics is described in the next Lemma.

Lemma 10. *Let* $p : N \to M$ *be a* **surjective submersion** *(smooth mapping that is onto and of constant rank) which is* **proper** *(i.e. such that all the compact subspaces $K \subset M$ have compact pre-image $p^{-1}(K) \subset N$) and let the manifold M be connected. Then, the triple (N, p, M) is the fibre bundle.*

See [70], Ch.III, Sec.9 for the proof of this statement.

Remark 81. Configurational bundles of a Field Theory $\pi : Y \to X$, introduced in Chapter 1, are fibre bundles (often paracompact).

Dynamical variables (in mechanics) and dynamical fields (in Field Theory) appear in geometrical formalism of Variational Calculus as the "**sections**" of *configurational bundles*. Thus, we define

Definition 37. *Let* $\pi : Y \to X$ *be a fibre bundle. Let $V \subset X$ be an open subset of X. A smooth (having infinitely differential components) mapping $s : V \to Y$ is called a* **section of the bundle** π *over V if $\pi(s(x)) = x$ for all points $x \in V$.*

Denote by $\Gamma_V(\pi)$ the set of all sections of bundle π over V. The topology of uniform convergence of mappings on the compact subsets of V endows $\Gamma_V(\pi)$ with the "compact-open topology", see [60].

Remark 82. In the domains of fibred charts $(W; x^i, y^\mu)$ sections $s : U \to Y$ $U = \pi(W)$ over U correspond to the m-tuples of functions $s = \{y^\mu(x) \in C^\infty(U)\}$.

Notice that a fibre bundle always has *local sections* (over the domains of fibred charts) but it might not have global sections defined on the whole base X. The most well known examples of fibre bundles having no global sections are tangent mappings of some manifolds (see [60]) and the Hopf bundle $S^3 \to S^2$ ([60], Ch.10).

65.1. **Tangent and Cotangent bundles.** Any differentiable manifold M is endowed with a family of vector bundles (see below) associated with M. The most well known are the tangent bundle $T(M) \to M$ and the cotangent bundle $T^*(M) \to M$. Fiber $T_x(M)$ of the tangent bundle over a point x is the vector space of tangent vectors at x to the manifold M.

Cotangent bundle is dual to the tangent one in the sense that the fiber $T^*_x(M)$ of cotangent bundle $T^*(M)$ at the point $x \in M$ is the vector space dual to the fiber $T_x(M)$ of the tangent bundle over the same point.

Let $\pi : Y \to X$ be a fibre bundle. Projection $\pi : Y \to X$ defines the epimorphism of tangent bundles $T(\pi) : T(Y) \to T(X)$ (tangent mapping of the mapping π) and the monomorphism (pullback of mapping π) $T^*(\pi) : T^*(X) \to T^{P*})Y$.

In local fibred coordinates (U, x^i, y^μ) vector fields $\{\partial_i = \partial_{x^i}, i = 1, \ldots, n\}$ form the basis for the (vector) space $X(U)$ of vector fields in the domain $U \subset X$. Vector fields $\partial_\mu = \partial_{y^\mu}$ form the basis of vector fields on the fibres Y_x of the bundle. Taken together, vector fields $\partial_{x^i}, \partial_{y^\mu}$ form the holonomic **frame** - the basis of tangent bundle in the domain $\pi^{-1}(U) \subset Y$.

For $y \in Y$, the tangent mapping $T(\pi)_y$ sends a tangent vector $\xi = \xi^i \partial_i + \xi^\mu \partial_\mu$ in Y at the point y to the tangent vector $\xi^i \partial_i \in T_{\pi(y)}(X)$ at the point $\pi(y) \in X$.

Differential 1-forms dx^i form the (coordinate) **coframe** basis of the vector space of 1-forms in the domain $U \subset X$ of the fiber chart, while 1-forms dx^i, dy^μ form the basis of 1-forms in the subspace $\pi^{-1}(U) \subset Y$.

Cotangent mapping $T^*(\pi)$ (pullback) sends the covector $\omega_i(x) dx^i$ at a point $x \in U$ to the same form considered as 1-form (covector) at any point y in the fiber Y_x over x.

In between the fibred bundles, the bundles of next three classes - vector bundles, affine bundles and the principal bundles are the most simple and most often appear in geometry and in different applications. **Tangent and Cotangent bundles** belongs to the class of **vector bundles**.

We specially mark the class of Jet bundles - principal objects of Geometrical Theory of differential equations, including Variational Calculus (see below Sec.69). Configurational 1-jet bundle $J^1\pi$ and its geometrical structures were the main objects of our investigation starting from Chapter I.

66. VECTOR AND AFFINE BUNDLES.

Let $\pi : Y \to X$ be a fibred bundle. Bundle π is called a **vector bundle** if the following two conditions are fulfilled:

(1) Typical fibre F of the bundle π and all the fibres $Y_x = \pi^{-1}(x)$, $x \in X$ are real vector spaces of dimension m;

(2) There is a bundle atlas $\Psi = \{U_\alpha, \phi_\alpha\}$ such that the trivialization morphisms $F \to Y_x$ and the transition mappings $\psi_{\alpha\beta}$ of this atlas are **linear isomorphisms**.

As a result, a vector bundle is endowed with the linear fibre coordinates y^i such that if e_i, $i = 1, \ldots, m$ is a fixed basis of the model fibre F and $e_i(x)$, $i = 1, \ldots, m$ (corresponding basis of the fibres Y_x), we have for $y \in Y$ the following vector representation

$$y = y^i e_i(\pi(y)).$$

Notice that the vector bundle has the canonical, **zero section** $0 : X \to Y$ sending a point $x \in X$ to the zero vector at the fiber Y_x. We refer to the source [122] for the proof of the following important theorem.

Theorem 32. *Let $\pi : Y \to X$ be a fibre bundle whose fibres Y_x are diffeomorphic to R^m. Then, any section of the bundle π over any closed embedded submanifold $S \subset X$ (for instance over a point) **can be extended to the global section**. In particular, such a bundle always has global sections.*

Tangent bundles $\tau : T(M) \to M$, cotangent bundles $\tau^* : T^*(M) \to M$, (p, q)-tensor bundles $T^p_q(M) \to M$ and bundles of k-forms $\Omega^k(M) \to M$ are examples of vector bundles.

Let $\bar{\pi} : \bar{Y} \to X$ be a vector bundle with the typical fiber $\bar{F}$. An **affine bundle** modeled on the vector bundle $\bar{\pi}$ is a fiber bundle $\pi : Y \to X$ whose typical fiber F is the affine space modeled on the vector space - typical fiber $\bar{F}$ of the bundle $\bar{\pi}$ and such that the following conditions hold:

(1) All the fibers Y_x are affine spaces modeled over corresponding fibres $\bar{Y}_x$ of the bundle $\bar{\pi}$,

(2) There is an "affine bundle atlas" $\Psi = \{(U_\alpha, \psi_x),\}$ which is an atlas of charts where **trivialization morphisms and transition mappings between domains of charts are affine isomorphisms**.

66.1. **Vertical bundle** $V(\pi)$.

Definition 38. *Let $\pi : y \to X$ be a fibre bundle. A tangent vector $v \in T_y(Y)$ is called vertical if $T\pi_y(v) = 0$.*

Lemma 11. *Linear subspaces $V_y \subset T_y(Y)$ of π-vertical tangent vectors form the subbundle of the tangent bundle $T(Y) \to Y$ This subbundle is called the **vertical bundle** $V(\pi) \to Y$ of fibre bundle π. The sequence of bundles over Y (i.e. with the base Y):*

$$0 \to V(\pi) \to T(Y) \to T(\pi)^*T(X) \to 0 \tag{66.1}$$

is exact.

*Here $T(\pi)^*T(X)$ is the **pullback** of the tangent bundle $T(X)/X$ to the space Y - vector bundle over Y with the fiber over a point $y \in Y$ being the tangent space $T_{\pi(y)}(X)$.*

In particular, at any point $y \in Y$, the sequence of vector spaces

$$0 \to V_y(\pi) \to T_y(Y) \to T_{\pi(y)}(X) \to 0 \tag{66.2}$$

is exact.

Proof. The proof can be done locally, in the domains of fibred charts and then the proof is straightforward. $\qquad\qquad\square$

The dual bundle to the vertical bundle is the bundle $V(\pi)^*$. Correspondingly, we have the exact sequence of bundles dual to (66.2)

$$0 \leftarrow V^*(\pi) \leftarrow T^*(Y) \leftarrow T^*(X) \leftarrow 0. \tag{66.3}$$

In the domain $W \subset Y$ of fibred coordinates (x^i, y^μ), tangent vectors $\{\partial_i = \frac{\partial}{\partial y^i}\}$ form the basis of vertical bundle $V(\pi)|_W$ while the 1-forms dy^μ form the basis of dual to vertical bundle $V^*(\pi)|_W$.

66.2. **Natural bundles. Natural bundles** (tensor bundles, bundles of tensor densities, etc., see [33]) often appear as configurational bundles in Classical Field Theory. If a bundle $\pi : Y \to X$ is *natural*, the metric G on the base X generates the metrics h_x on the fibers Y_x of the configurational bundle, smoothly depending on x.

Tensor bundles $T^{p,k}(X) \to X$ and the bundles of tensor densities $T^{p,k}(X) \otimes \Lambda^n(X) \to X$ ([33, 69]) are examples of natural bundles.

For the bundles of these classes, one can get more: to a pseudo-riemannian metric G on the base X there corresponds the Levi-Civita connection Γ^G on the tangent bundle $\tau_X : T(X) \to X$. The covariant metric $G^{-1} = G^{ij}$ on the fibers of the cotangent bundle $\tau_X^* : T^*(X) \to X$ generates the connection Γ^{*g} on the cotangent bundle. Metrics: G on the tangent bundle τ_X and G^{-1} on the cotangent bundle τ_X^* generate metrics on the fibers of all tensor and tensor density bundles. At the same time, connections Γ^G and Γ^{*G} generate the connections on all tensor and tensor densities bundles.

Using these connections one can define **natural metrics** on the **total spaces of tensor and tensor densities bundles** over X, see [33]. In Sec.9 we are using metrics on the total spaces of natural bundles defined in this way.

67. MAPPINGS (MORPHISMS) AND AUTOMORPHISMS OF BUNDLES.

Let $\pi : \pi : Y \to X$ and $\nu : Z \to U$ be two bundles (over different basis). A mapping (**bundle morphism**) of the bundle π to the bundle ν is couple of maps

$f' : X \to Z$ and $f : X \to U$ such that $\nu \circ f' = f \circ \pi$, or, what is the same, the commutative diagram

$$
\begin{array}{ccc}
Y & \xrightarrow{\;\hat{f}\;} & Z \\[4pt]
\pi \downarrow & & \downarrow \nu \\[4pt]
X & \xrightarrow{\;f\;} & U
\end{array}
$$

where $\hat{f}$ and f are smooth mappings of manifolds.

Remark 83. Automorphisms of the bundles plays a prominent role in the properties of Lie group actions on the differential equations, their solutions, Lagrangians of variational problems defining the "projectable actions" of Lie groups, see Sec. 75 below.

Definition 39. *An **automorphism** of a fiber bundle $\pi : Y^{n+m} \to X^n$ is a couple of diffeomorphisms $\phi : Y \to Y$; $\bar{\phi} : X \to X$ such that the diagram*

$$
\begin{array}{ccc}
Y & \xrightarrow{\;\phi\;} & Y \\[4pt]
\pi \downarrow & & \downarrow \pi \\[4pt]
X & \xrightarrow{\;\bar{\phi}\;} & X
\end{array}
$$

is commutative.

It is easy to check that automorphisms of a bundle π form the subgroup $Aut(\pi) \subset Diff(Y)$ of the group of all diffeomorphisms of the space Y. Automorphisms could be called the *projectable diffeomorphisms* of Y.

Example 39. Let $\phi : Y \to Y$ be a diffeomorphism of Y acting along fibers of the bundle π: $\pi\phi(y) = \pi(y)$ for all $y \in Y$. Then, ϕ is the automorphism of the bundle π. Such automorphisms form the subgroup of **(pure) gauge automorphisms** $Gau(\pi) \subset Aut(\pi)$ of the automorphisms group $Aut(\pi)$ of the bundle π.

Example 40. Let ν be a *zero curvature* connection on the bundle π (see bellow). Let $\bar{G} \subset Diff(X)$ be a subgroup of diffeomorphisms of the base space X. Then the ν-horizontal lifts of transformations $g \in \bar{G}$ to Y form the subgroup $G \subset Aut(\pi)$.

Let $X = R^4$ be the physical space-time and T^4 is the 4-dim group of translations of R^4. The lift of T to the bundle space Y via the trivial connection (if such one exists) and its further flow lift to the 1-jet bundle $\pi_{10} : J^1(\pi) \to Y$ is used for the formulation of the stress-energy-momentum balance (and conservation) laws.

67.1. One parametrical groups of automorphisms and the infinitesimnal automorphisms of fiber bundles. Let ϕ^t be a one-parameter group of automorphisms of the bundle π. Let ξ be the vector field (phase field of the group ϕ^t). In a fibred chart (x^i, y^μ) this vector field has the components

$$
\xi = \xi^i(x^\alpha)\partial_{x^i} + \xi^\mu(x, y)\partial_{y^\mu}.
$$

Components $\xi^i(x^\alpha)$ of the vector field ξ depend on the coordinates on the base X only. Such vector fields are called **projectable**. Vice versa, any **projectable vector field** in Y has automorphisms of the bundle π as the mappings ϕ^t of its phase flow -(locally and possible globally defined). These automorphisms form the one-parametrical group of diffeomorphisms of space Y covering some (local and possibly global) one-parametrical group of diffeomorphisms of X (see [60]).

68. Connections on the fibre bundles (FOR MORE DETAILS, SEE [46, 69, 70]).

Let $\pi : Y^{n+m} \to X^n$ be a fibre bundle with the n-dimensional base X and m-dim fibres Y_x, $x \in X$.

Definition 40.   (1) *A **connection** (called also an* Ehresmann connection*) ν on the bundle $\pi : Y \to X$ is a differentiable n-dimensional distribution (called the ν-horizontal distribution) $Hor \subset T(Y)$ such that for any point $y \in Y$ the restriction of the tangent projection $\pi_* : T(Y) \to T(X)$ to the subspace $Hor_y(\nu) \subset T_y(Y)$ - the mapping $\pi_{*y} : Hor_y(\nu) \to T_{\pi(y)}(X)$ **is the linear isomorphism**.*

    (2) *Let a connection ν be given. Let $y \in Y$ and $x = \pi(y)$. Linear mapping $\pi^{-1} : T_x(X) \to Hor_y(\nu)$, inverse to the projection of $Hor_y(\nu)$ to $T_x(X)$, is called the ν-**horizontal lift of tangent vectors** from X to Y.*

    (3) *Let $\xi \in \mathcal{X}(X)$ be a vector field in X. Vector field $\hat{\xi}$ defined by the conditions: For all $y \in Y$,*
       (a) *$\hat{\xi}(y) \in Hor_y(\nu)$,*
       (b) *$\pi_{*y}(\hat{\xi}(y)) = \xi(\pi(y))$,*
    *is called the ν-**horizontal lift** of the vector field ξ from X to Y.*

Some simple properties and two equivalent definitions of connections in fiber bundle, are presented in the next proposition.

Proposition 15. *Let ν be a connection in the bundle $\pi : Y \to X$.*

    (1) *Distribution $Hor(\nu)$ is complemental to the* vertical distribution $V(\pi) \subset T(Y)$ *(see Lemma 11),*
    i.e. the sum

$$T(Y) = V(\pi) \oplus Hor(\nu)$$

*is **the direct sum of vector bundles over** Y,*

    (2) *A connection ν **defines and is defined by the projection operator** $P_\nu :$ $T(Y) \to V(\pi)$ with $Hor(\nu) = Ker(P_\nu)$ or, equivalently, by the $V(\pi)$-valued 1-form on Y that is the identity on the vertical subbundle $V(\pi) \subset T(Y)$ and has the ν-horizontal subbundle $Hor(\nu) \subset T(Y)$ as its kernel,*

    (3) *There is a bijection between the connections on the bundle $\pi : Y \to X$ and the smooth sections $\Gamma : Y \to J^1(\pi)$ of the bundle $\pi_{10} : J^1(\pi) \to Y$.*

For the proof of this Proposition we refer to the monograph [69] or [70].

Let $\pi : Y \to X$ be a fibre bundle. Let (W, x^i, y^μ) be a fibred chart in the bundle π. Let ν be a connection in the domain $W \subset Y$. The ν-**horizontal lift of a vector field** $\bar{\xi} = \xi^i \partial_i$ in the base X has the form

$$\hat{\xi} = \xi^i \partial_i + \Gamma_j^\mu \xi^j \partial_{y^\mu} \tag{68.1}$$

for some smooth (C^∞) functions $\Gamma_i^\mu \in C^\infty(W)$. Functions Γ_i^μ are called the coefficients (or Christoffel coefficients) of the connection ν.

In terms of a local coordinate frame $\{\partial_i, \partial_\mu\}$ and corresponding coframe $\{dx^i, dy^\mu\}$ defined by a fibred chart (W, x^i, y^μ), the projector P_ν (see the last Proposition) has the form of the $(1,1) - tensor$ field

$$P_\nu = \partial_\mu \otimes (dy^\mu - \Gamma_j^\mu dx^j). \tag{68.2}$$

One-form ω of this connection has, in local fibred coordinates, the form

$$\omega = \partial_\mu \otimes (dy^\mu - \Gamma^\mu_k dx^k). \tag{68.3}$$

ν-horizontal lift of vector fields from X to Y is the **R**-linear monomorphism of bundles of vector fields $\mathcal{X}(X) \to \mathcal{X}(Y)$. Let $\xi = \xi^i \partial_i \in \mathcal{X}(X)$ be an arbitrary vector field in X. Horizontal lift of a vector field ξ has the form

$$\hat{\xi} = Hor_\nu(\xi) = \xi^i \partial_i + \Gamma^\mu_i \xi^i \partial_{y^\mu} \tag{68.4}$$

Notice that the ν-horizontal lift of vector fields from X to Y does not necessary preserve the Lie brackets of vector fields since there is an obstruction for the mapping $\xi \to Hor_\nu(\xi)$ to be the homomorphism of Lie algebras. This obstruction is called the **curvature of the connection** ν.

To introduce the tensorial presentation of the curvature, let $\xi = \xi^i \partial_i, \eta = \eta^k \partial_k$ be two vector fields in X and let $\hat{\xi} = \xi^i \partial_i + \Gamma^\mu_i \xi^i \partial_{y^\mu}$, $\hat{\eta} = \eta^k \partial_k + \Gamma^\nu_k \eta^k \partial_{y^\nu}$ are ν-horizontal lifts of these vector fields. We have

$$[\hat{\xi}, \hat{\eta}] - \widehat{[\xi, \eta]} = [\xi^i \partial_i + \Gamma^\mu_i \xi^i \partial_{y^\mu}, \eta^k \partial_k + \Gamma^\nu_k \eta^k \partial_{y^\nu}] - \widehat{(\xi^i \eta^k_{,i} - \eta^i \xi^k_{,i})} \partial_k =$$
$$= \xi^i \eta^k [(\Gamma^\mu_{k,i} - \Gamma^\mu_{i,k}) + (\Gamma^\nu_i \Gamma^\mu_{k,\nu} - \Gamma^\nu_k \Gamma^\mu_{i,\nu})] \partial_\mu = R^\mu_{ik} \xi^i \eta^k \partial_\mu. \tag{68.5}$$

The obtained object, $R^\mu_{ik} \xi^i \eta^k \partial_\mu$ represent the π-vertical vector field linearly depending (symmetrically) on the vector fields ξ, η. Coefficients of these vector fields valued bilinear forms of arguments ξ, η form the **curvature tensor** (tensor with respect to the change of fibred coordinates) of connection ν:

$$R^\mu_{ik} = (\Gamma^\mu_{k,i} - \Gamma^\mu_{i,k}) + (\Gamma^\nu_i \Gamma^\mu_{k,\nu} - \Gamma^\nu_k \Gamma^\mu_{i,\nu}) \tag{68.6}$$

Different classes of fibre bundles have curvature tensors of different structure or some complemental structural tensors. **Torsion** and **Contortion** tensors are two examples (see below, Sec.69, or [6],Chapter 6). Below we will describe specific structure of curvature tensor for the jet bundles.

Fiber bundles of special types are often endowed with the special classes of connections obligated by their properties to the specific properties of these bundles.

Most useful classes of bundles n are **vector** and **principal** bundles, see [60].

Example 41. Let $\pi : Y \to X$ be a vector bundle, and let ν be a connection on the bundle π. Then, the linear group $GL(m)$ acts on the fibres of the bundle π. If the horizontal distribution $Hor(\nu)$ is invariant under the action of $GL(m)$, the connection ν is called the **linear connection**.

Remark 84. If ν is a linear connection on a vector bundle, then the vertical valued 1-form (67.3) of its projector operator P_ν is invariant under the action of the linear group $GL(m)$.

Example 42. In the domain $(W \subset Y)$ (with $\bar{W} = \pi(W) \subset X$) of a fibred chart (x^i, y^μ) there is defined the **trivial connection** having $\Gamma^\mu_i = 0$ for all i, μ. Horizontal lift defined by this connection has the simple form $\hat{\partial}_i = \partial_i$, $i = 1 \dots, n$. The curvature of trivial connection is zero: $R^\mu_{ik} = 0$, for all μ, i, k.

68.1. Connections in the bundle tower.

Now, let a *tower of the bundles*

$$Z \to Y \to X$$

be given. In this situation we might have connections of three types $Y/X, Z/Y, Z/X$: connections in the bundle $\pi : Y \to X$, connections in the bundle X/Y and the connections in the composite bundle Z/X . There are some relations between these connections (see [45]) for a detailed study of such connections).

Let γ be a connection in the bundle Z/X. In the fibred chart of this tower, where $(\bar{U}, x^i)$ is the chart at X, $(U = \pi^{-1}\bar{U}, x^i, y^\mu)$ is the corresponding fibred chart at $U \subset Y$ and, finally, (U^1, x^i, y^μ, z^k) is the corresponding fibred chart in Z. Connection γ is defined by the 1-form

$$\gamma = (\partial_{x^i} + \gamma_i^\mu \partial_{y^\mu} + \gamma_i^\sigma \partial_{z^k}) \otimes dx^i. \tag{68.7}$$

Corresponding horizontal distribution in $T(Z)$ is the linear span of the horizontal lifts of ∂_i:

$$Hor(\gamma) = \langle \widehat{\partial^{x^i}} = \partial_{x^i} + \gamma_{x^i}^\mu \partial_{y^\mu} + \gamma_i^k \partial_{z^k} \rangle.$$

If the connection coefficients $\gamma_{x^i}^\mu$ do not depend on the variables z^k, then the projection of subspaces $Hor_z(\gamma)$ to Y - linear span of vectors $\langle \partial_{x^i} + \gamma_{x^i}^\mu \partial_{y^\mu} \rangle$ is defined correctly, that is it does not depend on a choice of the point $z \in \pi_{Z/Y}^{-1})$. Projection γ is called projectable if it defines the connection Γ in the bundle Y/X with the 1-form

$$\Gamma = (\partial_{x^i} + \gamma_i^\mu \partial_{y^\mu}) \otimes dx^i.$$

Vice versa, let Γ be a connection in the bundle Y/X with the 1-form $\Gamma = (\partial_{x^i} + \Gamma_{iu}^\mu \partial_{y^\mu}) \otimes dx^i$, and let K be a connection in the bundle Z/Y with the form

$$K = (\partial_{x^i} + K_i^k \partial_{z^k}) \otimes dx^i + (\partial_{y^\mu} + K_\mu^k \partial_{z^k}) \otimes dy^\mu. \tag{68.8}$$

combining connections of these two types, we get the **composite connection** in the bundle Z/X with the form

$$(\partial_{x^i} + \Gamma_i^\mu \partial_{y^\mu} + (K_i^k + \Gamma_i^\mu K_\mu^k)\partial_{z^k}) \otimes dx^i. \tag{68.9}$$

At the same time, we get the vertical connection (in the bundle Z/Y) defined by the part of connection K:

$$KV = (\partial_{y^\mu} + K_\mu^k \partial_{z^k}) \otimes dy^\mu. \tag{68.10}$$

69. LINEAR CONNECTIONS.

A **linear connection on a differentiable manifold** M is the connection on the tangent bundle $T(M) \to M$ of the manifold M. It can also be defined as the connection on the bundle of linear frames $L(M) \to M$ ([69], Vol.I, Ch.3). These connections have a number of special properties which general vector bundles and general (Ehresmann) connections are lacking in. That is why, in this section, we review specific facts concerning linear connections on the manifolds and related notions - curvature, torsion, non-holonomic frames, etc.

69.1. Linear connections. Let M^n be a connected, paracompact C^∞-manifold. Denote its tangent bundle by $\tau : T(M) \to M$ and its cotangent bundle by $\tau* : T*(M) \to M$.

If (U, x^i) is a local chart in M, denote by $\{\partial_i = \frac{\partial}{\partial x^i}\}$ corresponding local frame in $U \subset M$ and by $\{dx^i\}$ - corresponding coframe in U. Local coordinates x^i in $U \subset M$ generate corresponding coordinates (x^i, ξ^k) in the tangent bundle $T(M)$ where coordinates (x^i, ξ^k) correspond to a point $(x, \xi) \in T(M)$ where $\xi = \xi^k \frac{\partial}{\partial \xi^k}$.

We introduce linear connections first redefining definition of general connections on the tangent bundle $\tau : T(M) \to M$.

Definition 41.   (1) *A **connection** Γ on the tangent bundle $\tau : T(M) \to M$ (or, simply, "on the manifold M") is the rule associating with any point $y = (x, \xi) \in T(M)$ the linear monomorphism $L_{x,y} : T_x(M) \to T_y(T(M))$ (Γ-horizontal lift $v \to hor_y v$ of tangent vectors) smoothly depending on the point $(x, y) \in T(M)$. The image of mapping $L_{x,y}$ is denoted by $Hor_{y=(x,\xi)}$ and is called a "Γ-horizontal subspace" of $T_y(T(M))$.*

   (2) *A connection Γ **is called linear** (sometimes also called **affine connection**). More specifically, the horizontal lift of vectors and vector fields depends linearly on the fibre point $\xi \in T_x(M)$: if T_c is the linear translation by a vector $c \in R^n$ in the fibre $T_y(T(M))$ and $T_{c*y} : T_y(T(M)) \to T_{T_c y}(T(T(M))$ - corresponding tangent mapping at the point y, then*

$$T_{c*y}(hor_y v) = hor_{T_c y} v$$

for all $v \in T_x(M)$, for all $y \in T_x(M)$ and all $c \in R^n$.

In the domain U of a chart $\{x^i\}$ in M, a connection is defined by its coefficients which are smooth functions $\Gamma^i_{k}(x, \xi)$. Γ-horizontal lift of a vector $v = v^i \partial_i \in T_x(M)$ to the point $y = (x, \xi)$ is equal to

$$hor_y \xi = v^i \partial_{x^i} + \Gamma^k_{i}(x, \xi) v^i \partial_{\xi^k}.$$

For a *linear connection*, coefficients Γ^i_{k} are linear by the fibre variables ξ^k:

$$\Gamma^k_{i}(x, \xi) = \Gamma^i_{kj}(x)\xi^j,$$

so, the horizontal lift has the form

$$Hor_y(\xi = v^i \partial_{x^i}) = \xi + \Gamma^i_{jk}(x)\xi^j v^i \partial_{,\xi^k}. \tag{69.1}$$

Return now to the properties of the general and linear connections - a fact that the connection defines horizontal lifts of vector fields and curves and a related property of some connections called completeness.

Definition 42.   (1) *Let $v(x) = v^i(x)\partial_i$ be a vector field in M. The horizontal lift of a vector field v is the vector field $\hat{v}$ in $T(M)$ defined, in a local fibred chart with local coordinates (x^i on the base X and ξ^j) in the tangent spaces, as follows*

$$\hat{v}(x, y) = v^i(x)\partial_{x^i} + \Gamma^k_{i}(x, y)v^i(x)\partial_{\xi^k} = v^i(x)\partial_{x^i} + \Gamma^k_{ij}(x, y)\xi^j v^i(x)\partial_{\xi^k}.$$

   (2) *Let $\gamma : (a, b) \to U$ be a smooth curve in M, $x_0 \in (a, b)$ is a point on this curve and $y_0 \in T_{x_0}(M)$ a point in the tangent space at x_0. Horizontal lift of the curve γ at the point (x_0, y_0) is the curve $\hat{\gamma}(t)$ defined by the conditions: $\hat{\gamma}(0) = y_0$, $\hat{\gamma}'(t) = hor_{\gamma(t)}(\gamma'(t))$. The curve $\hat{\gamma}$ is (locally) uniquely defined by these conditions.*

(3) *A connection Γ is called **complete** if for any curve $\gamma : (a,b) \to M$, and any point $(\gamma(t_1), y) \in T(M)$ horizontal lift $\hat{\gamma}$ such that $\hat{\gamma}(t_1 0 = y$ extends to the whole interval (a,b) producing the horizontal curve over $\gamma(t)$.*

(4) *Let $p, q \in M$ and let $\gamma : (0,1) \to M$ be a curve such that $\gamma(0) = p, \gamma(1) = q$. The correspondence*

$$\xi \in T_p(M) \to \hat{\gamma}(1) \in T_q(M)$$

defines the parallel translation mapping $\Pi_\gamma : T_p(M) \to T_q(M)$. This, the parallel translation mapping depends on a choice of curve γ connecting points p, q. Parallel translation mapping Π_γ is linear on the fibres of tangent bundle if the connection Γ is linear.

The most often used example of a linear connection is the **Levi-Civita connection** defined by a **pseudo-Riemannian metric** g on a manifold M. The Levi-Civita connection Γ^i_{jk} is uniquely defined by two conditions:

(1) **Torsion** $T^i_{jk} = \frac{1}{2}(\Gamma^i_{jk} - \Gamma^i_{kj})$ of Levi-Civita-connection **vanish**, or, using another terminology, connection Γ is symmetric, i.e. $\Gamma^i_{jk} = \Gamma^i_{kj}$.

(2) Metric g_{ij} is invariant under parallel translation defined by Levi-Civita-connection (see below).

Levi-Civita connection is defined by the metric g as follows:

Theorem 33. *Let g be a pseudo-Riemannian metric on a connected smooth manifold M (see [69, ?]). Then, the Levi-Civita connection defined by the conditions 1),2) above exists and is unique.*

If, in a local chart (U, x^i), metric g is defied by the tensor $g_{ij}(x) = g_{ij}(x)dx^i dx^j$, then the Levi-Civita connection has the following coefficients:

$$\Gamma^i_{jk} = \frac{1}{2}g^{ih}\left(\frac{\partial g_{jh}}{\partial x^k} + \frac{\partial g_{kh}}{\partial x^j} - \frac{\partial g_{jk}}{\partial x^h}\right). \tag{69.2}$$

Example 43.

An equivalent Definition of a linear connection on the manifold M equivalent to the Definition 41 can be done in terms of covariation differentiation.

Definition 43. *A linear connection on the manifold M (on its tangent bundle $\tau(M)$) is the mapping $\eta \to \nabla_\eta$ associating with each vector field η in M, an operator of covariant differentiation of vector fields on M along η: $\nabla_\eta : \mathcal{X}(M) \to \mathcal{X}(M)$ in correspondence to each vector field $\eta \in \mathcal{X}(M)$ such that*

(1) *Operator ∇_η is linear over $\mathbb{R}$,*

(2) *(Leibniz formula) For any $f \in C^\infty(M)$,*

$$\nabla_\eta(f\xi) = \eta \cdot f\xi + f\nabla_\eta\xi,$$

(3) *Linearity by η:*

$$\nabla_{X+Y} = \nabla_X + \nabla_Y, \ \nabla_{fX}\eta = f\nabla_X\eta.$$

In terms of coefficients Γ^i_{jk} covariant differentiation is expressed as follows:

$$\nabla_X Y = \left(\frac{\partial Y^i}{\partial X^k} + \Gamma^i_{kj}Y^j\right)X^k\partial_{x^i}. \tag{69.3}$$

69.2. Curvature and Torsion. Linear connections posses two tensor characteristics: **torsion and curvature**.

(1) Torsion tensor T^i_{jk} is defined as a measure of non-symmetry of connection coefficients Γ^i_{jk} by lower indices:

$$T^i_{jk} = \frac{1}{2}(\Gamma^i_{jk} - \Gamma^i_{kj}).\tag{69.4}$$

Example 44. Let g be a pseudo-riemannian metric in M and ${}^g\Gamma$ be the corresponding Levi-Civita connection (see above ()). The connection coefficients are symmetric by low(covariant) indices: ${}^g\Gamma^i_{jk} = {}^g\Gamma^i_{kj}$, see below, and, therefore, **torsion of Levi-Civita connection is zero**.

(2) **The curvature of a connection** Γ^i_k can be defined as a measure non-preservation of brackets of vector fields in the horizontal lift. More specifically, let ∂_i, ∂_j be two basic vectors in $U \subset M$ and let $\widehat{\partial}_i = \partial_i + \Gamma^k_i \partial_{\xi^k}$, $\widehat{\partial}_j = \partial_j + \Gamma^s_j \partial_{\xi^s}$ denote their horizontal lifts. Then,

$$[\widehat{\partial}_i, \widehat{\partial}_j] = [(\partial_i\Gamma^s_j - \partial_j\Gamma^s_i) + (\Gamma^s_i\partial_{\xi^s}\Gamma^k_j - \Gamma^s_j\partial_{\xi^s}\Gamma^k_i)]\partial_{\xi^k}.$$

This equality delivers an expression for the curvature tensor of a connection Γ as the vertical (along the projection $\tau : T(M) \to M$) part (measure of non-horizontality) of the bracket of horizontal lifts of two basic vector fields:

$$R^s_{ij} = (\partial_i\Gamma^s_j - \partial_j\Gamma^s_i) + (\Gamma^s_i\partial_{\xi^s}\Gamma^k_j - \Gamma^s_j\partial_{\xi^s}\Gamma^k_i)\tag{69.5}$$

By linearity, the curvature tensor defines the non-horizontal part of the commutator of lifts of all vector fields $\eta\zeta$:

$$[\widehat{\eta}, \widehat{\zeta}] = hor([\widehat{\eta}, \widehat{\zeta}]) + R^k_{ij}\eta^i\zeta^j\partial_{\xi^k}.\tag{69.6}$$

Exercise1. Check calculations in (69.5) and (69.6).

Exercise 2. Prove that tensor R is antisymmetric by the lower (covariant) indices: $R^k_{ji} = -R^k_{ij}$.

For a linear connection, curvature tensor has the form

$$R^i_{j,kl} = \frac{\partial\Gamma^i_{lj}}{\partial x^k} - \frac{\partial\Gamma^i_{kj}}{\partial x^l} + \Gamma^i_{kp}\Gamma^p_{lj} - \Gamma^i_{lp}\Gamma^p_{kj}.\tag{69.7}$$

A useful contraction of this tensor (the one by indices i, k) defines the Ricci tensor of linear connection Γ (in the second equality a change of indices is used):

$$R_{ij} = R^k_{i,kj} = \frac{\partial\Gamma^k_{ji}}{\partial x^k} - \frac{\partial\Gamma^k_{ki}}{\partial x^j} + \Gamma^k_{kp}\Gamma^p_{ji} - \Gamma^k_{jp}\Gamma^p_{ki}.\tag{69.8}$$

Finally, lifting index in the Ricci tensor and contracting, we get the scalar curvature of metric g:

$$R(g) = g^{ij}Ric_{ij}.\tag{69.9}$$

Example 45. Let $g = g_{ij}$ be a a pseudo-riemannian metric in the manifold M and let $\omega = \omega_i dx^i$ be a 1-form in M. These data determines the **Weyl connection** in M:

$$\Delta^i_{jk} = {}^g\Gamma^i_{jk} + (\omega_j\delta^i_k + \omega_k\delta^i_j + g^{il}\omega_l g_{jk}).\tag{69.10}$$

The Weyl connection is *torsion free*, but metric g is not invariant under the Δ-parallel translation. In infinitesimal form we have:

$$\nabla_i g_{jk} = -\omega_i g_{jk}.\tag{69.11}$$

If $\gamma : [0, 1] \to M$ is a curve connecting points $p = \gamma(0)$ with $q = \gamma(1)$ and $\xi \in T_p(M)$ is a tangent vector in a starting point of this curve, then the parallel translation of vector ξ along γ is proportional to the vector ξ (see [21], Sec.30):

$$\Pi_q^p \xi = e^{-\int_\gamma \omega} \xi. \tag{69.12}$$

69.3. Metric connections and the Nonmetricity. Let M^n be a manifold endowed with a Riemannian metric g and the corresponding Levi-Civita connection $^g\Gamma$). Let Δ be *another linear connection* in M. Parallel translation defined by the connection Δ fails, in general, to preserve the metric tensor g_{ij}. **Nonmetricity tensor** $D^\Delta(g)$ of the connection Δ is defined as

$$D^\Delta(g) = \nabla_i^\Delta g_{jk} = \frac{\partial g_{jk}}{\partial x^i} = \Pi_{ij}^q g_{qk} - \Pi_{ik}^q g_{jq} \tag{69.13}$$

serves as the measure of failure of the Δ-parallel translation to preserve metric g.

Represent a linear connection Δ in the form

$$\Delta_{jk}^i =^g \Gamma_{jk}^i + R_{jk}^i,$$

where R_{jk}^i is a (1,2)-tensor field in M. Calculating the covariant derivative in (65.14) we find the g-nonmetricity of connection Δ tensor

$$D^\Delta(g)_{ijk} = -R_{ij}^q g_{qk} - R_{ik}^q g_{qj}. \tag{69.14}$$

A linear connection having zero nonmetricity is called the **metric connection** (relative to the metric g).

This expression for the nonmetricity tensor D is used in Section (40). In the same section we deal with the question: when do two linear connections define the same parallelism (i.e. the same geodesic lines (with fixed initial conditions: $c(0) = x_0$, $c'(0) = \xi \in T_x(M)$)). Next result answer this question.

Proposition 16. *(Friesecke, Thomas), see [21], Sec.12.*
Let L_{jk}^i be a linear connection in a manifold M^n. Another linear connection K_j^i define the same parallelism if and only if

$$K_{jk}^i = L_{jk}^i + 2\delta_j^i \psi_k, \tag{69.15}$$

for some 1-form $\psi = \psi_k dx^k$

Corollary 3. *If two symmetrical (torsionless) connections define the same parallelism (i.e. they define the same parallel directions along any curve), these connections coincide.*

70. Absolute parallelism.

Definition 44. *A connected manifold M^n with a linear (affine) connection Γ is called a **space with absolute parallelism** (or AP-connection), see [21, 69, 120] if for any two points $p, q \in M$, and for any path $\gamma : I \to M$ connecting them, the Γ-parallel translation*

$$\Pi_\Gamma : T_p(M) \to T_q(M), \ p = \gamma(0), q = \gamma(1),$$

does not depend on the choice of path γ.

Γ-parallel translation in a space (M, Γ) with absolute parallelism will be denoted by Π_q^p. Tangent vectors $\xi \in T_p(M)$ and $\eta \in T_q(M)$ in the manifold M are called Γ-**parallel** if $\eta = \Pi_q^p \xi$. A vector field $\xi(x)$ is called Γ-parallel or *covariantly constant* with respect to the connection Γ if vectors of this vector field at any two points of the manifold M are Γ-parallel: $\Pi_\Gamma(v(p)) = v(\Pi(p))$.

General considerations lead to the conclusion that **in a space with absolute parallelism curvature tensor vanishes. If M is simply connected, the reverse statement is also valid.**,[69], Ch.1.

In the geometry of spaces with absolute parallelism, the space $\mathcal{X}_{cc}$ of covariantly constant vector fields plays a very important role. The next result is explicit evidence of it:

Theorem 34. *Let (M^n, Γ) be an n-dimensional connected manifold with an affine connection Γ. The curvature tensor of connection Γ vanishes identically if and only if the space of Γ-covariantly constant vector fields $\mathcal{X}_{cc}$ is n-dimensional. In particular, for a space of absolute parallelism $\dim\mathcal{X}_{cc} = n$.*

For the proof, see [21], Sec.9.

Let (M^n, Γ) be a manifold with an absolute parallelism defined by a zero curvature-connection Γ. Let $\xi_i, i = 1, \ldots, n$ be a local frame of covariantly constant vector fields in M. In terms of this frame, parallel translation of tangent vectors from a point p to a point q looks as follows: for a vector $\xi(p) = \sum_{i=1}^n \xi^i \xi_i(p))$,

$$\Pi_q^p(\xi) = \sum_i \xi^i \xi_i(q), \tag{70.1}$$

i.e. the image at the point $q \in M$ of the vector $\xi \in T_p(M)$ has the same components with regard the basis $\xi_i(q)$ as the vector ξ has with respect to the basis $\xi_i(p)$.

70.1. Non-holonomic frame and absolute parallelism.

A holonomic frame in a domain $U \subset M$ is the basis $\partial_i = \frac{\partial}{\partial x^i}$ of tangent spaces $T_x(M)$ generated to a local chart (U, x^i). Corresponding holonomic coframe is the collection of 1-forms dx^i.

Definition 45. *A non-holonomic frame in a domain $U \subset M^n$ is a collection of n vector fields in U **linearly independent at all points** $x \in U$.*

Remark 85. Holonomic frames defined in a domain of local coordinate systems are special cases of the general, non-holonomic frames.

A situation when, on a manifold M^n, there exists a **global frame** - a collection of n globally defined and linearly independent at all points of M vector fields strongly depends on the topology of manifold M, see [122]. Yet, in a domain of every trivialization chart (U, x^i) there are lot of non-holonomic frames $\{\eta_i\}$ obtained by a nondegenerate linear point dependent transformations (nonholonomic transformation) of the holonomic frame $\{\partial_i\}$:

$$\eta_i = D_j^i \partial_{x^j}, \; \det(D_j^i) \neq 0.$$

Any **non-holonomic frame** (U, ξ_α) defines the corresponding (dual) non-holonomic **coframe** ω^k by the condition

$$\langle \omega^k, \xi_i \rangle = \delta_i^k = 0.$$

Let $(U, x^i, 1 = 1, \ldots, n)$ be a local chart in a domain $U \subset M$. Let, as above, $\{\xi_i, i = 1, \ldots, n\}$ be a local frame of vector fields in $U \subset M$ and, let, in the holonomic (chart defined) frame $\{\partial_i\}$,

$$\xi_j = \xi_j^i(x)\partial_i.$$

Then, the absolute parallelism connection Γ corresponding to the (non-holonomic) frame ξ_j is defined as follows (see [21, 120])

$$\Gamma_{ij}^k = \xi_s^k \frac{\partial \xi_i^{-s}}{\partial x^j} = -\xi_i^{-s} \frac{\partial \xi_s^k}{\partial x^j}, \tag{70.2}$$

where minus in the power means taking inverse matrix.

Curvature (65.7) of such connection vanishes: $R_{j,kl}^i = 0 \; \forall \; i, j, k, l$.

The **torsion** of connection Γ is the antisymmetric expression

$$T_{ij}^k = \frac{1}{2}(\Gamma_{ij}^k - \Gamma_{ji}^k) = \frac{1}{2}\left(\xi_s^k \frac{\partial \xi_i^{-s}}{\partial x^j} - \xi_s^k \frac{\partial \xi_j^{-s}}{\partial x^i}\right). \tag{70.3}$$

Geometrically, connection Γ is defined by the condition that the vector fields ξ_j of the defining frame are *covariantly constant*, i.e. invariant under the parallel translation defined by the connection Γ.

Now, let Γ be a zero curvature affine connection in a simply connected manifold M^n. Pick a point $m \in M$ and let $\xi_s, s = 1, \ldots, n$ be a basis of the tangent space $T_m(M)$. Let $l \in M$ be any other point and let $\gamma : [0, 1] \to M$ be a smooth curve connecting $m = \gamma(0)$ to $l = \gamma(1)$. Parallel translation of vectors ξ_s along γ to the point l defines the basis of tangent space $T_l(M)$. This basis does not depend on the choice of curve γ. As a result, we obtain a smooth, non-holonomic in general, frame in M. This proves the following

Proposition 17. *Let M^n be a linearly connected manifold endowed with an absolute parallelism Γ. There exists a bijection between*

(1) *Γ-parallel (covariantly constant) frames in M,*
(2) *Baseses of tangent space $T_p(M)$ at a fixed point $p \in M$.*

70.2. **Non-holonomic (pure gauge) transformations and induced connections.** Let $(U \subset M; x^i)$ be a coordinate chart in M^n, ∂_i be the corresponding (local) frame and $\{dx^i\}$ be the corresponding coframe.

Let D be a nonholonomic (gauge) automorphism of the tangent bundle $T(M)$. For any point $m \in M$, $D_m : T_m(M) \to T_m(M)$ is the linear automorphism of the tangent space $T_m(M)$ defined by

$$\eta_i = D_m \partial_{x^i} = D_i^j(x)\partial_j. \tag{70.4}$$

Vectors $\eta_i = D_i^j(x)\partial_j$ form the non-holonomic frame in the domain U. This frame defines the zero curvature connection

$$\Gamma_{ij}^k = D_s^k \frac{\partial D_i^{-1\,s}}{\partial x^j}, \tag{70.5}$$

with the torsion

$$T_{jk}^i = \frac{a}{2}\left(D_s^k \frac{\partial D_i^{-1\,s}}{\partial x^j} - D_s^k \frac{\partial D_j^{-1\,s}}{\partial x^i}\right).$$

This construction will be used in Sec.3, Chapter 5.

71. Automorphisms of the vertical bundle and their prolongation.

In this section we present some properties of automorphisms of vertical bundles of a fibre space $\pi : Y \to X$. These results are used in Ch.5, Sec.42.

Let $\pi : Y \to X$ be a fibre bundle and $V(\pi) \to Y \to X$ be its vertical (double) bundle. Let $\psi = (\chi, D)$ be a (non-holonomic) automorphism of the vertical bundle $\pi_V : V(\pi) \to Y$ **projecting to the identity diffeomorphism of** X (it can be called a gauge automorphism of the bundle $V(\pi) \to X$).

Automorphism ψ determines (and is determined by) the couple: an automorphism $\chi : (x, y) \to (x, w = \chi(x, y))$ of the bundle $\pi : Y \to X$ covering the identity diffeomorphism of X and the (pure) gauge automorphism D of the bundle $\pi_V : V(\pi) \to Y$.

More specifically, a diffeomorphism χ of Y acting on the points of Y: $w^i = \chi^i(y^k)$, canonically defines the automorphism $\chi_* : T(Y) \to T(Y)$ linear on the fibres. Due to the condition on χ, automorphism χ_* *leaves the vertical subbundle* $V(\pi) \subset T(Y)$ *invariant* and, therefore, defines the automorphism of the vertical (double) bundle $V(\pi) : V(\pi) \to Y \to X$.

Composition $\psi \circ \chi_{*V}^{-1}$ is the automorphism of the vertical bundle projecting to the identity on the base Y. Therefore, it is defined by the action of linear automorphisms D_y on the fibres V_y so that

$$\psi_y = D_y \circ \chi_{*y}. \tag{71.1}$$

Such fields of linear automorphisms are in one to one correspondence with the non-degenerate $(1,1)$-tensor field $y \to D_y$, i.e. fields $y \to D_y$ such that $Det(D_y) \neq 0$ for any $y \in Y$. This bijection is not canonical - it is defined by a choice of the vertical frame - (local or global basis $\eta_\alpha, \alpha = 1, \ldots, m$ of vertical vector fields in Y.

In a (local) vertical chart y^μ, automorphism D has the form

$$D : \xi = \xi^\mu(x, y)\partial_\mu \to \eta = D\xi = (D_\nu^\mu \xi^\nu)\partial_\mu. \tag{71.2}$$

Combining this representation of gauge automorphism D with the action of the automorphism χ we get the representation of automorphism ψ in local coordinates:

$$\{y^\alpha\} \to \{w^\beta = \chi^\beta(x, y)\}; \; \xi(y) = \xi^\alpha(y)\partial_{y^\alpha} \to \eta(w) = (D_\alpha^\beta \xi^\alpha)\partial_{w^\beta} = (D_\beta^\gamma \chi_{,y^\mu}^\beta \xi^\mu)\partial_{w^\gamma}. \tag{71.3}$$

To define a prolongation of an automorphisms (71.2) to the 1-jet bundle we will be using properties of 1-jet bundle $J^1(V(\pi)) \to X$ of the vertical bundle $V(\pi)$. In particular, we use the following basic result ([119], Thm.4.4.1, or [45],Sec.2.1).

Theorem 35. *There is the canonical isomorphism of the bundles $J^1V(\pi) \to V(\pi_1)$ over $\mathcal{V}(\pi)$:*

If ξ is an automorphism of the vertical bundle $V(\pi) \to Y$ covering the identity diffeomorphism of X, it defines the automorphism of the bundle $\nu(\pi_1) : V(\pi) \to X$ with the same property. Therefore, it canonically defines the automorphism of the bundle $J^1(V(\pi)) \to X$ and, by the Theorem above, the automorphism of the bundle $\nu(\pi_1) : V(J^1(\pi)) \to X$.

Corollary 4. *Any automorphism ψ of the bundle $V(\pi) \to X$ can be canonically prolonged to the automorphism ψ^1 of the bundle $V(\pi_1) : V_X(J^1(\pi)) \to X$ such that* *the diagram*

$$
\begin{array}{ccc}
V_X(J^1(\pi)) & \xrightarrow{\ \psi^1\ } & V_X(J^1(\pi)) \\
\pi_v \downarrow & & \pi_v \downarrow \\
V_X(Y) & \xrightarrow{\ \psi\ } & V_X(Y)
\end{array}
$$

is commutative.

Proof. This statement follows from the Theorem 31 since any automorphism of a bundle has a prolongation to the 1-jet bundle of this bundle, see [119], Sec.4.2. $\square$

Appendix II. Jet bundles and related geometrical structures.

72. Introduction.

Differential equations are functional relations between the dynamical variables (dependant variables) and their derivatives. These relations may also include independent variables. Functional relations may be linear as in Laplace or classical wave equations, may be simple nonlinear as in the simplest wave equation $u_[t+uu_x = 0$ or quite complicated as the Monge-Amper equation $det(\frac{\partial^2 h}{\partial x^i \partial x^j}) = h$. Classical works of S.Lie, E.Cartan and other mathematicians demonstrated the unsurpassed usefulness of using algebraic and geometrical methods for study of differential equations and the beauty of obtained results.

In 1948 S.Ehresmann had developed the machinery of jet bundles where differential equations and systems of such equations were realized as the submanifolds of the differential manifolds - **k-jet bundles** - spaces whose points codes the values of dynamical variables $y^\alpha(x)$ and their derivatives up to some order k,where $1 \leqq k \leqq \infty$.

We refer to the sources [119, 70] for detailed development of the theory of k-jet bundles and present here only basic notions and results necessary for us. We define k-jet bundles ($1 \leqq k \leqq \infty$) of a fiber bundle $\pi : Y \to X$, contact structure on jet bundles and the flow prolongation of vector fields from the space Y to the 1- and k-jet bundles.

73. Jet bundle $J^1(\pi)$.

We start with a simple example where $m = n = 1$ and the configurational bundle $\pi"Y \to X$ is the trivial bundle $\pi : R^2_{x,y} \to R_x$. Here x is the independent and y - dependent variable, a section s of this trivial bundle has the form $s(x) = (x, y(x))$. Space $J^k(\pi)$ that would carry information about derivatives of sections (functions) $y(x)$ up to the order k should have coordinates $(x, y; y_1, y_2, \ldots, y_k)$, where y_i is the jet coordinate defined by the condition: a smooth section $s(x) = (x, y(x))$ of the bundle π defines the point $j^k s(x) \in J^k(\pi)$ with

$$y_i(j^k s(x)) = \frac{\partial^i y}{\partial x^i}(x), \ i = 1, \ldots, k.$$

A point of the k-jet space $j^k s(x) \in J^k(\pi)$ corresponding to a smooth section $s(x)$ defined in a neighborhood of the point x contains information about the point x, possible values $y(x)$ of sections defined at the point x and derivatives of the function $y(x)$ AT THE POINT x (ONLY) In order to define correctly points of the k-jet space, we have to identify sections defined in a neighborhood of the point x and having the same derivatives of order $\leqq k$ AT THIS POINT. This leads to the following definitions of 1-jet bundles and, in the next subsection, of k-jet bundles, $k \geqq 2$.

73.1. **The 1-jet bundle of a fibre bundle** $\pi : Y \to X$. Let $\pi : Y^{n+m} \to X^n$ be a fibre bundle. We will be using local fibred charts (W, x^i, y^α) in this bundle. Here $W \subset Y$ is the domain of this chart. Denote by $\bar{W} \subset X$ the projection $\pi(W)$ - domain of the chart in X with local coordinates $x^i, i = 1, \ldots, n$. In terms of the fibre coordinates (x^i, y^α), a section $s : \bar{W} \supset U \to Y$ of the bundle π is the collection of fields $s = \{y^\alpha(x) \in C^\infty(U)\}$, $\alpha = 1, \ldots, m$.

© Springer International Publishing Switzerland 2016
S. Preston, *Non-commuting Variations in Mathematics and Physics*,
Interaction of Mechanics and Mathematics,
DOI 10.1007/978-3-319-28323-4_8

Definition 46. *(Definition-Proposition).*

(1) *Given a fibre bundle* $\pi : Y^{n+m} \to X^n$ *we say that two sections* $s, s' : U \to Y$ *defined in a neighborhood* U *of a point* $x \in X$ *are equivalent of order one* **(have the same 1-jet at this point)** *if*

$$\begin{cases} s(x) = s'(x), \\ s_{*x} = s_{*x} : T_x(X) \to T_{s(x)}(Y). \end{cases}$$

In a fibred chart (x^i, y^α), *the last condition has the form*

$$\frac{\partial s^i}{\partial x^\mu}(x) = \frac{\partial s'^i}{\partial x^\mu}(x), \quad i = 1, \ldots, m; \mu = 1, \ldots, n.$$

We will write $s_x \sim_1 s'_x$ *for a sections equivalent of order one at a point* x.

(2) *Relation* $s_x \sim_1 s'_x$ *is the* **equivalence relation** *on the space of all section of* π *defined in a neighborhood of* $x \in X$. *Classes of equivalence are called* **1-jets of sections** *at the point* x. *For a section* s *defined in a neighborhood of a point* $x \in X$ *its 1-jet at* x *is denoted by* $j^1 s(x)$. *Set of all classes of equivalence of local (defined in a neighborhood of a point* $x \in X$ *) sections is denoted by* $J^1_x(\pi)$.

(3) *Union* $J^1(\pi) = \bigcup_{x \in X} J^1_x(\pi)$ *is endowed with the epimorphic mapping* $\pi_{10} : J^1(\pi) \to Y$ *generated by the correspondence* $j^1 s(x) \to s(x)$.

(4) *Let* $(W, x^i, y^\alpha; i = 1, \ldots, n; \alpha = 1, \ldots, m)$ *be a fibred chart in* Y. *Then the induced local coordinate system* $(x^i, y^\alpha, y^\alpha_i; i = 1, \ldots, n; \alpha = 1, \ldots, m)$ *in* $J^1(\pi)$ *is defined by the condition*

$$y^\alpha_i(j^1_x s) = \frac{\partial y^\alpha}{\partial x^i}(x)$$

in the open subset $W^1 = \pi_{10}^{-1}(W) \subset J^1(\pi)$.

(5) *Set* $J^1(\pi)$ *is endowed with the structure of differentiable manifold such that*

　(a) *The mappings* $J^1(\pi) \xrightarrow{\pi_{10}} Y \xrightarrow{\pi} X$ *are fibrations.*,

　(b) *The fibration* $\pi_{10} : J^1(\pi) \to Y$ *is the* **affine bundle** *over* Y *modeled over the vector bundle* $\pi^*(T^*(X)) \otimes V(\pi) \to Y$, *where* $V(\pi) = Ker(\pi_*) \subset T(Y)$ *is the* **vertical bundle** *- subbundle of the tangent bundle* $\tau_Y : T(Y) \to Y$ *(see Appendix I).*

(6) *A section* $\sigma : V \to J^1(\pi)$ *of the 1-jet bundle* π_1 *over an open subset* $V \subset X$ *is called* **holonomic** *if* $\sigma(x) = j^1 s(x)$ *for a section* $s = \pi_{10} \circ \sigma : V \to Y$.

Proof. Using local (fibred) coordinates one can easily check that the relation introduced in the statement 1) of this definition is the equivalence relation. Formula for the change of partial derivatives under the change of variables proves the independence of this fact from a fibred chart (see Sec.71).

Statement 3) follows from the fact that $\pi : Y \to X$ is the **locally trivial bundle** (see Appendix I, Sec. 65). As a result, for any point (x, y) such that $y \in Y_x$, there exists a smooth section $s : U \to Y$ defined *and having its compact support* in a neighborhood U of the point x such that $s(x) = y$.

To prove the point 4) and the statement 5b we look at the change of jet variables y^α_i under a transformation of fibred chart, let $x' = x'(x), y' = y'(x, y)$ be a change of fibred charts in the intersection of their domains - open subsets of Y. Then, the action of this transformation $(x, y)' = \phi(x, y)$ on a section s is: $s^\phi = \phi(s(\bar{\phi}^{-1}(x))$, where $\bar{\phi} : x \to x'$ is the induced change of variables in the base X. In terms of components we have $s'^\alpha = \phi^\alpha(s^\beta(x(x')))$. Calculating derivatives by x^ν we find that

$$\frac{\partial y'^i}{\partial x'^\nu} = \frac{\partial y'^k}{\partial y^i}\frac{\partial x^\mu}{\partial x'^\nu}z^i_\mu + \frac{\partial y'^k}{\partial x^\sigma}\frac{\partial x^\sigma}{\partial x'^\nu}.$$

Thus, jet variables y^α_i transforms, under a change of fibred chart, **by affine transformations**.

It is easy to see now that the affine bundle $\pi_{10} : J^1(\pi) \to Y$ **is modeled** on the vector bundle $\pi^*T^*(X) \otimes V(\pi) \to Y$. $\qquad\qquad\qquad\qquad\square$

Remark 86. If $\pi : Y \to X$ is a vector bundle, 1-jet bundle $J^1(\pi) \to Y$ is the vector bundle as well. To see this we notice that in a vector bundle there exists special, globally defined, section - zero section s_0. For $x \in X$, value 0f section s_0 at the point x - $s_0(x)$ is the zero point of the (vector) space - fiber Y_x over x. One jet $j^1 s_0$ of section s_0 defines the point at each affine fiber $J^1(\pi)_{s_0(x)}$ over X. This endoves affine fibers $J^1(\pi)_x$ with the structure of vector spaces. Projection $\pi_{10} : J^1(\pi) \to Y$ becomes (over each point $x \in X$) the linear mapping of vector spaces. Kernel of this mapping over points $y \in Y$ are VECTOR spaces - fibers of the bundle $\pi_{10} : J^1(\pi) \to Y$.

74. Higher order jet bundles $J^k(\pi)$.

In this section we introduce higher order jet bundles $J^k(\pi) \to Y \to X, 1 < k < \infty$ of the configurational bundle $\pi : Y \to X$. For the proof of the properties of these bundles and for the extensive exposition of their properties we refer to the monographs [119], Sec.6.2.

We will use multiindex notations $\Lambda = \{\lambda_1, \ldots, \lambda_n\}, \lambda_i \in \mathbb{N}$ In particular, we denote by ∂^Λ the differential operator $\partial^\Lambda f = \partial^{\lambda_1}_{x^1} \cdot \ldots \partial^{\lambda_n}_{x^n}$ (acting, for instance, on the functions from $C^\infty(X)$.

Definition 47. *(Definition-Proposition). Let $1 < k < \infty$.*

(1) *Given a fibre bundle $\pi : Y \to X$ we say that two sections $s, s' : U \to Y$ defined in a neighborhood U of a point $x \in X$ **are equivalent of order** k at the point x (or, what is the same, **have the same k-jet at x**)) if*
   (a) *$s(x) = s'(x)$, and*

   (b) *At some (and, therefore, at all) fibred chart $(W, x^i, y^\alpha; \ldots, y^\alpha_I, |I| \leqq k)$, $\partial^J s_{*x} = \partial^J s_{*x}$ for all multiindices $J = (j_1, \ldots, j_n, |J| \leqq k$ (i.e. for all derivatives of order less or equal k).*
   We will write $s_x \sim_k s'_x$ for a sections having the same k-jet at a point x.

(2) *Relation $s_x \sim_k s'_x$ is the **equivalence relation** on the space of all section of π defined in a neighborhood of $x \in X$. Classes of equivalence are called k-jets of sections at the point x. For a section s defined in a neighborhood of a point $x \in X$ its k-jet at the point x is denoted by $j^k s(x)$. Set of all classes of equivalence of local (defined in a neighborhood of x) sections are denoted by $J^k_x(\pi)$.*

(3) *Union $J^k(\pi) = \bigcup_{x \in X} J^k_x(\pi)$ is endowed with the epimorphic mapping π_{k0} onto Y generated by the correspondence $j^k s(x) \to s(x) : J^k(\pi) \to Y$.*

(4) *Set $J^k(\pi)$ is endowed with a structure of differentiable manifold such that*
   (a) *The mappings $J^k(\pi) \to J^{k-1}(\pi) \to \ldots \to Y \to X$ are smooth surjective submersions.,*

   (b) *The fibration* $\pi_{k(k-1)} : J^k(\pi) \to J^{k-1}(\pi)$ *is the **affine bundle** modeled in the vector bundle* $\pi_{k-1}^*(S^k(T^*(X)) \otimes \pi_{(k-1)0}^* V(\pi)$, *where* $V(\pi)/Y \subset T(Y)/Y$ *is the vertical subbundle of the bundle* $T(Y) \to Y$, *see Ch.9, Sec.1.2.*

  (5) *Let* $(W, x^i, y^\alpha; i = 1, \ldots, n; \alpha = 1, \ldots, m)$ *be a fibred chart in* Y. *Then the induced fibred local coordinate system* $(x^i, y^\alpha, y_I^\alpha; |I| \leq k)$ *in* $J^k(\pi)$ *is defined by the conditions*

$$y_I^\alpha(j_x^k s) = \partial^I y^\alpha(x)$$

in the open subset $W^k = \pi_{k0}^{-1}(W) \subset J^k(\pi)$.

Below we will be using the notation:

  (1) Here and below for an open subset $W \subset Y$, the notation W^k is used for the open subset $W^k = \pi_{k0}^{-1}(W) \subset J^k(\pi)$.

  (2) Given a fibre bundle $\pi : Y \to X$ we will use the notation $J^k(\pi)$ both for the k-jet bundle $\pi_k : J^k(\pi) \to X$ of sections of the bundle $\pi,$[119, 45, 70] and for the space $J^k(\pi)$ of this bundle.

We denote by $\pi_{kr} : J^k(\pi) \to J^r(\pi), k \geq r \geq 0$ the natural projections in the tower between jet bundles of different order and by $\pi_k : J^k \to X$ the projection to the base manifold X.

Projection mappings $\pi_{k(k-1)} : J^k(\pi) \to J^{k-1}(\pi)$ form the **projective system -** the **tower of k-jet bundles**

$$\ldots \xrightarrow{\pi_{(k+1)k}} J^k(\pi) \xrightarrow{\pi_{k(k-1)}} J^{k-1}(\pi) \xrightarrow{\cdots} J^1(\pi) \xrightarrow{\pi_{10}} Y \xrightarrow{\pi} X. \qquad (74.1)$$

In this bundle, epimorphisms $\pi_{k(k-1)}$ are *affine bundles* modeled over the vector bundle $\pi_{(k-1)}^* S^k T^*(X) \otimes_{\to J^{k-1}(\pi)} \pi_{(k-1)0}^* V(\pi) \to J^{k-1}(\pi)$ (see Definition 32, part 4).

To shorten the notations **we denote by** $J^k(\pi)/X$ **the whole tower** (74.3). Correspondingly, denote by $Aut(\pi_k/X)$ the group of automorphisms of the tower $J^k(\pi)/X$ formed by the collection of diffeomorphisms ϕ_l of $J^l(\pi), 0 \leq l \leq k$ and $\bar{\phi}$ of X commuting with all the projection of the tower $J^k(\pi)/X$: $\pi_{ab} \circ \phi_a = \phi_b \circ \pi_{ab}$.

Denote by $\mathcal{X}(J^k(\pi)/X$ the Lie algebra of vector fields in $J^k(\pi)$ projectable to all $J^q(\pi), q < k$, to Y and X in the coherent way.

In studying symmetries and conservation laws of differential equations and systems of such equations we will be using the "characteristic of vector fields". So, it is convenient to define here this notion.

Definition 48. *Let*

$$\xi = \xi^i \partial_{x^i} + \xi^\mu \partial_{y^\mu} + \sum_{\mu, I \ |I| \leq k} \xi_I^\mu \partial_{y_I^\mu}$$

be a vector field in $J^k \pi$. *Characteristic of vector field* ξ *is the vertical vector field*

$$Q(\xi) = \sum_\mu (\xi^\mu - y_i^\mu \xi^i) \partial_{y^\mu}. \qquad (74.2)$$

74.1. Infinite jet-bundle $J^\infty(\pi)$. Recall [45, 71, 119] that the k-jet bundles $J^r(\pi)$ of a bundle $\pi: Y \to X$ form the projective system

$$\ldots \xrightarrow{\pi_{(k+1)k}} J^k(\pi) \xrightarrow{\pi_{k(k-1)}} J^{k-1}(\pi) \xrightarrow{\cdots} J^1(\pi) \xrightarrow{\pi_{10}} Y \xrightarrow{\pi} X. \tag{74.3}$$

whose **inverse limit** $J^\infty(\pi)$ is called the **infinite order jet bundle** of the bundle π. Projections $J^\infty(\pi) \to J^k(\pi)$ for $k > 0$ will be denoted by $\pi_{\infty k}$, the same for projections $\pi_\infty : J^\infty(\pi) \to X$ and $\pi_{\infty 0} : J^\infty(\pi) \to Y$.

A fibred chart $(W; x^i, y^\alpha)$ in Y determines the fibred chart $(W^\infty; x^i, y^\alpha, y_I^\alpha)$ in $W^\infty = \pi_{\infty 0}^{-1}$, where multi-index $I = (i_1, i_2, \ldots, i_n)$ is a collection of natural numbers modulo permutations.

Corresponding to the projective system (74.3) of bundles over X (and over Y), there is the inverse system of Lie algebras of vector fields $\{X_k \in \mathcal{X}(J^k(\pi))\}$ on the k-jet bundle spaces **projectable to all the bundles** $J^q(\pi)$. $q < k$, to Y and to X. Inverse limit of this sequence is the **Lie algebra $\mathcal{X}(J^\infty(\pi))$ of vector fields on the space $J^\infty(\pi)$**. In a domain of a fibred chart, vector fields from $\mathcal{X}(J^\infty(\pi))$ have the form of infinite series

$$\xi = \xi^i(x)\partial_i + \xi^\alpha(x, y)\partial_{y^\alpha} + \sum_{\alpha, I\ |I|>0} \xi_I^\alpha \partial_{y_I^\alpha} \tag{74.4}$$

where coefficients $\xi_I^\alpha \in C^\infty(J^k(\pi))$.

Dually, there is the direct system of exterior algebras of differential forms on the spaces $J^k(\pi)$

$$\Lambda^*(X) \xrightarrow{\pi^*} \Lambda^*(Y) \xrightarrow{\pi_{10}^*} \Lambda^*(J^1(\pi)) \xrightarrow{\cdots} \xrightarrow{\pi_{k(k-1)}^*} \Lambda^*(J^k(\pi))) \cdots \to \cdots \tag{74.5}$$

induced by the pullback of the forms from the lower order jet bundles to the higher order jet bundles.

Limit of this direct system is the exterior Z-graded algebra called the bundle $\Omega_\infty^* = \Lambda^*(J^\infty(\pi))$ of exterior forms on $J^\infty(\pi)$.

Thus, in a fibred chart $(x^i, y^\alpha, \ldots)$, a k-form on $J^\infty(\pi)$ is the pullback to $J^\infty(\pi)$ of a k-form $\psi^k \in \Lambda^k(J^r(\pi))$ on the r-jet bundle $J^r(\pi)$ for some $r \geq 0$, or from X.

It is obvious that the duality between the k-vector fields and exterior k-forms on the jet bundles of finite order is extended to the duality between the k-vectors and k-forms on $J^\infty(\pi)$.

In particular, smooth functions f from $C^\infty(J^\infty(\pi))$ depend on a finite number of variables $(x^i, y^\alpha, y_I^\alpha), |I| \leq k$ for some natural k.

74.2. Total derivatives. For each i, $1 \leq i \leq n$, the vector field (total derivative by x^i)

$$d_i = \partial_i + y_i^\alpha \partial_\alpha + \sum_{\alpha, I | |I| > 0} y_{I+i}^\alpha \partial_{,y_I^\alpha} \tag{74.6}$$

is defined correctly in $J^\infty(\pi)$. Action of these vector fields on the functions, Lie derivatives of these vector fields on the exterior forms on the infinite jet space $J^\infty \pi$ are defined correctly due to the fact that all coefficients of exterior forms, as well as the smooth functions, depend on a finite number of variables. It is easy to see that total derivatives d_i commute:

$$[d_i, d_j] = 0. \tag{74.7}$$

We also introduce, for each multiindex $I = (i_1^{k_1} \cdot \ldots \cdot i_n^{k_n})$, the total derivative d_I of higher order to be

$$d_I = \prod_{k=1}^{n} d_k^{i_k}. \tag{74.8}$$

Since total derivatives d_i commute between themselves, the order of factors in the expression for d_I is inessential.

Notice that the total derivatives map functions on the k-jet bundle $J^k(\pi)$ to the functions on the $(k+1)$-jet space $J^{k+1}(\pi)$. To remedy this in order to use total derivatives on individual k-jet bundle,s we introduce the **k-truncated total derivatives**:

$$d_i = \partial_i + y_i^\alpha \partial_\alpha + \sum_{\alpha, I | 0 < |I| < k} y_{I+i}^\alpha \partial_{,y_I^\alpha} : C^\infty(J^k(\pi)) \to C^\infty(J^k(\pi)).$$

For these derivatives we will be using the same notation as for the whole total derivatives. From the context if will always be clear, what total derivatives are used in each case.

Denote by $\mathcal{V}(J^k(\pi))$ where $i \geqq k \geqq \infty$, the sheaf of vector fields in $(J^k(\pi))$ whose projection to X vanishes (equal to 0). This sheaf is generated by the vector fields $\partial_{y_I^\mu}$ with arbitrary multiindices I. It is easy to see that the sheaf of vector fields $\mathcal{X}(J^k(\pi))$ on the k-jet space is the direct sum of the sheaf of vertical vector fields and the space of k-truncated total derivatives:

$$\mathcal{X}(J^k(\pi)) = \langle d_i, i = 1, \ldots, n \rangle \oplus \mathcal{V}(J^k(\pi)). \tag{74.9}$$

Remark 87. If it is allowed to use coefficients $f \in C^\infty(J^\infty(\pi))$ for the vector fields in $J^k(\pi)$, these vector fields in $J^k(\pi)$ are called **generalized vector fields**, see [106].

75. CONTACT STRUCTURE ON THE K-JET BUNDLES.

Definition 49. *An exterior form $\alpha \in \Omega^*(J^k(\pi))$ is called* **contact** *if for any section $s : U \to Y$ of the bundle π, where $U \subset X$ is an open subset of X,*

$$(j^k s)^* \alpha = 0.$$

. Denote by $C\Omega^*(J^k(\pi))$ the space of contact forms on $J^k(\pi)$.

In a local fibred chart (x, y, y_I^α) of $J^k(\pi)$, define the **basic contact forms**

$$\omega^\alpha = dy^\alpha - y_i^\alpha dx^i, \ldots \omega_I^\alpha = dy_I^\alpha - y_{I+1_i}^\alpha dx^i, \ 0 \leqq |I| < k. \tag{75.1}$$

In the next Lemma we collected some simple properties of contact forms.

Lemma 12.   (1) *For $m > n = Dim(X)$ all the m-forms are contact on $J^k\pi$.*

    (2) *"Basic 1-forms"*

$$\omega^\alpha = dy^\alpha - y_i^\alpha dx^i, \ldots \omega_I^\alpha = dy_I^\alpha - y_{I+1_i}^\alpha dx^i, \ 0 \leqq |I| < k. \tag{75.2}$$

    are contact.

    (3) *Contact forms form the* **differential ideal** *$C\Omega^*(J^k(\pi)) \subset \Omega^*(J^k(\pi))$ in the algebra of exterior differential forms in $J^k(\pi)$.*

Proof. The first statement follows from the fact that on the base space X there are no nonzero forms of order grater then $n = dim(X)$. Second statement can be checked by taking the pullback of basic forms ω_I^α with respect of the k jet $j^k s$ of any section $x \ni s : U \to Y$ with $|I| \leq k$. The last statement follows from the fact that the pullback of exterior forms commutes with the exterior product of forms and with the exterior differential d of the forms. $\qquad\square$

For more details about contact structure, see [70, 71, 119].

Definition 50. *The contact ideal $C\Omega^*(J^k(\pi))$ defines in the k-jet-space $J^k(\pi)$, $k = 1, \ldots, \infty$, the **Cartan distribution**:*

$$Ca_{(k)}^k = \{\xi \in T_{(k)}(J^k(\pi)) | i_\xi \omega = 0, \ \forall \omega \in C\Omega^*(J^k(\pi))\}, z^{(k)} \in J^k(\pi). \qquad (75.3)$$

Proposition 18. $\quad$ (1) *Differential contact ideal is algebraically generated by the basic contact forms (75.1) and the forms dy_J^α, $|J| = k$, see [68].*
$\quad$ (2) *If $(W; x^i, y^\alpha)$ is a fibred chart in Y, then the 1-forms*

$$dx^i, \omega^\alpha, \omega_I^\alpha, |I| \leq k - 1, dy_J^\alpha, \ |J| = k \qquad (75.4)$$

form the local coframe - a linear basis of the cotangent space $T_{z^{(k)}}^*(J^k(\pi))$ at every point $y^{(k)} \in W^k$.

Cartan distribution and the contact forms defining it plays a major role in the specifying the *holonomic* sections σ of jet-bundles $\pi_k : J^k(\pi) \to X$ - sections σ of the bundles $\pi_k : J^k(\pi) \to X$ generated by the sections $s : X \to Y$ of the configurational bundle as follows: $\sigma_{hol} = j^k s$. Notice that the non-holonomic sections of the bundle $\pi_k : J^k(\pi) \to X$ are sections of the the general type of of this bundle. More specifically, we have

Proposition 19. *Let $\sigma : U \to J^k(\pi)$ be a section of the bundle π_k with the domain $U \subset X$. Then the following statements are equivalent*

$\quad$ (1) $\sigma = j^k s$ *for some section $s : U \to Y$ of section π,*

$\quad$ (2) *The pullback to U of any contact form $\nu \in Con(J^k(\pi))$ defined by the section σ - $\sigma^* \nu$ vanishes,*

$\quad$ (3) *For any $x \in U$, $\sigma_*(T_x(X)) \subset Ca_{\sigma(x)}$. In other words, the image of the tangent space of the base space X, under the tangent mapping σ_* of the section mapping σ, belongs to the Cartan distribution.*

Proof. Definition 7 shows that statement 2) follows from 1). Statement 3) follows from 1) because the pullback of any form by a section $\sigma : x \to J^k(\pi)$ is defined by restrictions of these forms to the image $\sigma_* T_x(X)$) of the tangent bundle of the base space. By property 1) all the restrictions of contact forms to this image vanish. Since the Cartan distribution is defined as the kernel of the ideal of contact forms, image $\sigma_* T_x(X)$ has to be a subspace of the space $Ca(\sigma(x))$ of Cartan distribution for each point $x \in X$.

Statement 2) follows from 3) by the same arguments. Finally, to prove that 1) follows from 2) or3) it is sufficient to write the condition $\sigma^* \omega_I^\alpha = 0$ for a section $\sigma : X \to J^k(\pi)$ in the components σ_I^μ. We get the family of equalities

$$\sigma_I^\mu = \sigma_{I+1_i}^\mu dx^i$$

valid for all i, I, μ. From this it follows that $\sigma_I^\mu = \partial^I \sigma^\mu$. As a result, section σ is the k-jet of the section $s : Wx \to Y$ with the components σ^μ. $\square$

Differentials of basic contact forms ω_Λ^i on $J^k(\pi)$ are given by the formulas

$$\begin{cases} d\omega_I^\alpha = -\omega_{I+j}^\alpha \wedge dx^j, \ for \ |I| \leq k - 2, \\ d\omega_I^\alpha = -dy_{I+j}^\alpha \wedge dx^j, \ for \ |I| = k - 1. \end{cases} \qquad (75.5)$$

In the case where $k = \infty$ the basic contact forms dy_Λ^i are absent form the list of generators of the ideal $C^k \Omega^*(J^\infty(\pi)))$.

For $k = \infty$, Cartan distribution Ca^∞ splits the tangent bundle of $J^\infty(\pi)$: tangent bundle of the $J^\infty(\pi)$ is the direct sum of the subbundle of π^∞-vertical vector fields and the n-dimensional Cartan distribution Ca^∞ generated by the **operators of total derivative**

$$T(J^\infty(\pi)) = Ca^\infty \oplus V(\pi^\infty), \qquad (75.6)$$

where

$$\begin{cases} Ca^\infty = \langle d_i = \partial_{x^i} + \sum_{|I| \geq 0} y_{I+i}^\alpha \partial_{y_I^\alpha} \rangle, \\ V(\pi_\infty) = \langle \partial_\alpha, \partial_{y_I^\alpha}, |I| > 0 \rangle. \end{cases} \qquad (75.7)$$

As a result, $Hor^T = Ca^\infty$ defines the connection T on the bundle $\pi_\infty : J^\infty(\pi) \to X$ It is easy to prove that

Lemma 13. *Connection T on the bundle $\pi_{\infty 0} : J^\infty(\pi) \to Y$ (see Sec...) is flat, in other words its curvature is zero.*

Proof. Total derivatives of vector fields d_i generating horizontal subbundle Hor^T commute. Thus, the curvature of connection T vanishes. $\square$

Dually, the algebra of exterior forms $\Omega^*(J^\infty(\pi))$ has the decomposition

$$\Omega^*(J^\infty(\pi)) = \Omega_X^*(J^\infty(\pi)) \oplus C\Omega^*(J^\infty(\pi)) \qquad (75.8)$$

into the sum of the subalgebra of π^∞-horizontal (basic) forms and the ideal of contact forms.

Remark 88. Decomposition of the tangent bundle, similar to the decomposition 75.6 exists on finite jet bundles $J^k(\pi)$ as well provided we are using truncated total derivatives:

$$T(J^k(\pi)) = Ca^k \oplus V_k, \qquad (75.9)$$

where

$$\begin{cases} Ca^k = \langle d_i = \partial_{x^i} + \sum_{k \geq |I| \geq 0} y_{I+i}^\alpha \partial_{y_I^\alpha} \rangle, \\ V(\pi_k) = \langle \partial_\alpha, \partial_{y_I^\alpha}, k \geq |I| > 0 \rangle. \end{cases} \qquad (75.10)$$

Now we introduce the class of vector fields and their phase flows closely related with the Cartan distribution Ca in the 1-jet space $J^1(\pi)$ introduced above.

Definition 51.   (1) *A diffeomorphism $\phi : J^k(\pi) \to J^k(\pi)$ is called a **contact transformation** or **Lie transformation** if it leaves invariant the Cartan distribution: if for any point $z \in J^k(\pi)$,*

$$\phi_{*z} Ca^k(z) = Ca^k(\phi(z).$$

(2) *A vector field $\zeta \in \mathcal{X}(J^k(\pi))$ is called an infinitesimal contact vector field or a **Lie field**, if its (local or global if it exists) phase flow consists of canonical transformations or, what is equivalent, if vector field ζ is tangent to the Cartan distribution at any point $z \in J^k(\pi)$.*

Proof of the following Proposition follows immediately from the relation between the Cartan distribution and the ideal of contact forms on the k-jet bundle $J^k(\pi)$ (see also [71], Ch.3, Sec.3).

Proposition 20. *Let ζ be a vector field in the k-jet space $J^k(\pi)$. Then the following properties of ζ are equivalent:*

(1) *ζ is the Lie vector field,*
(2) *For any contact form $\theta \in C\Omega^*(J^k(\pi))$, Lie derivative $\mathcal{L}_\zeta \omega \in C\Omega^*(J^k(\pi))$ i.e is contact too.*

*If these conditions are fulfilled for a vector field ζ we say that ζ **preserves the contact structure in** $J^k(\pi)$.*

To resume the introduction into the contact structure of the tower of jet bundles, we recall that the Cartan distribution Ca^k in the k-jet bundle $J^k(\pi)$ is generated by the 1-forms $\omega_I^\alpha = dy_I^\alpha - y_{I+i}^\alpha dx^i$ and dy^I, $|I| = k$. This means that being restricted to the Cartan distribution, contact exterior forms vanish. Vice versa, if a tangent vector $\xi \in T_z(J^k(\pi))$ at a point $z \in J^k(\pi)$ is annulated by all contact forms, then this tangent vector belongs to the Cartan distribution at the point z.

Contact forms generate the differential ideal (i.e. ideal of the exterior algebra $\Omega^*(J^k(\pi))$ closed under the action of exterior differential d). This differential ideal of exterior form in $J^k(\pi)$ was denoted $C\Omega(J^k)$.

76. Prolongation of vector fields to the jet bundles.

In this section we describe (following [106]) the procedure of the "**flow**"-prolongation of vector fields from the space Y of the bundle $\pi : Y \to X$ to the jet bundles $J^k(\pi)$, k=1,2,..., ∞. To do this we introduce the "**characteristic vector field** $Q = Q^\alpha \partial_{y^\alpha}$" associated with a vector field $\xi \in \mathcal{X}(Y)$ and describe the prolongation of ξ to the vector field in $J^k(\pi)$ in terms of the characteristic Q.

Definition 52. *Let $\xi = \xi^i \partial_i + \xi^\alpha \partial_\alpha$, be a vector field defined in an open subset $W \subset Y$.*

(1) *"The **characteristic**" of vector field ξ is the π-vertical vector field*

$$\xi_Q = Q^\alpha \partial_\alpha = \omega^\alpha(\xi)\partial_\alpha = (\xi^\alpha - y_i^\alpha \xi^i)\partial_\alpha \tag{76.1}$$

where $\omega^\alpha(\xi)$, $\alpha = 1,\ldots,m$ is the value of the basic contact form $\omega^\alpha = dy^\alpha - y_i^\alpha dx^i$ on the vector field ξ.

(2) *Let $1 \leq k \leq \infty$. **The flow prolongation of order** k of a vector field $\xi = \xi^i \partial_i + \xi^\alpha \partial_\alpha$ in W is the vector field $\xi^{(k)}$ in the space $J^k(\pi)$ defined (in the domain $W^k = \pi_{k1}^{-1}(W)$) as follows.*

$$Pr^k \xi = \xi^i d_i + \xi_Q + \sum_{k \geq |I| > 0} (d_I Q^\beta)\partial_{y_I^\beta}, \tag{76.2}$$

(3) *The flow prolongation of vector field $\xi \in \mathcal{X}(Y)$ to the **infinite jet bundle** $J^\infty(\pi)$ has the form*

$$Pr^\infty \xi = \xi^i d_i + Q^\alpha \partial_\alpha + \sum_{|I|>0} (d_I Q^\beta) \partial_{y_I^\beta}. \tag{76.3}$$

The first term in this representation is the Cartan vector field that belongs, at every point of its
    domain to the Cartan distribution in $J^\infty(\pi)$, while the second and third terms form the π_∞-vertical part of the prolongation.

Another way to write the prolongation formula is the following:
Let $\xi = \xi^i \partial_{x^i} + \xi^\mu \partial_{y^\mu}$, then

$$Pr^k(\xi) = \xi + \sum_{\mu=1}^k \sum_{I,|I|1\leqq k} \xi_I^\mu(x, y, y_i^\mu, \ldots, y_I^\mu) \partial_{y_I^\mu}, \tag{76.4}$$

where coefficients ξ_I^μ are given by the following formula:

$$\xi_I^\mu = d_I Q^\mu + \sum_{i=1}^k \xi^i y_{I,i}^\mu. \tag{76.5}$$

Here $y_{I,i}^\mu = \frac{\partial y_I^\mu}{\partial x^i}$.

Finally, one can use the **iterative formula** for the coefficients ξ_I^μ of the prolongation ([106],Sec.2.3). We present this formula here in order to be able to compare it with the iterative definition of twisted prolongations in Chapter 3.

$$\xi_{I,k}^\mu = d_k \xi_I^\mu - y_{I,m}^\mu d_k \xi^m. \tag{76.6}$$

Exercise 1. A nice Exercise for the readers is to check the equivalence of these three definitions for prolongation of vector fields from Y to, say, $J^3\pi$.

Example 46. It is easy to see that for all $k > 1$ $Pr^k \partial_i = \partial_i$ and $Pr^k \partial_\mu = \partial_\mu$.

Proposition 21. *([106], Thm.2.39) Prolongation of order k, where $1 \leq k \leq \infty$ is the homomorphism of Lie algebras of vector fields: $\xi \to Pr^k \xi^{(k)} : \mathcal{X}(Y) \to \mathcal{X}(J^k(\pi))$: Let $\xi, \eta \in \mathcal{X}(Y)$, then*

$$\begin{cases} Pr^k(c_1\xi + c_2\eta) = c_1 Pr^k\xi + c_2 Pr^k\eta, \\ Pr^k[\xi, \eta] = [Pr^k\xi, Pr^k\eta]. \end{cases} \tag{76.7}$$

The next Proposition gives the explicit description of the prolongations of vector fields from Y to $J^1(\pi)$ (see [?] for the proof).

Proposition 22. *Let $\xi = \xi^i(x,y)\partial_i + \xi^\mu(x,y)\partial_\mu \in \mathcal{X}(Y)$ be an arbitrary vector field in a domain $W \subset Y$ of a fibred chart (W, x^i, y^μ). Then,*

(1) *There is unique vector field $Pr^1\xi \in \mathcal{X}(J^1(\pi))$ (**1-prolongation of a vector field** ξ) defined by the conditions:*
    (a) *Vector field $Pr^1\xi \in \mathcal{X}(J^1(\pi))$ is projectable to Y and*

$$\pi_{10*}(Pr^1\xi) = \xi,$$

(b) *Local phase flow of the vector field $Pr^1\xi$ preserves the Cartan distribution Ca^1 (i.e is the **Lie vector field**, see Definition 51 and the monograph [71]) for more about Lie vector fields.*

(2) *The 1-prolongation $Pr^1\xi$ of a vector field $\xi = \xi^i(x,y)\partial_i + \xi^\mu(x,y)\partial_\mu$ has, in local fibred coordinates, the form*

$$Pr^1\xi = \xi^\mu(x,y)\partial_\mu + \xi^i(x,y)\partial_i + \left(d_\mu\xi^i - y_\nu^i d_\mu\xi^\nu\right)\partial_{y_\mu^i} =$$
$$= (\xi^\mu\partial_\mu - d_\mu\xi^\nu y_\nu^i \partial_{y_\mu^i}) + (\xi^i\partial_i + d_\mu\xi^i\partial_{y_\mu^i}), \quad (76.8)$$

where $d_\mu\xi^i = \frac{d\xi^i}{dx^\mu} = \frac{\partial\xi^i}{\partial x^\mu} + y_\mu^j\frac{\partial\xi^i}{\partial y^j}$ is the total derivative of the function ξ^i and similarly for $\xi^\mu(x,y)$.

(3) *Vice versa, any Lie vector field in the 1-jet space $J^1(\pi)$ is the flow prolongation of the form 76.8 of some vector field $\xi \in \mathcal{X}(Y)$ ([71], Ch.2).*

(4) *The mapping $\xi \to Pr^1\xi$ is the homomorphism of Lie algebras:*

$$Pr^1[\xi,\eta] = [Pr^1\xi, Pr^1\eta]$$

for all $\xi,\eta \in \mathcal{X}(Y)$.

(5) *For a π-projectable vector field $\xi \in \mathcal{X}(Y)$ with $\xi = \xi^i(x)\partial_{x^i} + \xi^\mu(x,y)\partial_{y^\mu}$, its **1- prolongation** $Pr^1\xi$ coincides with the **flow prolongation**: Specifically, let $y \in Y$ and let ϕ^t be the flow of vector field ξ in a neighborhood of a point $y \in Y$. Let $\bar{\phi}^t$ be the phase flow of the vector field $\bar{\xi}$ - projection $\pi_*\xi$ of the vector field ξ in the corresponding neighborhood of the projection $\pi(y) \in X$. Then, the lift $\hat{\phi}^t$ of local diffeomorphisms ϕ^t to the 1-jet space $J^1(\pi)$ is the **local phase flow of the 1-prolongation** $Pr^1\xi$ in a neighborhood of any point z in the fibre $\pi_{10}^{-1}(y)$ over y.*

Proof. For a proof of Statement 5), see the proof of Theorem 4.4.11 in [119] . $\square$

The following equivalent conditions formulated in terms of contact forms, commuting with total derivatives, are used for determining **if a given vector field is the Lie field**.

Proposition 23. *Let $n > 1$. Let ζ be a vector field in $J^k(\pi)$ projecting to the vector field $X \in \mathcal{X}(Y)$. Following properties of vector field ζ are equivalent*

(1) *ζ is the Lie field (see Def. 51),*

(2) *For any contact form $\theta \in C\Omega(J^k(\pi))$, one has $\mathcal{L}_\zeta\omega \in C\Omega(J^k(\pi))$,*

(3) *$\zeta = Pr^k(\xi)$ for some vector field $\xi = \xi^i\partial_i + \xi^\alpha\partial_\alpha \in \mathcal{X}(Y)$ (if $n = 1$, for a vector field $\xi \in \mathcal{X}(J^1(\pi))$),*

(4) *For any contact form $\theta \in C\Omega(J^k)$ and for $i = 1,\ldots,n$, $i_{[d_i,\zeta]}\theta = 0$,*

(5) *$[d_i,\zeta] = c_i^k(z)d_k + V_i$, where $c_i^k(z) \in C^\infty(J^k(\pi))$ and V_i is the $\pi_{k(k-1)}$-vertical vector field (i.e. vector field V has the form $V = \sum_{\alpha,I||I|=k} v_i^\alpha(z)\partial_{y_I^\alpha}$.*

Proof. Equivalence of properties 1) and 2) is proved in Lemma 10 above. Equivalence of 2) and 3) is the classical result (the Backlund theorem), see, [71], Ch.3, Thm.3.3 or [106].

Equivalence of properties 3) and 4) follows from the fact that k-truncated vector fields d_m span the set of non-(k(k-1))-vertical vector fields in the annulator of the contact forms.

To prove the equivalence of properties 2) and 4) we notice that $d_i = \partial_i + \sum_{\alpha, I} y^\alpha_{I+i} \partial_{y^\alpha_I}$. The standard computation using the explicit form of vector field $\zeta = \zeta^k \partial_k + \sum \ldots$ will lead us to

$$[d_i, \zeta] = (d_i \zeta^k)\partial_k + (d_i \zeta^\alpha_J - \zeta^\alpha_{J+j}\partial_{y^\alpha_I});$$

hence we get $i_{[d_i,\zeta]}\omega^\alpha_J = -\zeta^\alpha_{J+i} + (d_i\zeta^\alpha_J - y^\alpha_{J+k}d_i\zeta^k)$. Obtained expression vanishes if and only if ζ satisfies to the prolongation formula (76.6).

To prove equivalence of 2) and 5) we use the property 4) and notice that the total derivatives d_i span the space of non-vertical vector fields for the fibration $J^k(\pi) \to J^{k-1}(\pi)$ in the **annulator of contact forms**. Another way to prove this is to apply to the function $c^k_i = d_i\xi^k$ the decomposition of differential of any function $f \in C^\infty(\pi)$ into "horizontal" and "contact" parts

$$df = (d_i f)dx^i + \hat{\theta}[f],$$

where explicit form of the second term is irrelevant (see [38]). . $\qquad\square$

Remark 89. For the (finite) contact transformations there exists a description similar to the one given in statement 2) of the last Proposition for Lie fields, see [71], Thm.3.1.

77. Connections on the 1-jet bundle $\pi_{10} : J^1(\pi) \to Y$.

In Sec.68 of Appendix I, we introduced connections on the fiber bundles, and defined their curvature tensors fields. In this Section we specialize these notions for the jet bundles and for the tower of these bundles.

We recall the double bundle

$$J^1\pi \xrightarrow{\pi_{10}} Y \xrightarrow{\pi} X \tag{77.1}$$

of the constitutive bundle $\pi : Y \to X$ and the exact sequence of tangent bundles

$$0 \to V(J^1\pi) \to T(J^1\pi) \to T(Y) \to T(X) \to 0 \tag{77.2}$$

defining the vertical subbundle $V(J^1\pi) \to J^1\pi$ of the tangent bundle $T(J^1\pi) \to J^1\pi$ of the bundle $\pi_{10} : J^1\pi \to Y$.

In the domain of the chart induced in $J^1\pi$ by a fibred chart (x^i, y^μ) in Y, *vertical tangent bundle* $V(J^1\pi) \subset T(J^1\pi)$ of the bundle $\pi_{10} : J^1\pi \to Y$ is generated by the vector fields $\partial_{y_i^\mu}$.

Definition 53. *An **Ehresmann connection** (or simply "**connection**") is the rule associating with any point $z = (x^i, y^\mu, y_i^\mu) \in J^1(\pi)$ a subspace $H_z \subset T_z(J^1(\pi))$ complemental to the vertical subspace $V_z(\pi_{10}) \subset T_z(J^1(\pi))$ and smoothly depending on the point z.*

Let a connection K be defined in the bundle $\pi_{10} : J^1(\pi) \to Y$. In the domain of a local fibred chart $(W^1, x^i, y^\mu, y_i^\mu)$ in $J^1(\pi)$ this connection is defined by the $T(J^1(\pi))$-valued 1-form on Y, (see the Section above and [45, 70]:

$$K = dx^l \otimes (\partial_l + K_{lj}^\nu \partial_{y_j^\nu}) + dy^\mu \otimes (\partial_\mu + K_{\mu j}^\nu \partial_{y_j^\nu}) \tag{77.3}$$

with the connection coefficients $K_{lj}^\nu, K_{\mu j}^\nu \in C^\infty(J^1(\pi))$.

A **K-horizontal lift** of a vector field $\xi = \xi^i \partial_i + \xi^\mu \partial_\mu$ in Y to the vector field in $J^1(\pi)$ has the form

$$Hor_K(\xi) = (\xi^i\partial_i + K_{ij}^\nu\xi^i\partial_{y_j^\nu}) + (\xi^\sigma\partial_\sigma + K_{\mu j}^\nu\xi^\mu\partial_{y_j^\nu}) = \xi + (K_{ij}^\nu\xi^i + K_{\mu j}^\nu\xi^\mu)\partial_{y_j^\nu}. \tag{77.4}$$

Example 47. $\lambda\mu$-connections.

A special class of connections in the bundles $J^1\pi \to Y$ appears in the works of C.Muriel, J.L.Romero, G.Gaeta, G.Cicogna and P.Morando, (see [13, 39, 40, 98, 102],etc.). $\lambda-$ and $\mu-$ connections are defined by the condition that the tensor K_{lj}^ν depends linearly on the components of tensor $K_{\mu j}^\nu$:

$$K_{lj}^\nu = y_l^\mu K_{\mu j}^\nu. \tag{77.5}$$

Cited authors are using tensor $K_{\mu j}^\nu$ in the form of matrix valued horizontal 1-form $\nu = \Lambda_{i\mu}^\nu dx^i$, where, of course, $\Lambda_{i\mu}^\nu = K_{\mu j}^\nu$. We will call these connections - $\lambda\mu$-connections by the name of key objects (function λ and 1-form μ)) as was done in the work of these authors.

The curvature of a connection (75.3) is defined, as in Section (68), by comparing commutators of K-horizontally lifts of vector fields in Y with the lift of commutators of the same fields:

$$R(\xi, \eta) = [Hor_K(\xi), Hor_K(\eta)] - Hor_K([\xi, \eta]). \tag{77.6}$$

Let $\xi = \xi^i \partial_i + \xi^\mu \partial_\mu$, $\eta = \eta^j \partial_j + \eta^\nu \partial_\nu$ be two vector fields in Y.

Then,

$$\begin{cases} Hor_K(\xi) = \xi + (K^{\nu}_{ij}\xi^i + K^{\nu}_{\mu j}\xi^{\mu})\partial_{y^{\nu}_j} = \xi + \Phi^{\nu}_j \partial_{y^{\nu}_j}, \\ Hor_K(\eta) = \eta + (K^{\sigma}_{kl}\eta^k + K^{\sigma}_{\lambda l}\eta^{\lambda})\partial_{y^{\sigma}_l} = \eta + \Psi^{\sigma}_l \partial_{y^{\sigma}_l}. \end{cases} \tag{77.7}$$

Here we introduce the notation $\Phi^{\nu}_j = K^{\nu}_{ij}\xi^i + K^{\nu}_{\mu j}\xi^{\mu}$ and $\Psi^{\sigma}_l = K^{\sigma}_{kl}\eta^k + K^{\sigma}_{\lambda l}\eta^{\lambda}$ for vertical components of horizontally lifted vector fields.

The commutator of obtained vector fields in $J^1\pi$ has the form

$$[Hor_K(\xi), Hor_K(\eta)] = [\xi + \Phi^{\nu}_j \partial_{y^{\nu}_j}, \eta + \Psi^{\sigma}_l \partial_{y^{\sigma}_l}] =$$
$$= [\xi, \eta] + [\xi, \Psi^{\sigma}_l \partial_{y^{\sigma}_l}] + [\Phi^{\nu}_j \partial_{y^{\nu}_j}, \eta] + [\Phi^{\nu}_j \partial_{y^{\nu}_j}, \Psi^{\sigma}_l \partial_{y^{\sigma}_l}]. \tag{77.8}$$

On the other side,

$$[\xi, \eta] = [\xi^i \partial_i + \xi^{\mu}\partial_{\mu}, \eta^j \partial_j + \eta^{\nu}\partial_{\nu}] =$$
$$= [\xi^i \eta^j_{,i} + \xi^{\mu}\eta^j_{,\mu} - \eta^k \xi^j_{,k} - \eta^{\nu}\xi^j_{,\nu}]\partial_j + [\xi^i \eta^{\nu}_{,i} + \xi^{\mu}\eta^{\nu}_{,\mu} - \eta^i \xi^{\nu}_{,j} - \eta^{\mu}\xi^{\nu}_{,\mu}]\partial_{\nu} = q^j \partial_j + q^{\nu}\partial_{\nu}, \tag{77.9}$$

where notation q^j and q^{ν} are introduced.

Horizontal lift of this commutator of vector fields is

$$Hor_K[\xi, \eta] = (q^j \partial_j + q^{\nu}\partial_{\nu}) + (K^{\mu}_{ij}q^i + K^{\mu}_{\nu j}q^{\nu})\partial_{y^{\mu}_j} =$$
$$= [\xi, \eta] + (K^{\mu}_{ij}q^i + K^{\mu}_{\nu j}q^{\nu})\partial_{y^{\mu}_j}. \tag{77.10}$$

Notice that the expression in the first bracket is $[\xi, \eta]$.

Curvature $R(\xi, \eta)$ in the direction of vector fields ξ, η is defined as the difference

$$R(\xi, \eta) = [Hor_K(\xi); Hor_K(\eta)] - Hor_K[\xi, \eta] = [\xi, \Psi^{\sigma}_l \partial_{y^{\sigma}_l}] + [\Phi^{\nu}_j \partial_{y^{\nu}_j}, \eta] +$$
$$+ [\Phi^{\nu}_j \partial_{y^{\nu}_j}, \Psi^{\sigma}_l \partial_{y^{\sigma}_l}] - (K^{\mu}_{ij}q^i + K^{\mu}_{\nu j}q^{\nu})\partial_{y^{\mu}_j} =$$
$$= [\xi \cdot \Psi^{\sigma}_l \partial_{y^{\sigma}_l} - \eta \cdot \Phi^{\nu}_j \partial_{y^{\nu}_j}] + [\Phi^{\nu}_j \partial_{y^{\nu}_j}, \Psi^{\sigma}_l \partial_{y^{\sigma}_l}] - (K^{\mu}_{ij}q^i + K^{\mu}_{\nu j}q^{\nu})\partial_{y^{\mu}_j} = I + II - III. \tag{77.11}$$

In the last equality we have used the fact that coefficients of vector fields ξ, η do not depend on the jet variables y^{μ}_i.

In simplifying obtained expression it is easy to see that the part in term I containing derivatives of components of vector fields ξ and η will be canceled by part III. As a result, only terms containing products of components of of vector fields ξ, η (i.e. bilinear by ξ, eta and the part II which has similar structure, are left:

$$R(\xi,\eta) = I + II - III =$$
$$= \xi^i \eta^k [(\partial_i K^\nu_{kj} - \partial_k K^\nu_{ij}))] + \xi^i \eta^\lambda [(\partial_i K^\nu_{kj} - \partial_k K^\nu_{ij})] +$$
$$+ \xi^\nu \eta^k [(\partial_\nu K^\nu_{kj} - \partial_k K^\nu_{\mu j})]] + \xi^\nu \eta^\lambda [(\partial_\nu K^\nu_{\lambda j} - \partial_\lambda K^\nu_{ij})] +$$
$$+ \xi^\mu \eta^\lambda [K^\sigma_{\mu l}(\partial_{y^\sigma_l} K^\nu_{\lambda j}) - K^\sigma_{\lambda l}(\partial_{y^\sigma_l} K^\nu_{\mu j})] +$$
$$+ + \xi^\mu \eta^k [K^\sigma_{\mu l}(\partial_{y^\sigma_\lambda} K^\nu_{kj}) - K^\sigma_{kl}(\partial_{y^\sigma_l} K^\nu_{\mu j})]$$
$$+ \xi^i \eta^\lambda [K^\sigma_{il}(\partial_{y^\sigma_j} K^\nu_{\lambda j}) - K^\sigma_{\lambda l}(\partial_{y^\sigma_l} K^\nu_{ij})] +$$
$$+ \xi^i \eta^k [K^\sigma_{il}(\partial_{y^\sigma_l} K^\nu_{kj}) - K^\sigma_{kl}(\partial_{,y^\sigma_l}) K^\nu_{ij}]. \qquad (77.12)$$

Recombining the terms in the last formula we get the following for the curvature tensor:

$$R(\xi,\eta) = I + II - III =$$
$$= \xi^i \eta^k \{[(\partial_i K^\sigma_{kl} - \partial_k K^\nu_{ij}))] + [(K^\sigma_{il}(\partial_{y^\sigma_l} K^\nu_{kj}) - K^\sigma_{kl}(\partial_{,y^\sigma_l}) K^\nu_{ij})]\} +$$
$$+ \xi^i \eta^\lambda \{[(\partial_\lambda K^\nu_{ij} - \partial_\lambda K^\nu_{ij})] + [K^\sigma_{il}(\partial_{y^\sigma_j} K^\nu_{\lambda j}) - K^\sigma_{\lambda l}(\partial_{y^\sigma_l} K^\nu_{ij})]\} +$$
$$+ \xi^\mu \eta^k \{[(\partial_\mu K^\sigma_{kl} - \partial_k K^\nu_{\mu j})] + [K^\sigma_{\mu l}(\partial_{y^\sigma_\lambda} K^\nu_{kj}) - K^\sigma_{kl}(\partial_{y^\sigma_l} K^\nu_{\mu j})]\} +$$
$$+ \xi^\mu \eta^\lambda \{[(\partial_\mu K^\sigma_{\lambda l} - \partial_\lambda K^\nu_{\mu j})] + [K^\sigma_{\mu l}(\partial_{y^\sigma_l} K^\nu_{\lambda j}) - K^\sigma_{\lambda l}(\partial_{y^\sigma_l} K^\nu_{\mu j})]\}. \qquad (77.13)$$

Notice that all terms of the curvature are coefficients of $\partial_{y^\nu_j}$.

For basic *vertical* vector fields $\partial_\mu, \partial_\nu$ we have

$$[Hor_K(\partial_\mu), Hor_K(\partial_\nu)] = [\partial_\mu + K^\lambda_{\mu j}\partial_{y^\lambda_j}, \partial_\nu + K^\sigma_{\nu k}\partial_{y^\sigma_k}] =$$
$$= [(\partial_\mu K^\nu_{\lambda j} - \partial_\lambda K^\nu_{\mu j})]\partial_{y^\nu_j} + [K^\sigma_{\mu l}\partial_{y^\sigma_l} K^\nu_{\lambda j} - K^\sigma_{\lambda l}\partial_{y^\sigma_l} K^\nu_{\mu j}]\partial_{y^\nu_j}, \qquad (77.14)$$

so that the corresponding (vv or vertical-vertical) component of the curvature tensor has the form

$$R^\nu_{\mu\lambda j} = [(\partial_\mu K^\nu_{\lambda j} - \partial_\lambda K^\nu_{\mu j})] + [K^\sigma_{\mu l}\partial_{y^\sigma_l} K^\nu_{\lambda j} - K^\sigma_{\lambda l}\partial_{y^\sigma_l} K^\nu_{\mu j}]. \qquad (77.15)$$

Remark 90. This part of the curvature appears as the tensor $\tilde{R}$ in the calculation of obstructions for the K-twisted prolongation of vertical vector fields, see Chapter 3, Sec.20.

For basic horizontal vector fields ∂_i, ∂_j we have

$$[Hor_K(\partial_i), Hor_K(\partial_j) = [\partial_i + K^\mu_{ik}\partial_{y^\mu_k}, \partial_j + K^\nu_{js}\partial_{y^\nu_s}] =$$
$$= [(\partial_i K^\mu_{jk} - \partial_j K^\mu_{ik}) + (K^\nu_{is}\partial_{y^\nu_s} K^\mu_{jk} - K^\nu_{js}\partial_{y^\nu_s} K^\mu_{ik})]\partial_{y^\mu_k}. \qquad (77.16)$$

Thus, the hh (horizontal-horizontal) component of the curvature tensor is

$$R^\sigma_{ijk} = (\partial_i K^\sigma_{jk} - \partial_j K^\sigma_{ik}) + (K^\nu_{is}\partial_{y^\nu_s} K^\sigma_{jk} - K^\nu_{js}\partial_{y^\nu_s} K^\sigma_{ik}). \qquad (77.17)$$

Finally, the hv (horizontal-vertical) component of curvature obtained by calculating

$$[Hor_K(\partial_i), Hor_K(\partial_\mu)]$$

has the form

$$R^\nu_{k\sigma j} = [\partial_k(K^\nu_{\sigma j}) - \partial_\sigma K^\nu_{kj}) + [K^\mu_{ki}\partial_{y^\mu_i} K^\nu_{\sigma j} - K^\lambda_{\sigma l}\partial_{y^\lambda_l}(K^\nu_{kj}]. \qquad (77.18)$$

Combining components of the curvature tensor we get the following matrix

$$\mathcal{R}(K) = \begin{pmatrix} R^\sigma_{\mu\nu k} & R^\sigma_{i\mu k} \\ R^\sigma_{i\mu k} & R^\sigma_{ijk} \end{pmatrix} \qquad (77.19)$$

77.1. Vertical connections. In order to define the lift $\xi \to Pr^1_K\xi$ of π-vertical vector fields it is sufficient to have the restriction of a connection (7.1) to the vertical subbundle $\mathcal{V}(\pi) \subset T(Y)$ or, more specifically, the operation of lift of π-vertical vector fields $\xi = \xi^\mu\partial_\mu$ on Y to the π_1 vertical vector fields $\hat\xi = \xi^\mu\partial_\mu + K^\mu_i\partial_{y^\mu_i}$ on $J^1(\pi)$. Notice that the associating of the vertical tangent bundle $V(\pi) \to Y \to X$ with the bundle $\pi : Y \to X$ is the covariant functor defined on the category of fiber bundles over X and taking values in the category of double vector bundles over X. The following commutative diagram illustrate this association:

$$
\begin{array}{ccc}
\mathcal{V}(\pi_{10}) & \longrightarrow & J^1(\pi) \\
{\scriptstyle \mathcal{V}(p_{10})}\downarrow & & \| \\
\mathcal{V}(\pi_1) & \longrightarrow & J^1(\pi) \\
{\scriptstyle \mathcal{V}(\pi_{10})}\downarrow & {\scriptstyle \pi_{10}}\downarrow & \cdot \\
\mathcal{V}(\pi) & \longrightarrow & Y \\
\downarrow & {\scriptstyle \pi}\downarrow & \\
\cdot \quad\longrightarrow & & X
\end{array}
$$

Here $p_{10} : \mathcal{V}(\pi_{10}) \hookrightarrow \mathcal{V}(\pi_1)$ is the inclusion of the π_{10}-vertical tangent vectors to $J^1\pi$ to the space of π_1-vertical vectors on $J^1(\pi)$.

This *"vertical connection"* lifts π-vertical vector fields ξ in Y to the π_1-vertical vector fields in $J^1(\pi)$: $\xi^\mu\partial_\mu \to \xi^\mu\partial_\mu + K^\nu_{i\mu}\xi^\mu\partial_{y^\nu_i}$.

Commutator expression (77.6) determines the curvature of this connection.

Remark 91. Recall the bijection between the connections ν on the bundle $\pi :$ $Y \to X$ and the sections $\rho_\nu : Y \to J^1(\pi)$, see [70] or [75]. Section ρ_ν determined the point $\rho_\nu(y)$ in the affine fiber $\pi_{10}^{-1}(y)$ over any point $y \in Y$. This determines the identification of the 1-jet bundle $\pi_{10} : J^1(\pi) \to Y$ with the vector bundle $\pi^*(T^*(X)) \otimes V(\pi) \to Y$.

Remark 92. Notice that a choice of a vertical connection $\xi \to Hor_{K_v}(\xi)$ defines the direct splitting of the tangent bundle $T(J^1(\pi))$:

$$T(J^1(\pi)) = Ca^1 \oplus Hor_{K_v} \oplus V(\pi_{10}) \qquad (77.20)$$

into the sum of Cartan distribution $Ca^1 = \langle d_i \rangle$, K_ν-horizontal distribution, $Hor_{K_v} = \langle Pr^1_K(\partial_{y^\mu}) \rangle$ and the π_{10}-vertical distribution $V(\pi_{10}) = \langle \partial_{y^\mu_i} \rangle$. This decomposition can not be projected to Y.

In the next diagram we collect the relations between tangent bundles of all three participating spaces: $X, Y, J^1(\pi)$, their vertical subbundles and present the basic tangent vectors in vertical subbundles

$$
\begin{array}{ccccc}
\partial_{y_i^\alpha} & & \partial_{y^\alpha}, \partial_{y_i^\alpha} & & \partial_{y^\alpha} \\
V(\pi_{10}) & \xrightarrow{\ \pi_{10v}\ } & V(\pi_1) & \xrightarrow{\ \pi_v\ } & V(\pi) \\
\downarrow & & \downarrow & & \downarrow \\
T(J^1(\pi)) & =\!=\!= & T(J^1(\pi)) & \xrightarrow{\ \pi_{10*}\ } & T(Y) \\
\pi_{10*}\downarrow & & \pi_{1*}\downarrow & & \downarrow \pi_* \\
T(Y) & \xrightarrow{\ \pi_*\ } & T(X) & =\!=\!= & T(X)
\end{array}
\tag{77.21}
$$

77.2. Connections in the infinite tower $J^\infty/\ldots/X$. . Consider the infinite tower of jet bundles

$$
J^\infty(\pi) \to \ldots J^k(\pi) \to \ldots \to J^1(\pi) \to Y \to X.
\tag{77.22}
$$

A connection C on the bundle $\pi_{\infty 0} : J^\infty(\pi) \to Y$ delivers the ("**C-horizontal**") lift of vector fields $\xi = \xi^i \partial_i + \xi^\mu \partial_\mu$ to the infinite jet bundle:

$$
\xi^{(\infty)} = \xi + \sum_{\mu, I\mid\, |I|>0} (C_{Ii}^\mu \xi^i + C_{I\nu}^\mu \xi^\nu) \partial_{y_I^\mu}
\tag{77.23}
$$

Connection coefficients $C_{Ii}^\mu, C_{I\nu}^\mu$ are, in general, arbitrary smooth functions in $J^\infty(\pi)$, i.e. they belongs to the spaces $C^\infty(J^s(\pi))$ where s may depend on μ, I.

In most cases (except of the so called *generalized vector fields*, see definition 79 below), $s \leqq |I|$. Let this be true for a given connection C.

If we drop terms with $|I| > s$, in the series (77.23), we get the connection on the bundle $\pi_{s0} : J^s(\pi) \to Y$ with the (C)-horizontal lift of vector fields $\xi \to \xi^{(s)}$.

Remark 93. Curvature of a connection C in the higher order k-jet bundles $J^k \pi \to Y$ (here $1 < k \leqq \infty$) is defined in the same way as in the case $k - 1$. For the information about connections on the higher order jet bundles we refer to the monograph [70].

Remark 94. Notice that the tangent bundle to the space $J^\infty(\pi)$ naturally splits into the π_∞-vertical component (kernel of the tangent mapping $\pi_{\infty*} : T(J^\infty(\pi)) \to T(X)$) and the complemental (n)-dim subbundle Hor (linear span of the vector fields d_μ)

$$
T(J^\infty(\pi)) = Hor \oplus V(\pi_\infty).
\tag{77.24}
$$

This splitting depends smoothly on the point $z^\infty \in J^\infty(\pi)$ and defines the **canonical flat connection** S_0 on the bundle $\pi_\infty : J^\infty(\pi) \to X$.

Appendix III. **Symmetry groups of system of differential equations.
and the Noether Theorem**.

Introduction.

Noether Theory of Symmetries and Conservation laws is a beautiful chapter in mathematical physics. Here we present the form this theory take place in the presence of non-commuting variations. Shortly speaking, in the presence of non-commutative variations, Noether laws becomes balance (not conservation) laws. This does not harm their usefulness. Even more, in some cases, these balance laws are related with the II law of thermodynamics and carries information about entropy production (and the energy dissipation).

Symmetry groups are usually Lie groups. These are smooth manifolds G having a group structure - group product $G \times G \to G$, smooth by each argument, unit element $e \in G$ such that $ge = eg = g$ for all $g \in G$ and the operation of taking inverse - $g \to g^{-1}$ that is also smooth mapping (infinitelly differential).

Vector spaces R^k with operation of addition, Complex nombers with the zero removed and operation of product; Unit circle S^1 with addition of angles; General linear group $GL(k, h)$ - space of non-degenerate (with $det(M) \neq 0$) $k \times k$ matrices M over a field h (cases where $k = R$ or $k = C$ are most useful), subgroups of the group $GL(k, R)$ preserving Euclidian scalar product in R^k or Lorentz scalar product in R^4 are some examples of Lie groups that could met often in Mathematics, Physics and Engineering.

We do not want to enlarge the text more and, instead, we refer reader to known sources [71, 106] about symmetries and conservation laws. We also suggest several sources about Lie groups, Lie algebras (see below) and about the actions of Lie groups and their Lie algebras on the manifolds as the Transformation Groups: [51, 106].

Action of a Lie group G **on a manifold** M is a smooth mapping $\mu : G \times M \to M$ such that, for any element of the group $g \in G$ and any $x \in M$,

$$\begin{cases} \mu(e, x) = x, \\ \mu(g, (g'x)) = (gg')x. \end{cases}$$

In particular, any element of the group defines the diffeomorphism of the manifold M.

A Lie group has a variety of subgroups. Most useful of these subgroups are one-dimensional subgroups H. Such subgroups are also called one-parametrical - being a one dimensional submanifold of M such subgroups H is the one-dimensional manifold.

Such one-parametrical subgroups of a Lie groups G define the algebraic structure at the tangent space $T_e(G)$ of the Lie group G. Shortly speaking, this tangent space becomes the Lie algebra $\mathfrak{g}$ - vector space having the same dimension as the Lie group G and the operation of Lie bracket $(x, y) \to [x, y]$ for all couples of elements $(x, y \in \mathfrak{g}$. We refer to the cited sources for the information about Lie algebras, Lie groups and their action on the manifolds including the infinitesimal acting of Lie algebra $\mathfrak{g}$ generated by the action of corresponding Lie group.

© Springer International Publishing Switzerland 2016

S. Preston, *Non-commuting Variations in Mathematics and Physics*,
Interaction of Mechanics and Mathematics,
DOI 10.1007/978-3-319-28323-4_9

78. **Lie groups actions on the jet bundles and the symmetry groups of (systems of) differential equations.**

Introduction of jet bundles, and the geometrical structures in the tower of jet bundles (contact structure, connections, metrics) allows us to use powerful geometrical tools for the study of differental equations and systems of such equations (shortly DE). This geometrical theory of differential equations developed by Pfaff, Frobenius, S.Lie, E.Cartan, Kahler and, later, by Kuranishi, Guillemin, Sternberg, Goldschmidt, Griffiths, Anderson, Olver, Bryant etc. present a variety of powerful algebraic and geometrical tools for studying differential equations. We refer to the several important sources for the results of this theory (see references in the monographs[7, 71, 106]). Here we present some simple notions and results of this Theory.

Let $\pi : Y \to X$ be a configurational bundle of a Field theory and let

$$J^\infty \pi \ldots \to \ldots J^k(\pi) \ldots J^1(\pi) \ldots Y \to X$$

be the jet bundles tower corresponding to this configurational bundle.

Any system of DE $\Delta_\mu = 0, \mu = 1, \ldots, k$ of order k ($k \geqq 1$) (we take $n = dim(X) = 1$ for a system of ODE and $n = dim(X) > 1$ for a system of PDE) can be invariantly defined as the subset $\Sigma_\Delta \subset J^k(\pi)$ of the k-jet space defined by m equations

$$\Delta_\mu = 0, \mu = 1, \ldots, m. \tag{78.1}$$

Here, Δ_μ are smooth functions in $J^k\pi$: $\Delta_\mu \in C^\infty(J^k\pi)$. Order k here is the highest order of derivatives (jet variables) entering the equations 76.1.

Vice versa, any system of p equations $\{f_\beta = 0, \ \beta = 1, \ldots, p\}$ in a k-jet space $J^k\pi$ of a bundle $\pi : Y \to X$ defines the system of differential equations of order k for the sections $s : D \to Y$ (D being an open subset of the base X) of the bundle π. The case of a system of ordinary differential equations is specified by the condition that the space of independent variables is one-dimensional: $dim(X) = 1$.

A solution of a system Δ is a section $s : D \to Y$ with $D \subset X$ being an open subset of the base X such that the image $j^k s(D)$ of the domain D under the $k-jet$ of a section s (mapping $D \to J^k(\pi)$) lays in the subset $\Sigma_\Delta : j^k s(D) \subset \Sigma_\Delta$.

Now let Δ be a system of differential equations of order k and $\nu : G \times J^k(\pi) \to J^k(\pi)$ be a smooth action of a Lie group G in the k-jet space.

Definition 54. *Let* $\nu : G \times J^k(\pi) \to J^k(\pi)$ *be an action of a Lie group* G *on the k-jet space* $J^k(\pi)$. *This group is called the **symmetry group of the system** Δ **of differential equations** if the subset* $\Sigma_\Delta \subset J^k(\pi)$ *is invariant under the action of* G: $g(\Sigma_\Delta) \subset \Sigma_\Delta$ *for all* $g \in G$.

It is easy to see that a symmetry group G of a system Δ maps solutions of DE system Δ to other solutions of this system.

79. Symmetries of Lagrangian and the first Noether Theorem.

In this section we present the Emily Noether Theory of conservation laws associated with the Lie Groups of symmetries of a Lagrangian Action. Our presentation of this topic will be a slight modification of the approach of P.Olver in Chapter 4 of the monograph [106]. Our modification concerns the Euler-lagrange equations with the sources (see below) where the Noether formalism leads, in general, to the balance laws instead of conservation laws. In addition, we pay somewhat more attention to the origin and thr form of different terms in Noether **balance laws**, relating them to the sources and to the NC-tensors K.

79.1. **Symmetries and infinitesimal symmetries of the Lagrangian Action.**

Consider an Euler-Lagrange system of equations with a Lagrangian $L \in C^\infty(J^k(\pi))$ of order k and arbitrary sources in the right side

$$E_\alpha(L)(y) = f_\alpha, \ \alpha = 1\ldots, m. \tag{79.1}$$

Let

$$\mathcal{A}(y) = \int_D L(x^i, y^\alpha(x), \ldots, y^\alpha_{,I}(x), \ldots)dv \tag{79.2}$$

be an action functional defined by a Lagrangian L of order k, so that in the expression (79.2) for action, $|I| \leqq k$.

In this section we remind the notions of **variational** and **divergent symmetries** of the Lagrangian.

Let G be a Lie group of smooth transformations acting on an open subset $W \subset Y$. Let $s = \{y^\alpha(x)\} : D_1 \to W$ be a section of the configurational bundle $\pi : Y \to X$ defined in the subset $D_1 \subset D$ and such that *its graph* $\Gamma_s = \{(x, s(x)y)|x \in D_1\}$ *lays in* W.

Each element $g \in G$ that is close enough to the unit element $e \in G$ defines the transformation of section s: $s \to s^g$ as follows: $s^g(gx) = gs(x)$.

Remind now the notion of projectable transformation. Let $\phi : Y \to Y$ be a diffeomorphic transformation of the space Y of a bundle $\pi : Y \to X$. Transformation ϕ is called projectable if there exists the diffeomorphism $\bar{\phi} : X \to X$ such that the couple $(\phi, \bar{\phi})$ define the automorphism of the bundle $\pi : Y \to X$, see Definition 39, Section 67.

In a case of *projectable transformations* g, the transformed section can be defined by the equivalent formula $s^g(x) = gs(\bar{g}^{-1}x)$. Here $\bar{g}$ is the diffeomorphisms of X - projection of g to X (see Sec.) In this case transformed section s^g is defined in the domain $\tilde{D}_1$.

Remark 95. If the transformations of the group G are not projectable, domain $\tilde{D}_1$ of transformed section s^g depends both on g *and on the section s*.

Definition 55. *A local group G of smooth transformations acting on $W \subset Y$ is a* **variational symmetry group** *of the action functional $\mathcal{A}(\{y^\alpha\})$ if whenever $D_1 \subset D$ is a sub-domain such that $\bar{D}_1 \subset D$ and $s : D_1 \to W$ be a section of the bundle π over D_1, and, finally, g is an element such of the group G that $\tilde{s}(gx) = gs(x)$ is a single-values section defined in $\widetilde{D}_1$, the following equality is valid:*

$$\int_{\tilde{D}_1} L(\tilde{x}, j^k(s^g(\tilde{x}))d\tilde{v} = \int_{D_1} L(x, s(x), j^k(s)(x))dv. \tag{79.3}$$

Remark 96. In a case of a Lie group G of projectable transformations, this definition simplifies. In particular, the transformed section is defined by the formula $s^g(x) = gs(g^{-1}x)$ and the equality (79.3) takes the form

$$\int_{\tilde{D}_1} L(\tilde{x}, j^k(s^g(\tilde{x})))d\tilde{v} = \int_{D_1} L(x, s(x), j^k(s)(x))dv. \qquad (79.4)$$

Notice that we use the same letter ξ both for an element of Lie algebra $\mathfrak{g}$ of the Lie group G and for the vector field generated by this element in the domain W.

A relation of an action of a Lie group G and corresponding infinitesimal action of its Lie algebra $\mathfrak{g}$ on a manifold M (in our case, on the bundle $\pi : Y \to X$), allows us to formulate a condition of variational invariance (i.e. of the symmetry) of the action functional, in an infinitesimal form - in terms of the prolonged action of the Lie algebra $\mathfrak{g}$ on the lagrangian L.

Theorem 36. *Infinitesimal Criteria of variational Invariance.* *([106], Sec.4.2.)*
A connected Lie group G of transformations acting on the domain $W \subset Y$ is a variational symmetry group of an action functional (75.2) (i.e. for the action with the Lagrangian L of order k) if and only if for any infinitesimal generator of the group G a vector field $\xi \in \mathcal{X}(Y))$,

$$\xi = \xi^i(x,y)\partial_{x^i} + \xi^\alpha(x,y)\partial_{y^\alpha} \in \mathcal{X}(W),$$

corresponding to an element ξ in the Lie algebra $\mathfrak{g}$ of the group G and for all points $(x^i, y^\alpha, y^\alpha_I, |I| \leqq k) \subset W^k$, the following relation is fulfilled

$$Pr^k\xi \cdot L + LDiv(\xi) = 0. \qquad (79.5)$$

Here $Div(\xi) = \sum d_i\xi^i$ is the total divergence of the n-tuple of functions $\{\xi^i\}$.

In the case of variational formalism, it is natural to introduce a more general definition of the symmetries of the action A_D related to the fact that adding a total divergence term of the form $\sum_i d_i K^i$, $K^i \in C^\infty(J^k(\pi))$ to a Lagrangian L, does not change the Euler-Lagrange equations.

Definition 56. *Let* $\mathrm{A}(s) = \int Ldv$ *be an action defined by a Lagrangian of order k. A vector field $\xi = \xi^i(x,y)\partial_i + \xi^\alpha(x,y)\partial_\alpha$ defined in a domain $W \subset Y$ is an **infinitesimal divergent symmetry** of the action A (of Lagrangian L) if there exists an n-tuple of functions $\{B_i \in C^\infty(J^k(\pi)), \; i = 1, \ldots, n\}$ such that in the domain W,*

$$Pr^k\xi \cdot L + Ldiv(\xi) = Div(B). \qquad (79.6)$$

Remark 97. Compare this definition to the infinitesimal criteria of the variational invariance (79.3). If in the situation of last definition $B = 0$, one recover the notion of variational symmetry.

We refer to [106], Chapter 5 for the proof of the following

Theorem 37. *Let ξ be an **infinitesimal divergent symmetry** of a Lagrangian problem with Lagrangian L, then the (possibly local) one-parameter group g^t corresponding to the vector field ξ **is the symmetry group of the Euler-Lagrange equations** $E_\alpha(L) = 0$, $\alpha = 1, \ldots m$.*

For the proof of this result, see [106], Section 5.3.

79.2. Generalized vector fields and the symmetries of systems of differential equations. Now we introduce the notion of generalized vector field, [106]:

Definition 57. *A **generalized vector field** in Y is the (formal) vector field*

$$\xi = \xi^i \partial_i + \xi^\mu \partial_\mu, \tag{79.7}$$

where $\xi^i, \xi^\mu \in C^\infty(J^\infty(\pi))$.

Characteristic of a generalized vector field $\xi = \xi^i \partial_i + \phi^\alpha \partial_\alpha + \sum_{\mu,I} \phi \partial^\mu I$ is defined by the same formula as the conventional vector fields

$$Q^\mu = \xi^\mu - \xi_i^\mu \phi^i.$$

Definition 58. *Generalized vector fields of the form $\xi = \xi^\alpha \partial_\alpha$ are called the **evolutional vector field**.*

The **characteristic** of an evolutional vector field ξ is the collection (m-tuple of functions) $\{Q^\alpha = \xi^\alpha\}$ or the vector field $Q = Q^\mu \partial_\mu$.

Definition 59. *Let $\xi = \xi^i \partial_i + \xi^\alpha \partial_{al}$ be a vector field in Y. Its characteristic is the m-tuple $Q^\alpha = (\xi^\alpha - y_i^\alpha \xi^i,\ \alpha = 1, \ldots, m$ introduced above, see Definition 52. Then, the evolutional vector field*

$$\xi_Q = Q^\alpha \partial_\alpha \tag{79.8}$$

*is called the **evolutional representative** of vector field ξ.*

The importance of introducing of evolutional representatives is explained by the following result.

Proposition 24. *A **generalized vector field** ξ **is a symmetry of a system of differential equations if and only if its evolutional representative** ξ_Q **is**.*

For the proof, see [106], Proposition 5.5.

80. Symmetries and Noether conservation laws.

In this section we present the proof of the first Noether Theorem establishing the relation between the infinitesimal variational (and divergent) symmetry vector fields ξ of a system of differential equations (ordinary or partial) and the conservation laws of the Euler-Lagrange system of this system of differential equations. We import this proof from the monograph of P.Olver, see [106], Sec.4.4. We present this proof in details because part of this proof is used in next Section for the proof of the analog of the first Noether Theorem for the Euler-Lagrange systems with the sources induced by an Ehresmann connection K.

Theorem 38. *(**First Noether Theorem**) Let $L \in C^\infty(J^k(\pi))$ be a Lagrangian of order k of a configurational bundle $\pi : Y \to X$. Let G be a (local) one dimensional Lie group of variational symmetries of variational problem with the Lagrangian L. Let $\xi = \sum_{i=1}^{i=n} \xi^i(x,y)\frac{\partial}{\partial x^i} + \sum_{\mu=1}^m \xi^\mu(x,y)\frac{\partial}{\partial y^\mu}$ be the infinitesimal generator of G and $Q^\mu(x,y) = \xi^\mu - \sum_1^n \xi^i y_i^\mu,\ y_i^\mu = \frac{\partial y^\mu}{\partial x^i}$ be corresponding characteristic of*

*vector field ξ (see Sec. 79 above). Then, there exists an n-tuple $P = (P_1, \ldots P_m)$
of functions on the k-jet space $J^k(\pi)$ such that*

$$Div(P) = Q \cdot E(L) = \sum_\alpha Q^\mu \cdot E_\mu(L). \tag{80.1}$$

*This relation defines the Noether conservation law in characteristic form for the
system of the Euler-Lagrange Equations $E_\mu(L) = 0$, $\mu = 1, \ldots, m$. In other words,
for all solutions $y = \{y^\mu, mu = 1, \ldots, m\}$ of the Euler-Lagrange system $\frac{\delta L}{\delta y^\mu} =
0, \mu = 1, \ldots, m$, $Div(P)(y) = 0$.*

Proof. Substitute the prolongation formula (76.2) into the infinitesimal invariance
criteria (79.5). We get

$$0 = Pr^k v(L) + L \cdot Div(\xi) = Pr^k v_Q(L) + \sum_{i=1}^n \xi^i d_i L + L \sum_{i=1}^n d_i \xi^i = Pr^k v_Q(L) + Div(L\xi),$$
$$\tag{80.2}$$

where $L\xi$ is the n-tuple with the components $(L\xi^1, \ldots, L\xi^n)$.

First term in the last expression can be integrated by parts:

$$Pr^k v_Q(L) = \sum_{\alpha, J} d_J Q^\alpha \frac{\partial L}{\partial y^\alpha_J} = \sum_{\alpha, J} Q^\alpha \cdot (-d)_J \frac{\partial L}{\partial y^\alpha_J} + div(A) =$$

$$= \sum_{\alpha=1}^m Q^\alpha E_\alpha(L) + Div(A), \quad (80.3)$$

where $A = (A_1, \ldots, A_m)$ is some m-tuple of functions depending on Q, L and their
derivatives whose precise form is not essential for the proof. We have proved that

$$Pr^k v_Q(L) = Q \cdot E(L) + Div(A) \tag{80.4}$$

for some n-tuple A. Moving the expression in the left side of the last equality to
the right and using first equality in (79.2) giving the expression for $Pr^k v_Q(L)$ we
obtain the equality

$$0 = Q \cdot E(L) + Div(A + L\xi). \tag{80.5}$$

As a result, equality (79.1)holds with the n-tuple $P = -(A + L\xi)$. As a result,
conservation law $Div(P) = 0$ holds for all the solutions of Euler-Lagrange system
$\frac{\delta L}{\delta y^\mu} = 0$, $\mu = 1, \ldots, m$. $\qquad\square$

In a case of a first order Lagrangian, the next result delivers the explicit expression for density-flux components P^i.

Proposition 25. *([106], Corollary 4.30).*
*Let $L \in C^\infty(J^1(\pi))$ be a first order Lagrangian for the configurational bundle π :
$Y \to X$. Let a vector field*

$$\zeta = \sum_{i=1}^n \zeta^i(x, y)\partial_{x^i} + \sum_{\alpha=1}^m \zeta^\alpha(x, y)\partial_{y^\alpha}$$

*be a variational symmetry of the Lagrangian, i.e. let $Pr^1\zeta \cdot L + LDiv(\bar\zeta) = 0$. Then,
the functions*

$$P^i = \sum_\alpha \zeta^\alpha \frac{\partial L}{\partial y^\alpha_i} + \zeta^i L - \sum_{\alpha, j} \zeta^j y^\alpha_j \frac{\partial L}{\partial y^\alpha_i} \tag{80.6}$$

are components of the conservation law $Div(P) = 0$ for the solutions of Euler-Lagrange equations $E_\alpha(L) = 0$.

In the case of Divergence symmetry group, the statement of the first Noether Theorem remains the same. The only change in the proof of this theorem is the inclusion of the $Div(B)$ in the proper places. For instance, formula (79.5) is replaced by the equality

$$Q^\mu E_\mu(L) + Div(A + L\xi) = Div(B). \tag{80.7}$$

As a result, final conclusion (80.1) holds now with $P = B - A - L\xi$.

81. NOETHER BALANCE LAWS.

Following the proof of the First Noether Theorem we establish the relation between infinitesimal variational symmetry - vector field $\xi \in \mathcal{X}(Y)$ and the corresponding **balance law** that is satisfied by the solutions of the Euler-Lagrange systems with the sources (79.1):

$$E_\mu(L) = f_\mu, \ \mu = 1, \ldots, m. \tag{81.1}$$

In particular, it includes the systems of Euler-Lagrange equations with the NC-variations (6.6).

We were using these balance laws in Chapter 2 and 6 above.

Following the proof above we get the equality (74.5)

$$Div(A + L\xi) = -Q^\mu E_\mu(L).$$

On the solutions $y = \{y^\mu, \mu = 1, \ldots, m\}$ of system (79.1), $E_\mu(L)(sol) = f_\mu$.

Substituting this to the equality (74.5) we get for solutions of the system (81.1), the **Noether balance law**

$$Div(A + L\xi) = -Q^\mu f_\mu(y). \tag{81.2}$$

Thus, we've proved the following

Theorem 39. *Let L be a Lagrangian of order k and let*

$$E_\mu(L) = f_\mu, \ \mu = 1, \ldots, m \tag{81.3}$$

be an Euler-Lagrange system with the Lagrangian L of order k and the sources $f_\mu, \mu = 1, \ldots, m$. Let ξ be a vector field in Y - infinitesimal generator of variational symmetry of the Lagrangian L (see Sec. 79) and let $Pr^k(\xi)$ be its prolongation of order k. Then there exist an n-tuple of the smooth functions A^i such that for all solutions of the system (81.1) the following equality is fulfilled

$$Div(A + L\xi) = -Q^\mu \cdot f_\mu \tag{81.4}$$

Here $Q = \{Q^\mu = \xi^\mu - y_i^\mu \xi^i\}$ is the characteristic *of the vector field ξ.*

Remark 98. When the right side of this balance law is zero, (examples are: material homogeneity, isotropy, absence of body forces) the corresponding balance law becomes the conservation law.

Next Corollary follows directly from the last Theorem.

Corollary 5. Let ξ be an infinitesimal divergence symmetry of a variational problem of order k with a Lagrangian L. *Then*

$$Pr^k\xi \cdot L + LDiv(\xi) = Div(B) \qquad (81.5)$$

*for a π-horizontal (n-1)-form $B^i\eta_i$ on $J^k(\pi)$ (see [106], Sec.4.4). As a result,*on the solutions of system *(77.1) the balance law 81.4 takes the form*

$$Div(P) = -\sum_\alpha Q^\alpha f_\alpha, \ P = A + L\xi - B. \qquad (81.6)$$

In particular, in a case where $f_\alpha = 0, \alpha = 1, \ldots, m$, previous balance law becomes the conventional Noether conservation law.

<h1 style="text-align:center">82. Conclusion.</h1>

<h2 style="text-align:center">References</h2>

[1] Anderson, A. and Choquet-Bruhat, Y and York, J., *Einstein Equations and Equivalent Hyperbolic Dynamical Systems*, preprint arXiv, gr-qc/990/099 V2., 1999.

[2] Arnovitt, R., Deser, S., Misner, C.W., *Gravitation: An introduction to Current Research*, W.H.Freeman and Co.,New York, 1970.

[3] , S.Preston, *Analytical solution in the models of Aging Ch.VI.*, manuscript (unpublished).

[4] T.M.Atanackovic, *A note on a Variational Principle for Simple Materials*, Acta Mechanica, 26, pp.331-335, 1977.

[5] T.M.Atanackovic, *The Sufficient Conditions for an Extremum in the Variational Principle with Non-Commuting Variational Rules*, Proc. IUTAM-ISSIMM Symp. on Modern Developments in Analytical Mechanics, 1983, pp.463-467.

[6] R.Bishop, R. Crittenden, *Geometry of Manifolds*, Academic Press, N.Y,London, 1964.

[7] R.Bryant, S.S.Chern, R.B.Gardner, H.L.Goldschmidt, P.A.Griffiths, *Exterior Differential systems*, Springer-Verlag,N.J.,Berlin, 1991.

[8] Boltzmann, L *Uber die Form der Lagrangschen Gleichungen fur nichtholonome,generalisierte Coordinaten*, S.-B. Acad. Wiss. Wien. Abt. IIa Math-Naturwiss Cl.111 (1902), 1603-1614.

[9] J.T. Boyle, J.Spence, *Stress Analysis for Creep*, Butterworths, London,1983.

[10] E.Binz,H. Fischer, J. S'niatycki, *Geometry of Classical Fields*, Amsterdam, North-Holland, 1988.

[11] Carter, B. and Quintana, *Foundations of general relativistic high-pressure elasticity theory*, Proceedings of Royal Society London,1972,Ser.A 331, pp. 57-83.

[12] V.Ciancio, M.Francaviglia, *Non-Euclidian Structures as Internal Variables in Non-Equilibrium thermodynamics*, Balcan J. of Geometry ans its Applications, v.8,No.1,2003, pp.33-43.

[13] , Cicogna G., Gaeta G. and Morando P., *On the relation between standard and μ-symmetries for PDEs*, J. Phys. A 37 (2004) 9467-9486.

[14] , Cicogna G., Gaeta G.,*Noether Theorem for μ-symmetries.*, arXiv; 07-83144v1, [math=ph], 23Aug 2007.

[15] A. Chudnovsky, S. Preston, "4D Geometrical Modeling of A Material Aging", International Journal of Geometric Methods in Modern Physics, v.3, N8 (2006), pp.1-30,

[16] A. Chudnovsky, S. Preston, *Variational Formulation of a Material Ageing Model*, in "Configurational Mechanics of Materials", ed. G.Maugin, R. Kienzler, Springer, Wien, 2001, pp.273-306.

[17] , A. Chudnovsky, S.Preston, "Geometrical Modeling of Material Aging, Extracta Mathematicae 11(1) (1996) 1-15.

[18] , B-HoChoi, Z.Zhou, a. Chudnovsky, S. Stivala, K.Sehanobish, C.Bosnyak,*Fracture Initiation associates with chemical degradation: observation and modeling, manuscript*, 2004.

[19] B. Dimityrov, *A modified variational principle in relativistic hydrodynamics*, arXiv,gr-qc/9908032v2, 1999.

[20] Dj, Djukich, B. Vujanovic, *On some geometrical aspects of classical nonconservative mechanics*, J. Math. Phys., Vol.16, No.10, 1975, pp. 2099-2102.

[21] L. Eisenhart, *Non-Riemannian Geometry*, AMS, New York, 1927.

[22] *M.Epstein, M. Elzanowski, Material Inhomogeneities and their Evolution*, Springer, Berlin, 2007.

[23] *M.Elzanowski, S.PrestonS. (Prishepionok), On homogeneous configurations of Uniform Elastic Bodies*, in "Matherials Sciences Forum", vol.123-125 (1993), p.155-164.

[24] *M. Elzanowski, S. Prishepionok, Connections on Higher Order Frame Bundles*, in: Tamassy and J.Szenthe (eds), New Developments in Differential Geometry, Debrecin, 1994, pp. 131-142., Kluver Acad. Publishers, 1996.

[25] M. *Elzanowski,S. Prishepionok,*, *Higher Grade Material Structures*, in D.Parker, A.England (eds), IUTAM Symposium on Anisotropy, Inhomogeneity and Nonlinearity, Kluver Acad.Publishes, pp.63-68,,1995.

[26] Epstein, M. and Maugin, G., *The Energy-Momentum tensor and material uniformity in finite elasticity*, Acta Mechanica, vol.83,1990, pp.127-133.

© Springer International Publishing Switzerland 2016

S. Preston, *Non-commuting Variations in Mathematics and Physics,*
Interaction of Mechanics and Mathematics,
DOI 10.1007/978-3-319-28323-4

[27] A. Echeverria-Enriquez, M.Munoz-Lecanda, N.Roman-Roy, *Non-standard connections in classical mechanics*, dg-ga/9503001 2Mar, 1995.

[28] M. Epstein, *Self-driven continuous dislocations and growth*, Private comm., 2004.

[29] J. D. Eshelby,*The elastic energy-momentum tensor*, Journal of Elasticity, vol. 5, Nos. 3-4, 1975, pp. 321-335.

[30] J. D. Eshelby, *Energy relations and the energy-momentum tensor in continuum mechanics*, in *Inelastic behaviour of solids*. ed. M.Kannien, W.Adler,

[31] J. D. Eshelby, *The force of an elastic singularity*, Phil. Trans. Roy. Soc. London, A244, 1951, pp. 87-.

[32] G.Fedorovskii, *On the time in Mechanics of deformed bodies*,Preprint (in Russian), 1990. http://www.shaping.ru/congress/russian/fedorovsky/fedorovsky.asp

[33] L.Fatibene, M.Francaviglia, *Natural and Gauge Natural Formalism for Classical Field Theory*, Kluwer Academic Publ., 2003.

[34] Fisher, A. and Marsden, J., *The Einstein Equations of Evolution - a Geometrical Approach*, Journal of Mathematical Physics, vol.13, No.4,1972, pp. 546-568.

[35] P. Fiziev and H. Kleinert,*Anholonomic Transformations of Mechanical Action Principle*. Proceedings of the Workshop on the Variational and Local Methods in the Study of Hamiltonian Systems, editors A.Ambrosetti and G.F.Dell'Antonio, World Scientific, 1995, pp.166-177, arXive:gr-qc/9605046v1 21May 1996.

[36] A. Golebiewska-Hermann, *On conservation laws of continuum mechanics*, Int. J. Solids and Structures, 17, 1981, pp. 1-9.

[37] G.Cicogna, G.Gaeta, S. Walcher, *A generalization of λ-symmetry reduction for systems of ODEs: σ-symmetries*, J.Physics, A: Math. Theor. 45 (2012) 355205 (29pp.).

[38] G. Gaeta and P. Morando, *On the geometry of lambda-symmetries, and PDEs reduction*, J. Phys. A 37 (2004), 6955-6975.

[39] Gaeta G., *Lambda and mu-symmetries*, in Symmetry and perturbation theory (SPT2004) (G. Gaeta et al. eds.), World Scientific, Singapore 2005

[40] Gaeta G., *Frame invariance for differential equations and ?-prolongations*, Int. J. Geom. Meth. Mod. Phys., to appear

[41] , *Twisted symmetrries of differential equations*, J.Nonlinear Math.Phys., 16 (2009),Suppl1., 107-136.

[42] *A gauge-theoretic description of μ-prolongations, and μ-symmetries of differential equations*, Journal of Geometry and Physics, 59 (2009), 519-539.

[43] , *Twisted symmetries and integrable systems*, arXiv: 1002.1489vl. [math-ph] 7 Frb 2010.

[44] I.Gelfand, Fomin, *Variational Calculus*, Prentice-Hall,1963.

[45] G. Giachetta, L. Mangiarotti, G. Sardanashvily, *New Lagrangian and Hamiltonian Methods in Field Theory*, World Scientific, 1997.

[46] G. Giachetta, L. Mangiarotti, G. Sardanashvily,*Advances classical field theory*, world Scientific, New Jersey, 2009.

[47] M.Giaquinta, S. Hildebrandt, *Calculus of Variations, vol.I*, Springer, Berlin, 2004.

[48] V.L. Ginzburg, *Theoretical Physics and Astrophysics*,Oxford ; New York : Pergamon Press, 1979.

[49] Y.Grabovsky, A.Kuchar, L.Truskinovsky, *Weak variations of Lipschitz graphs and stability of phase boundaries*, continuum mechanics and Thermodynamics. (2011),v.23, pp.87-123.

[50] H. Goldschmidt, Sternberg,*The Hamilton-Cartan formalism in the Calculus of Variations*, Ann.Inst.H.Poincare, 23 (1973),(203-269.

[51] , V.Gorbatsevich, A.Onishchik, E.Vinberg, *Foundations of Lie Theory and Lie Transformation Groups*, Springer Verlag, 1997.

[52] A.Green,W. Zerna, *Theoretical Elasticity*, ed.2, Clarendon Press, Oxford, 1968.

[53] A.Green, P.Naghdi, *A re-examination of the basic postulates of thermomechanics*, Proc. Roy. Sos. Lomd., A432 (1991), pp.171-194.

[54] A.Green, P.Naghdi, *Thermoelasticity without energy dissipation*, Journal of Elasticity 31, pp.189-208, 1993.

[55] Groot, de, S., Mazur, P., *Non-equilibrium Thermodynamics*, Dover, N.Y., 1974.

[56] , E.T.Hall.*The Dance of Life*,Anchor Prtess/Doubleday, Garden City, New York,1984.

[57] G.Hamel, *Die Lagrange-eilerschen Glieichungen der Mechanik*, Z.Math. und Phys., Vol.50,1904.

[58] H.v. Helmholtz, Ges. Abh.;vgl. auch M.Plank, 8 Vorlesungen uber theoretische Physik. Leibzig 1910.

[59] Herrmann, A. G.,*On physical and material conservation laws* , Proc.IUTAM Symp. on Finite Elasticity, 1981, Martinus Nijhof, Boston, pp.201-209.

[60] Hjuzmoller.*Fibre bundles*, Springer-Verlag, New York, Berlin, 1966.

[61] R.Houwink, H.de Decker, *elasticity, plasticity and structure of matter*,3rd ed., Cambridge University Press, Cambridge, 1971

[62] Kaduchevich, *Thermodynamical time and microstress tensor in Vakulenko form*, in "Strength of Materials", 1991, pp.552-555.

[63] J.Kijowski,, A.Smolski, A.Gornicka, *Hamitlonian Theory of self-gravitating perfect fluid and a method of effective deparametrization of einstein's theory of gravitation.*, Physics Review D, Vol.41, No.9, 1990, pp.1875-1883.

[64] Kijowski, J. and Magli, G., *Relativistic elastomechanics as a lagrangian field theory*, Journal of Geometry and Physics, vol. 9,no.3,1992,pp. 207-233.

[65] H. Kleinert, A. Pelster, *Autoparallels From a New Action Principle*, Gen. Relativity Gravitation 31 (1999), no. 9, 1439–1447.

[66] H. Kleinert, *Gauge fields in Condensed Matter, vol.II Stresses and Defects*, World Scientific, Singapure, 1989.

[67] H.Kleinert, *Path Integrals in quantum Mechanics, Statistics, Polymer Physics and financial Markets*,3rd Ed. World Scientific, New Jersey,2004.

[68] H. Kleinert, *MULTIVALUED FIELDS in Condensed Matter, Electromagnetism, and Gravitation*, manuscript, 2007.

[69] S. Kobayashi, K.Nomizu, *Foundations of Differential Geometry*, Intersience Publ., N.Y,London, 1963.

[70] I.Kolar, P.Michor, J.Slovak, *Natural Operations in Differential Geometry*, Springer-Verlag, 1996.

[71] I. Krasil'shchick, A. Vinogradov, eds. *Symmetries and Conservative Laws for Differential Equations of Mathematical Physics*, AMS, 1999.

[72] H. Kraus, *Creep Analysis*, John Wiley and Sons, New York,1980.

[73] E. Kroner, Kontinuumstheorie der Versetzungen und Eigenspannungen, Springer Verlag, Berlin, 1958.

[74] D.Krupka, *Introduction to Global Variational Geometry*, Atlantis Press, 2015.

[75] D.Krupka, D.Saunders, *Handbook of Global Analysis*, Elsevier, Amsterdam, Boston, 2008.

[76] O.Krupkova, *The Geometry of Ordinary Differential Equations*, Lect. Notes in Math., v.1678,Springer Verlag, Berlin, 1991.

[77] L.Truskinovsky, *Nonequilibrium phase boundaries in the Earth's mantle*, Dokl.Acad. Nauk SSSR, 303(6), 1988, 1337-1342.

[78] L.Truskinovsky, *Kinks versus schocks*, in "Shocks Induced Transitions and Phase Structures in General Media", (Comment preceeding Eq(3.3), Ed. R.Fosdisk,E.Dinn and M.Semrod, IMA 52, Springer-Verlag, 1993, 185-229.

[79] Landau, L. and Lifshitz, E., *Elasticity Theory, 2nd ed.*, Pergamon Press, Elmsford, N.Y.,1970.

[80] L.Landau, S.Lifshetz, *Field Theory*, .

[81] Landa, P., *Nonlinear Oscillators and Waves*, Kluver Academic Press, 1996.

[82] Lee, E.H., *Elastic plastic deformation of finite strain*, ASME, Trans. J. Appl. Mech., 54,1-6, 1969.

[83] T. Levi-Civita, U. Amaldi,*Lezioni di Meccanica Razionale*, Vol.II, Bologna, 1926.

[84] M.de Leon, D.M.de Diego, M.Vaquero, *A Hamilton-Jacoby Theory for singular Lagrangian systems in the skinnerand Rusk setting.*, arXive:1205.0168v1 [math-ph] 1 May 2012.

[85] M.de Leon, J.C.Marrero, S.M.de Diego, *A Geometyric Hamilton-Jacoby Theory for Classical Field Theories*, arXiv:0801.181v1 [math-ph] 8 Jan 2008.

[86] P.Libermann, C-M Marle *Symplectic Geometry and Analytical Mechanics*, D.Reidel Publ. Co., Dordrecht, 1987.

[87] J.Powengrub and L.Truskinovsky *Quasi-incompressible Chan-Hillard fluids and topological transitions*, Proc.R.Soc. Lond. A(1998) 454,2617-2654.

[88] A. Lurie, *Analytic Mechanics*,Kluwer,1991.

[89] S. Manoff, *Geometry and Mechanics in different models of Space-Time*, Nova Science, New York, 2002.

[90] J. Marsden, T. Hughes,*Mathematical foundations of Elasticity*, Dover, 1994.

[91] Marsden, J., Simo, J., *On the Rotated Stress Tensor and the Material version of the Doyle-Ericksen Formula*, Arch. Rational Mech. Analysis 86, 1984, pp. 213-231.

[92] G.A. Maugin, V. Kalpakides. *Toward a variational mechanics of dissipative continua*, Ch.9 in "Variational and Extremum Principles in Macroscopic Systems", Elsevier, 2005, pp.187-205.

[93] Maugin,G., *Internal varaibles and dissipative structures*, J. Non-Equilib. Thermodyn., Vol.15 (1990), pp. 173-192.

[94] Maugin, G., *The thermomechanics of nonlinear irreversible behavior.*, World Scientific, Singapur, 1999.

[95] Maugin, G.,*Material Inhomogeneities in Elasticity*, Chapman Hall, London, 1993.

[96] Muschik, W., Papenfuss, C., Ehrentraut, E.*Concepts of Continuum Thermodynamics*, Kielce University of technology, Kielce, 1996.

[97] Misner, C. and Thorne, K. and Wheeler, J., *Gravitation*, Freeman, N.Y.,1973.

[98] P Morando, *Deformation of Lie derivative and mu-symmetries*, J. Phys. A: Math. Theor. 40 (2007) 11547-11559.

[99] I. Muller,*Thermodynamics*, Pitman Advanced Publ., Boston, 1985.

[100] , I.Muller, T.Ruggeri,*Extended Thermodynamcis*,Springer, 1993.

[101] C. Muriel and J.L. Romero, *New method of reduction for ordinary differential equations*, IMA Journal of Applied Mathematics 66 (2001), 111-125.

[102] C.Muriel, J.Romero, P.Olver, *Variational C^∞-symmetries and Euler-Lagrange Equations*, J.Diff. Equations., 222 (2006), pp.164-184.

[103] C.Muriel, J.Romero, *Nonlical symmetries, Telescopic Vector fields and λ-symmetries of Ordinary Differential Equations*, SIGMA 8 (2012) 106, 21p.

[104] Ju. Neimark, *On transpositional relations in mechanics*, Gorkov gos. Univers. Uchen. zap.,(35), 1957, 100-103.

[105] Ju. Neimark, N. Fufaev, *Dynamics of Nonholonomic systems*,AMS, Providence, 1972.

[106] P. Olver, *Applications of Lie Groups to Differential Equations*, 2nd ed., Springer-Verlag,New York, 1993.

[107] A.L.Onishchik, E.B.Vinberg(eds). *Lie Groups and Lie algebras III*, Encyclopedia of Mathematical Sciences, v.41. Springer Verlag, 1990.

[108] J.-P. Ortega, V.Planas-Bielsa, *Dynamics on Leibniz manifolds*, J. of geometry and Physics, 52 (2004), 1-27.

[109] vol Laue,M, *Die Relativitatstheorie, Erster Band*, Brainschweig, 1921.

[110] Paolo Podio-Guidugli, *Thermal Displacement, From Helmholtz's to the Present Day*, in Progress in the Theory and numerics in Configurational Mechanics, Erlangen-Nurenberg, 20-24 October 2008.

[111] Roberto Percacci, *Geometry of Nonlinear Field Theory*, World Scientific, 1986.

[112] S. Preston, Noncommuting variations in the Lagrangian field theory and the dissipation, IL NUOVO CIMENTO Vol. 32 C, N. 1 Gennaio-Febbraio 2009, pp. 159-172.

[113] S.Preston, *Variational Theory of Balance Systems*, International Journal of Geometrical Methods in Modern Physics, Vol.7, no.5 (2010), pp745-795.

[114] S.Preston, *Forms of Lepage Type and the balance systems*, Differential geometry and ins Applications, 29 (2011), pp. 196-206.

[115] Pucci, E., Saccomandi, G., *On the reduction methods for ordinary differential equations*, J.Phys. A 35 (2002), 6145-6155.

[116] Rayleigh, lord, *Theory of Sound*, Vol.1,(Reprint), Dover, N.Y., 1945.

[117] Hanno Rund, *The Hamiltonian-Jacoby theory in the Calculus of Variations*, Van Nostrand Co., Londing, 1966.

[118] H. Sagan, *Introduction to the Calculus of Variations*, McGRAW-HILL BOOK CO., New York, 1969.

[119] D. Saunders, *The geometry of jet bundles*, CUP, Cambridge, 1989.

[120] J.A. Schouten, *Ricci-Calculus*, Springer-Verlag, Berlin, 1954.

[121] R.W. Sharpe, *Differential Geometry*, Springer-Verlag, Berlin, 1997.

[122] Steenrod, N., *The Topology of Fibred Bundles*, Princeton Univ. Press, Princeton, 1972.

[123] Suslov, G.K., *Theoretical Mechanics*, (In Russian) gostechizdat, 1946.

[124] A.Vakulenko, *Theory of irreversible processes*, Vestn. Leningr. Gos. Univ., No.7, 84-90, 1969.

[125] A.Vakulenko, *Superposition in the Reology of continuum media*,Izvestia of UN USSR (in Russian), 1970, n.1, pp.69-73.

[126] A. Vakulenko, *Thermodynamical time and repeated loading of an elastoplastic medium*, Vopr. Mekh. Protsess. Upravl., No.13, 55-61, 1990.

[127] K.C.Valanis, *A theory of viscoplasticity without a yield surface*,Parts I,II, Archves of Mechanics, 2-3,517,535, 1977.

[128] K.C.Valanis, *fundamental consequences of a new intrinsic time measure. Plasticity as a limit of the endochronic theory*,Archiv of Mechanics, v.32, pp.171-191, 1980.

[129] Vlasov, A.A., *Distribution Functions*,(In Russian) Moskow, 1966.

[130] V. Volterra, Atti dell'Ace. di Torino, v.XXXIII, 1897, pp. 451-475, 542-558.

[131] V. Volterra, *Sopra uno classe di equazione dinamiche*, Atti Reale Accad., Science Torino, No.33, 1898.

[132] P. Voronets, *On the equations of motion for Non-holonomic systems*, (In Russian), Matem.Sbornik, vol.22,No.4, 1901.

[133] B.D. Vujanovic, S.E.Johns, *Variational Methods in Nonconservative Phenomena*, Academic Press, Boston, 1989.

[134] B.D. Vujanovich, *A Variational Principle for Non-Conservative Dynamical Systems*, ZAMM 55, pp.321-331, 1975

[135] D. Van Dantzig, *On the phenomenological thermodynamics of moving matter*, Physica VI, no. 8, 1939, pp. 673-704.

[136] von Laue,M. *Die Relativitatstheorie, Erster Band*, Brainschweig, 1921.

[137] Wang, C.C., Truesdell,C., *Introduction to rational elasticity*, Leyden, Noordhoff International Pub. Co., 1973.

[138] Frank W. Warner, *Foundations of differentiable Manifolds and Lie Groups*, Springer-Verlag, New York, Berlin, Heidelberg, 1983.

[139] Weiss, B., *Abstract vibrating systems*, J. Math. and Mech. 17,1967, pp.241-255.

[140] Whittaker,E.T., *A treatise of the Analyical Dynamics of particles and solid bodies*, Dover. Publ., New York, 1936.

[141] Ziegler, H., *An introduction to thermomechanics*, North-Holland, Amsterdam, 1977.

[142] Ziegler,H., Wehrli,C. *The derivation of constitutive Relations from the Free Energy and the Dissipative Function*, Adv. in Applied Mechanics, vol.25, 1987.

Index

DEPARTMENT OF MATHEMATICS AND STATISTICS, PORTLAND STATE UNIVERSITY, PORTLAND, OR, U.S.

E-mail address: `serge@mth.pdx.edu`

Printed by Printforce, the Netherlands